Edward Nichols

The outlines of physics

An elementary text-book

Edward Nichols

The outlines of physics
An elementary text-book

ISBN/EAN: 9783337275662

Printed in Europe, USA, Canada, Australia, Japan

Cover: Foto ©berggeist007 / pixelio.de

More available books at **www.hansebooks.com**

THE

OUTLINES OF PHYSICS

AN ELEMENTARY TEXT-BOOK

BY

EDWARD L. NICHOLS
PROFESSOR OF PHYSICS IN CORNELL UNIVERSITY

New York
THE MACMILLAN COMPANY
LONDON: MACMILLAN & CO., Ltd.
1897

Norwood Press
J. S. Cushing & Co. — Berwick & Smith
Norwood Mass. U.S.A.

PREFACE

I HAVE attempted in this book to outline a short course in physics which should be a fair equivalent for the year of advanced mathematics now required for entrance to many colleges. Such a course, especially if it is to be accepted as an alternative entrance subject, should possess at least an equally great disciplinary value.

To possess this disciplinary value, physics must be taught by laboratory methods, and the experiments should be, as far as possible, of a quantitative nature. The student should be given ample practice in the measurement of *length*, *mass*, and *time;* and he should be taught how to arrange and interpret his results and how to express them graphically by means of curves. In this way only can the science be adequately taught, and the younger the student the more imperative is the adoption of such methods.

In this book, accordingly, experimental work is given a prominent place. The experiments have been chosen with a view to the illustration of the leading principles of physics. In the selection of the methods and of the apparatus used, I have always had in view the greatest

possible directness and simplicity, rather than the highest
degree of accuracy. The inexperience and the immatu-
rity of the reader, and the necessarily inadequate equipment of school laboratories, have been likewise borne in
mind.

In spite of the place accorded to experimental study,
the volume is not to be regarded simply as a laboratory
manual of physics. It is intended to fill the place of a
text-book and laboratory guide combined, in those cases
where the amount of time allotted to the study of physics
does not make it advisable to use both a text-book and
a laboratory manual.

The successful use of such a book implies a teacher whose
knowledge of physics has been gained through practical
work. It likewise implies the existence of a laboratory
which, although it may be homely in its appointments,
must yet contain a reasonable amount of well-made apparatus. The list of instruments essential to the carrying
on of an effective course of laboratory practice in physics
is not a large one. It includes certain standard apparatus, such as the balance, the air pump, the thermometer,
the projecting lantern, the electrical machine, and the
galvanometer ; together with such accessories as are necessary to the use of these instruments. Without such
an equipment no college or school should attempt to give
instruction in physics.

The present work has been written with this method of
teaching in view. It contains statements of the most

important principles, and descriptions of the experiments
to be performed in illustration of them. I have endeav-
ored to make these descriptions sufficiently full to enable
the student to see what the solution of each problem
demands and to work it out intelligently. The book
has purposely been shorn of all but the briefest mention of the countless applications of physics to the arts,
with the details of which too many of our modern elementary text-books are encumbered.

To my wife, who made the drawings for nearly all the
illustrations in this volume, I desire to express my grati-
tude. But for her skillful and devoted aid, my task
would have been a much more difficult one.

E. L. N.

PHYSICAL LABORATORY OF CORNELL UNIVERSITY,
February 27, 1897.

TABLE OF CONTENTS

PART I

MECHANICS

CHAPTER | PAGE

I. INTRODUCTION 1

II. THE LAWS OF MOTION 10

III. THE LAWS OF FALLING BODIES 19

IV. GRAVITATION IN COMBINATION WITH OTHER FORCES . 31

V. THE SIMPLE PENDULUM 37

VI. THE PHYSICAL PENDULUM 51

VII. KINETIC ENERGY; POTENTIAL ENERGY; WORK . . 58

VIII. MACHINES 65

IX. THE BALANCE 75

X. COHESION, ADHESION, AND FRICTION 86

XI. ELASTICITY 95

XII. THE PROPERTIES OF LIQUIDS 108

XIII. DENSITY 122

XIV. PROPERTIES OF THE SURFACE FILM OF LIQUIDS . . 130

XV. PROPERTIES OF GASES 137

PART II

HEAT

XVI. NATURE AND EFFECTS OF HEAT 148

XVII. CALORIMETRY 161

XVIII. PHENOMENA ACCOMPANYING FUSION AND LIQUEFACTION . 174

ix

CHAPTER   PAGE
XIX. RELATIONS BETWEEN HEAT AND WORK   .   .   . 187
XX. TRANSMISSION OF HEAT   194

PART III

ELECTRICITY AND MAGNETISM

XXI. INTRODUCTION TO ELECTROSTATICS   .   211
XXII. CONDUCTORS AND NON-CONDUCTORS; ELECTRICAL MA-
CHINES   221
XXIII. DISTRIBUTION OF THE ELECTRIC CHARGE UPON CON-
DUCTORS   235
XXIV. CONDENSERS   245
XXV. THE ELECTRIC SPARK   255
XXVI. THE ELECTRIC CURRENT   264
XXVII. THE MAGNETIC EFFECTS OF THE CURRENT   .   . 271
XXVIII. MAGNETISM   280
XXIX. THE MEASUREMENT OF CURRENT, ELECTROMOTIVE
FORCE AND RESISTANCE   291
XXX. THE HEATING EFFECT OF THE ELECTRIC CURRENT   . 302
XXXI. ELECTROLYSIS   307
XXXII. THERMO-ELECTRICITY   315
XXXIII. ELECTROMAGNETIC INDUCTION   321

PART IV

SOUND

XXXIV. THE PROPAGATION OF SOUND   334
XXXV. VIBRATING BODIES; PITCH AND TIMBRE   .   .   . 341
XXXVI. EXPERIMENTS WITH TUNING FORKS; THE MEASURE-
MENT OF PITCH   348

CONTENTS   xi

CHAPTER   PAGE
XXXVII. THE VIBRATION OF STRINGS 355
XXXVIII. WIND INSTRUMENTS AND RESONATORS 362

PART V

LIGHT

XXXIX. REFLECTION AND REFRACTION 370

XL. DISPERSION 382

XLI. LENSES 390

XLII. POLARIZATION, DOUBLE REFRACTION AND INTERFER-

ENCE 403

XLIII. VISION AND THE SENSE OF COLOR 413

APPENDICES

I. TABLE OF THE RELATION OF BRITISH MEASURES TO THE

METRIC SYSTEM 427

II. THE USE OF CROSS-SECTION PAPER 428

III. BALANCES AND WEIGHTS 429

IV. READING MICROSCOPES 432

V. FILTERING PUMPS 434

VI. THE CONSTRUCTION OF A SENSITIVE GALVANOMETER . . 435

VII. A SIMPLE FORM OF GLASS CELL 439

VIII. THE CONSTRUCTION OF AN ELECTROSCOPE 440

IX. THE USE OF THE LANTERN 441

THE OUTLINES OF PHYSICS

———◦◦⦂◉⦂◦◦———

PART I — MECHANICS

———◆———

CHAPTER I

INTRODUCTION

1. Physics Defined. — Physics deals with the properties of matter, and with the various changes and motions which matter undergoes. All phenomena due to the presence of life, however, are generally excluded from the domain of physics. There are several branches of physics which are so distinct and so important that they are generally treated as separate sciences. Such are:

Astronomy, the physics of the outer universe.
Geology, the physics of the crust of the earth.
Meteorology, the physics of weather and climate.
Chemistry, the physics of the atom.

2. Physical Measurements. — In ascertaining the properties of matter and studying its behavior, the student of physics must make measurements of three kinds:

(1) Measurements of **length.**
(2) Measurements of **mass.**
(3) Measurements of **time.**

3. The Metric System. — It is of the greatest convenience to have some uniform system of measures in which the various quantities are related to each other in the simplest possible manner. The metric system, which fulfills these requirements more nearly than any other, has been adopted all over the world for use in scientific work.

4. The unit of length in physical measurements is the centimeter (cm.). What a centimeter is, expressed in familiar terms, will be seen from Fig. 1, in which inches and centimeters are shown side by side.

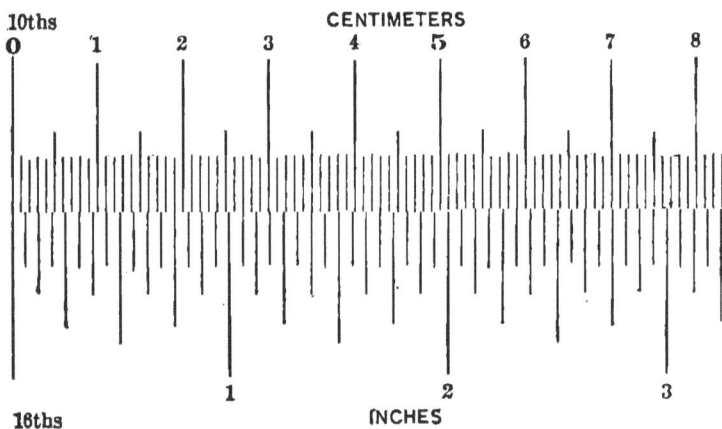

FIG. 1.

The centimeter, precisely defined, is one hundredth part of the distance between two parallel transverse lines ruled upon a bar of platinum, which is carefully preserved in the Archives of Paris. It was the intention in constructing this standard to have the distance (a meter) precisely one ten-millionth of the distance from the equator of the earth to the north pole measured along a meridian. It was soon found that the meter did not correspond very exactly to the intended length; but it was deemed much more satisfactory to use the bar as a standard, since it could be accurately copied, than to retain a standard of reference which could never be directly measured, and the precise length of which would always remain in dispute.

5. The unit of mass is the gram. It is one thousandth of the mass of a piece of platinum which is preserved in the Archives in Paris. This piece of metal, which is called the *standard kilogram*, was given as nearly as possible the same mass as that of one thousand cubic centimeters of pure water at 4° of the centigrade scale. The gram corresponds therefore, very nearly indeed, to the mass of one cubic centimeter of water at that temperature.

Water is chosen for the purpose of this definition instead of a metal such as platinum or gold or any solid material, because it is easily obtained in a state of purity, and because the mass of a given volume at a given temperature is always precisely the same. The density of a solid, however, depends upon its physical structure and the process by which it is prepared. A given volume of any metal does not therefore always possess the same mass.

A cubic centimeter of water would be a very inconvenient standard to handle in the actual comparison of masses. Platinum weights were therefore prepared, and these were adjusted until they corresponded in mass to the definition. These constitute the actual comparison standards.

Figure 2 shows a gram weight in brass, the true size; also, for comparison, a kilogram of the same material.

The precise relation of the centimeter and of the other units of the metric system to each other and to the British units, which are still largely used in the operations of everyday life, may be found in the tables of Appendix I. The names of the *more important* units of the metric system, viz.:

millimeter (mm.), centimeter (cm.), meter (m.), kilometer (km.);
milligram (mg.), centigram (cg.), decigram (dg.), gram (g.), kilogram (kg.); liter (l.);

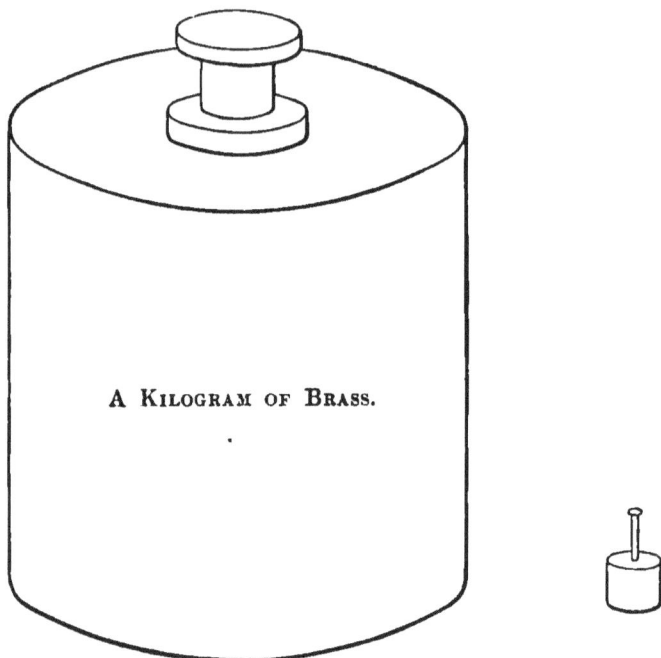

A KILOGRAM OF BRASS.

FIG. 2.

also their relations to one another, should be committed to memory.

The following approximate relations to the units of other systems, which can also readily be memorized, are very useful:

Two and a half centimeters make an inch (nearly).
A meter is about $39\frac{4}{10}$ inches.
A kilometer is about $\frac{5}{8}$ of a mile.
A gram is about $15\frac{4}{10}$ grains.
A kilogram is about $2\frac{2}{10}$ pounds.
A liter is about $1\frac{3}{4}$ pints (imperial) or rather more than 1 quart (1·056)
    U. S. standard.

6. The unit of time is the second. It is $\frac{1}{86400}$ of the mean time from noon to noon. The instrument by which time is measured is the clock, which is a machine for keeping a pendulum in motion, and for counting its vibrations. The pendulum is of such a length, usually, that it makes one vibration per second (see further, Chapters V and VI).

A simple apparatus by means of which short intervals of time are indicated to the ear, but not counted, is the

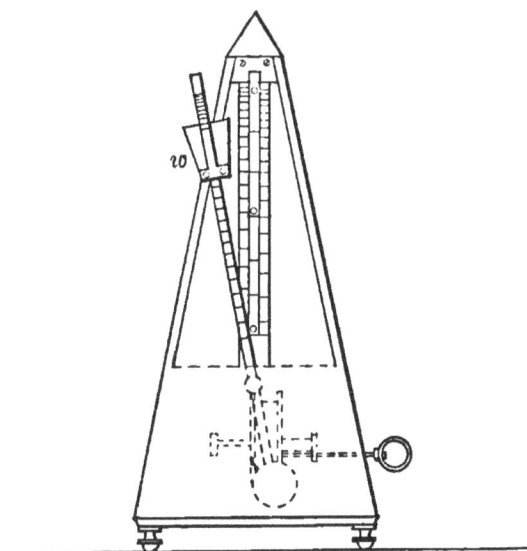

FIG. 3.

metronome (Fig. 3). This consists of a short pendulum driven by clockwork. The rate of the pendulum is adjustable by means of the sliding weight w, between 40 vibrations and 208 vibrations a minute. At each beat it ticks loudly. The metronome is used chiefly by students of music, but it is likewise a valuable apparatus in the physical laboratory. For illustrations of its application, see Arts. 38, 40, 43, 44, etc.

7. Measurements of length are always made by the use, directly or indirectly, of a divided scale, usually a copy, more or less exact, of a part or the whole or some multiple of the standard meter.

For measurements of precision the divisions of the scale are ruled with a diamond upon metal or glass, giving lines too fine for use with the naked eye. In such cases observations are made with a microscope.

8. The Estimation of Tenths. — In comparing any linear distance with a scale, the boundaries rarely (one may say never) coincide precisely with the scale divisions. The measurement, therefore, consists of two operations.

(1) The identification of the two divisions lying nearest to the boundaries of the distance to be measured.

(2) The estimation in each case of the distances between the boundary and the nearest division. Convenience of notation and computation make it desirable that the estimate should be expressed in tenths, and the second operation is called *the estimation of tenths.*

In Fig. 4, for example, *ab* and *cd* are lines the distance between which is to be measured.

FIG. 4.

The lines 3, 4, and 65, 66, are neighboring divisions of a scale, by means of which the measurement is to be made. The observer notes that *ab* is nearest to division 3, and he estimates its position as 0·2 beyond that mark. In the same way the position of *cd* is noted as 65·3.

These observations are recorded thus:

Position of *ab* 3·2 scale divisions
Position of *cd* 65·3 scale divisions
_____
Distance *ab* ··· *cd* 62·1 scale divisions

The required distance is thus determined to within 0·1 of a scale division (s. d.).

If each scale division is a tenth of one centimeter (= 1 mm.), a very common case because that is the smallest division of which tenths can be readily estimated by the unaided eye, the result is reduced to centimeters by moving the decimal point one place to the left, thus:

$$62·1 \text{ s. d.} = 6·21 \text{ cm.} \cdot$$

The estimation of tenths is an operation requiring a certain amount of practice. It affords excellent training for the eye, and is an essential attainment for all who have occasion to use the decimally divided scales which are commonly employed in science.

The necessary practice may be acquired by means of the following experiment.   ·

9. EXPERIMENT 1. — **The Estimation of Tenths.**

Apparatus:

(1) A scale at least 10 cm. long and divided to millimeters.
(2) A piece of cross-section paper. (See Appendix II.)

Procedure:

(*a*) Lay the scale upon the paper, its lines forming an oblique angle with those of the latter.[1]

_____

[1] Note that the estimation of tenths of the smallest parts of scales divided into eighths, twelfths, or sixteenths, does not in the least facilitate notation or computation of the results. It is on account of the use of such scales in everyday life that most persons remain unfamiliar with so simple an operation as the estimation of tenths.

(b) Letter every tenth line of those which lie within the limits of the scale a, b, c, etc., from left to right. (See Fig. 54.)

(c) Observe the position of the lines a, b, c, etc., at the edge of the scale, counting to the nearest division to the left, and estimating

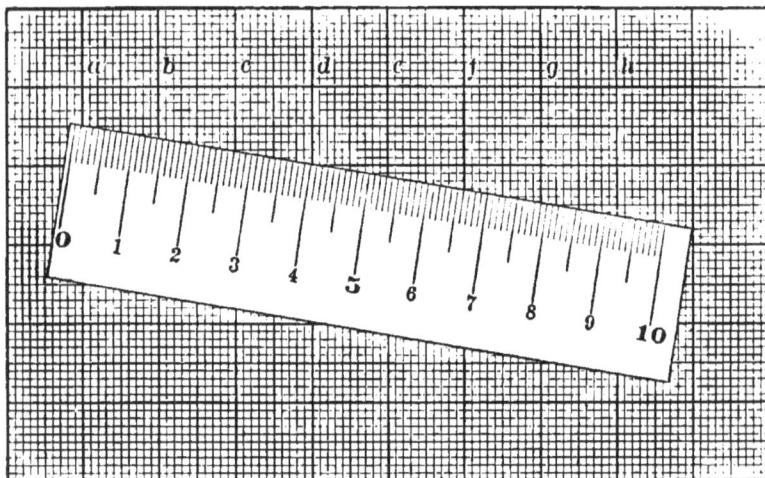

FIG. 5.

the tenths of a division. Tabulate the observations as below, and compute from them the most probable value of the distance between the lines upon the paper. Taking this as the true value, compute the positions which a, b, c, etc., should occupy upon the scale, and thus determine the error of each observation.

Note that the mean value of an interval, computed by dividing the sum of the distances from a by the sum of the numbers contained in the column marked " number of intervals, counting from a," is slightly different from the average of the successive differences. The former gives a more probable value, because it gives each observation due weight. The actual size of the error in reading the position of line h, for example, is not likely to be greater than in reading that of line b. The observed distance between a and h (90·6), however, is that of seven intervals, and the length of an interval computed by dividing that distance by 7, gives $\frac{90\cdot6}{7} = 12\cdot943$. In the same way $\frac{77\cdot7}{6} = 12\cdot950$, and $\frac{64\cdot6}{6} = 12\cdot920$, etc.

TABLE.—ESTIMATION OF TENTHS.

Line.	Scale reading in millimeters or scale divisions.	Distances from line "a" in millimeters or scale divisions.	Number of intervals included (counting from line "a").	Successive differences.	Error of reading (obtained by subtracting each "distance from a" from the product of the "mean value of an interval" multiplied by "number of intervals included.")
a	1·2	0·0			
b	14·2	13·0	1	13·0	− ·061
c	27·1	25·9	2	12·9	+ ·039
d	40·0	38·8	3	12·9	+ ·039
e	52·9	51·7	4	12·9	+ ·039
f	65·8	64·6	5	12·9	+ ·039
g	78·9	77·7	6	13·1	− ·161
h	91·8	90·6	7	12·9	+ ·039

$$\text{Sum } 362{\cdot}3 \qquad 28 \qquad 7\,|\,\underline{90{\cdot}6}$$
$$|\,12{\cdot}943 = \text{Av. interval.}$$

$$\text{Mean value of interval} = \frac{\text{sum of distances from ``a''}}{\text{sum of intervals included}} = \frac{362{\cdot}3}{28} = 12{\cdot}939 +$$

These values agree among themselves and with the mean value to within two digits in the fourth place, whereas the values of the interval obtained by considering the single intervals a to b, b to c, etc., separately (see column of successive differences) differ among themselves by two in the third place.

The example illustrates an important principle in the computation of the results of physical measurements.

CHAPTER II

THE LAWS OF MOTION

10. The fundamental characteristics of motion are that it occupies time and that it takes place along a path. The three questions to be answered in describing a motion are:

(1) In what direction?

(2) How fast?

(3) How much matter is there in the moving body or system of bodies?

Uniform motion is motion along a perfectly straight path and with unvarying speed. For such motion the answers to the three questions stated above would afford a complete description.

Uniform motion, however, does not exist in nature, except for very brief intervals of time, and when we inquire into the reasons, we find that it is because neighboring bodies act upon the moving body, tending to change its motion either as to direction, drawing it out of its straight path, or as to speed. Every such action of matter upon matter we call **force**.

11. The Usual Definition of Force. — *Force is anything that tends to produce or to modify motion.* Whenever a force exists, investigation shows that it is matter which is acting; one or more bodies acting upon the moving body. There are no exceptions to this statement.

Nothing further than the above can be said to be known about the nature of force; but two perfectly definite things can always be stated about any given force, the existence of which is observed:

(1) *Its direction.*
(2) *Its size.*

12. Graphical Representation of Forces. — Any force, the point of application of which is known, can be represented by a straight line, as *AB* (Fig. 6), one end of which is at

FIG. 6.

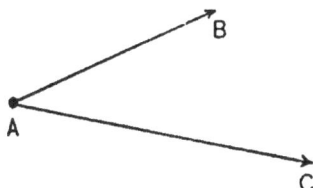

FIG. 7.

the point of application *A*. The direction of the line indicates the direction in which the force acts, its length, the size of the force. In the same way *AB* and *AC* (Fig. 7) represent two forces acting upon the point *A*.

13. The Composition and Resolution of Forces. — Frequently a number of forces may be considered as acting upon a single point. In such cases a single force can

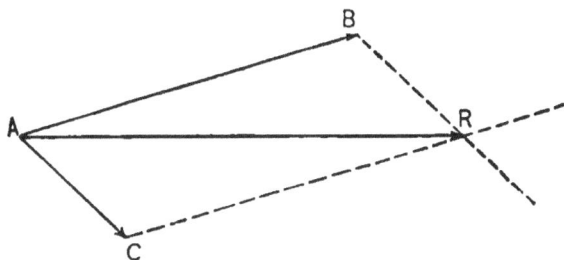

FIG. 8.

always be found which would take the place of all of them. The effect of this force, which is called the *resultant*, is precisely the same as the combined effects of all the original forces (or *components*) which it supplants.

In Fig. 8, *AB* and *AC* are forces acting upon the point *A*. *AR* is their resultant.

14. The Parallelogram of Forces. — The direction of the resultant of two forces and its size may be found by means of the principle of the parallelogram of forces; viz.:

The resultant of two forces which act upon a single point is found by completing the parallelogram of which the forces in question form adjacent sides, and drawing the diagonal which passes through the point of application. This diagonal represents, both as to size and direction, the resultant of the two forces.

EXAMPLE. — In Fig. 8, *AB* and *AC* are two forces acting upon the point *A*. They form adjacent sides of a parallelogram which may be completed by drawing *BR* parallel to *AC*, and *CR* parallel to *AB*, until they cut one another at *R*. If *A* and *R* be joined, we shall have the diagonal through *A*, and this line represents by its size and direction the resultant of *AB* and *AC*.

15. Useful Extensions of the Parallelogram of Forces. — (1) Any number of forces acting upon a single point may be combined, and their resultant may be found by taking them pairwise successively until only a single pair remains.

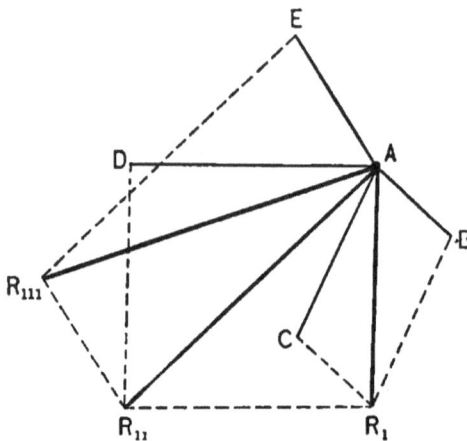

FIG. 9.

The resultant of this pair of forces will be the resultant of all the original forces.

EXAMPLE. — Let AB, AC, AD, AE (Fig. 9), be forces. AB and AC combined furnish the resultant AR_1 (Fig. 9). AR_1 may now be combined with AD, giving a resultant AR_{11}, and this in turn taken with AE, gives AR_{111}, which is the resultant of all the original forces.

(2) The only lines in the diagram (Fig. 9) which are essential to the construction are AB, BR_1, R_1R_{11}, $R_{11}R_{111}$.

Instead of drawing the various parallelograms and their diagonals, we may, knowing the sizes and directions of the forces AB, AC, AD, and AE (Fig. 9), draw AB as in Fig. 10, then BR_1 from B (having the direction and length of AC), R_1R_{11} from R_1 with the direction and length of AD, and $R_{11}R_{111}$ from R_{11}

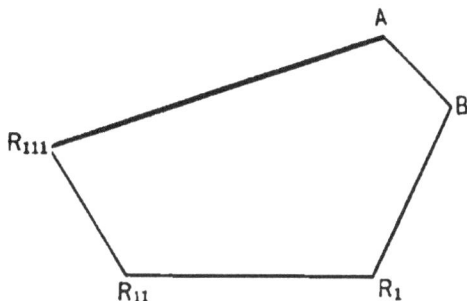
FIG. 10.

with the direction and length of AE. If A and R_{111} be then joined, a polygon will be formed (called **the polygon of forces**). The side AR_{111} will be the resultant of AB, AC, AD, and AE. It will be seen that AR_{111} is identical with the corresponding line in Fig. 9, which was obtained by means of the parallelogram of forces.

16. Newton's Laws of Motion. — Sir Isaac Newton expressed the fundamental facts with reference to motion in the form of three laws, as follows:

THE FIRST LAW. — *A body maintains its condition, whether of rest or of uniform motion, in a straight line, excepting as it is compelled by the action of forces to alter that condition.*

This law is sometimes called the *law of inertia.* It describes an important negative property of matter: its complete passivity or freedom from tendency to change its condition as regards the amount or direction of its motion. It indicates further that all changes of motion must be regarded as being brought about by the action of forces from without.

Illustrations. — (1) Imagine a block of smooth metal or wood sliding on a plane horizontal surface such as ice. It moves more and more slowly, and finally comes to rest. Newton's law tells us that we are not to look to any property of the block, by which it strives or tends to come to rest, but to forces acting from without. We find these forces in the friction between the surface of the block and that of the ice, and in the resistance of the atmosphere. The first of these would commonly be much the more important cause of retardation. If we polish the surfaces, thus reducing the friction more and more, the block will move further and further, with a given initial motion, before coming to rest. If this process could be carried on indefinitely, and finally, if the atmosphere were removed, the block would move on indefinitely without coming to rest. The motion of the block would approach uniformity as the opposing forces were removed.

(2) Imagine the above experiment to be tried in a wind blowing in the direction in which the block is moving. If we reduce the friction, we reach a point where the block will travel faster and faster under the action of the force due to the wind. To produce uniform motion, either the friction and the action of the atmosphere must be annihi-

lated as in the former example, or these must be adjusted so as to balance one another.

(3) Imagine the experiment to be tried in a transverse wind. The block will then suffer both retardation and deviation from its course. The forces of friction and of the wind cannot be balanced, and uniformity of motion is approached only as friction and the action of the atmosphere are annihilated.

The processes imagined in these three illustrations cannot be rigorously carried out, hence the statement made in Art. 10, that uniform motion does not exist in nature excepting during brief intervals of time. Experiments involving the first law will be found in subsequent sections.

17. THE SECOND LAW. — *Change of motion is always proportional in amount to the applied force (or to the resultant of the applied forces), and takes place in the direction in which the force acts.*

The second law leads us to consider the precise meaning of the term, *amount of motion.* To put a very large body and a small one into motion, such that they travel at the same speed, requires different amounts of force. In point of fact the force in each case will be proportional to the mass. The expression, *amount of motion*, as used in the second law, takes into consideration both the speed and the mass of the moving body. It is measured by the product of the two.

18. THE THIRD LAW. — *Every action (of a force) is accompanied by an equal and opposite reaction.*

This law is really a very important statement concerning the nature of force, viz. that it always consists of an *interaction* between two bodies or masses of matter.

What it was that Newton desired to express may, perhaps, be most readily shown by means of a number of familiar illustrations.

19. *Illustrations.* — (1) Imagine two precisely similar boats each containing one man. The masses of the two, including the occupants, are the same. A rope connects the two boats. If this is drawn in, the boats will approach each other by equal amounts, whether drawn by both men, or by either one singly, the other end being fastened. The force always takes the form of a strain, drawing the boats equally.

(2) If for one of the boats a vessel of large size be substituted, there will still be a *mutual* approach of the boat and the larger vessel, whatever be the size of the latter; and the *amount of motion* (taken in the sense in which that term is used in Art. 17) imparted to the two will be equal, from whatever end the rope be drawn in.

(3) The thrust of the shaft of a steamship is equally great in both directions, forwards against the hull and backwards against the water. The forward thrust, by virtue of which the ship is driven onwards, must be suitably met to prevent a movement of the shaft itself towards the bow of the vessel.

20. Reaction of a Jet of Water. — Reaction shows itself where the medium of force-transmission is a liquid or a gas, exactly as in the cases where solid matter intervenes. The reaction of water jets is a familiar example. It may be illustrated in the following simple manner.

A rubber tube firmly attached to the water faucet leads to a forked tube *T*, and thence by means of two flexible arms to a glass tube, as shown in Fig. 11. When the water is turned on, it finds exit at the opening O in the side of the glass tube. The re-action of the issuing jet thrusts the tube forcibly backwards, and in spite of the attractive force of

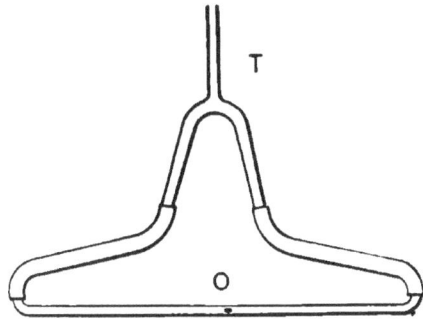

FIG. 11.

the earth it is supported in the position shown in Fig. 12.

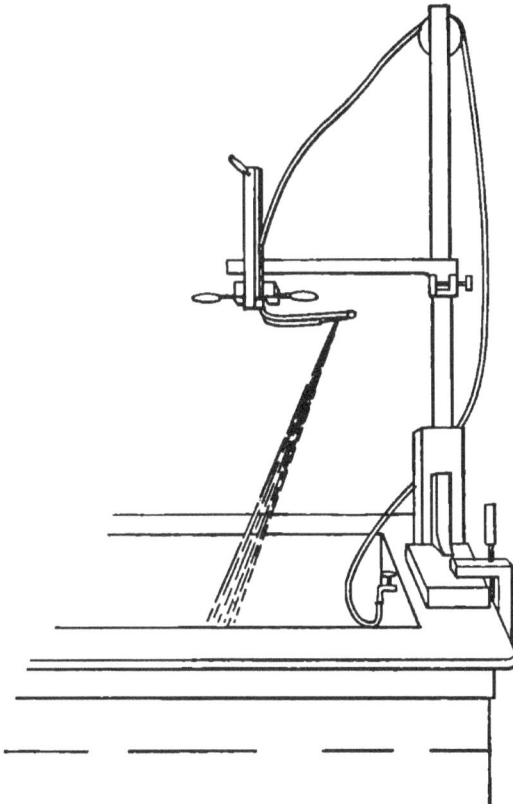

FIG. 12.

C

A familiar example of the same effect is the water tourniquet used for the sprinkling of lawns. Similar apparatus may be driven by means of the jets from a receiver of compressed air. One of the earliest of all heat engines, Hero's engine (Fig. 13), depended upon the reaction of steam jets, and one of the most successful

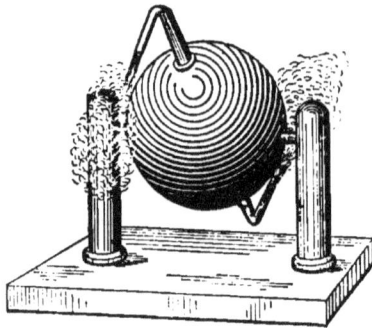

FIG. 13.

of modern steam engines, the steam turbine, is based upon the same principle.

CHAPTER III

THE LAWS OF FALLING BODIES

21. Galileo's Experiments. — The direct observation of the flight of bodies falling freely through space is difficult on account of the high speeds which they acquire in a short interval of time.

The first accurate studies of this subject were made by the Italian physicist Galileo (1604), who made use of the leaning tower at Pisa for his experiments. This building, a picture of which is shown in Fig. 14, is admirably adapted to such a purpose. It consists of a series of open galleries, one above another, reaching a total height of 55 meters, or about 179 feet. By dropping bodies, differing in mass and form and composed of various materials, from these galleries to the ground, and noting the time taken in falling, Galileo succeeded in ascertaining the following important facts with reference to the motions of matter falling freely through space under the attractive force of the earth: (1) Large bodies and small require precisely the same time to fall from a given height to the ground. In other words, *the time of fall is independent of the mass.* (2) Bodies made of different substances, as iron, copper, stone, wood, etc., fall in the same time. In other words, *the time of fall is independent of the material of which the falling body is composed.*

To this second statement Galileo found many apparent exceptions. A sheet of paper, a leaf, or a feather, for

example, would fall much more slowly than a bullet.   He
was able to show, however, that this difference was due
to the resistance of the air.   By gathering bodies, which

FIG. 14.

from their form offered a large surface to the action of the atmosphere, into compact pellets, he found that they approached the denser bodies more and more nearly in their rate of falling. Heavy metals like gold, on the other hand, when beaten into foil, fluttered slowly downwards.

22. Gravitation. — The general conclusion to be drawn from these experiments is that the earth exerts, upon each particle of matter separately, a force which is proportional to the mass of the latter, but is independent of the material of which it is composed. This force is called *gravitation*.

23. The Guinea and Feather Experiment. — At the time when Galileo made his investigations, the knowledge of the properties of the atmosphere was very vague and incomplete. No instrument for the production of a vacuum had as yet been devised. Nearly half a century later, when Otto von Guericke invented the air pump (Cologne, about 1650), one of the uses to which the new apparatus was put was the verification of Galileo's statement that, but for the resistance of the air, all bodies would fall to the earth with equal rapidity.

The form of the experiment, which has come down to the present time under the name of the *guinea and feather experiment*, is as follows:

A glass tube about a meter and a half in length and from five to ten centimeters in diameter (Fig. 15), is closed at the ends by means of brass caps. In one of these is inserted a stopcock threaded to fit the pump.

The tube contains a metal coin and a feather, or sometimes some disks of tissue paper. If these be brought to one end by holding the tube in a vertical position, and

FIG. 15.

then be made to fall freely the length of the tube, by suddenly turning the latter end for end, the difference in the rate of falling will be very noticeable. If the air be exhausted from the tube, however, and the observation be then repeated, the denser and lighter bodies will fall with equal rapidity.

24. Method of the Inclined Plane. — To determine the precise law followed by bodies falling freely under the action of a constant force like gravitation, Galileo made use of the inclined plane. His method depends upon the following principle.

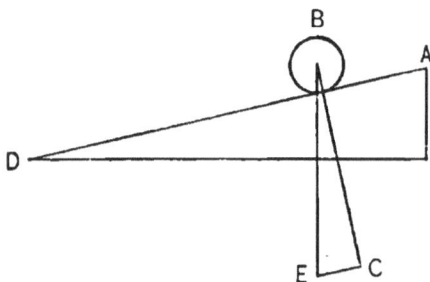

Let B (Fig. 16) be a ball resting upon an inclined plane AD.

FIG. 16.

BE is the force between it and the earth, under the action of which it would fall vertically downwards were it not partially supported by the plane.

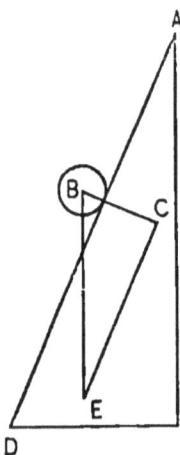

Owing to the presence of the latter, BE is in part expended in producing pressure. In point of fact, we must regard that force (see Art. 12) as resolved into two components: BC perpendicular to the plane, and CE parallel to it. The former of these components produces pressure, but no motion; the latter, CE, alone urges the ball down the inclined plane. If the grade or pitch of the plane be slight, CE will be small as compared with the original force BE, and the

FIG. 17.

ball will move slowly, but will follow the same *laws of motion* as though it were falling freely under the action of *BE*. If the plane becomes horizontal, *CE* vanishes and the whole of *BE* is consumed in the production of pressure. If the plane be very steep, as in Fig. 17, *CE* approaches *BE* in size, and the pressure-producing component, *BC*, diminishes. Finally, if the plane be vertical, *CE* equals *BE*, pressure against the plane ceases, and the ball falls freely under the action of gravitation.

25. EXPERIMENT 2. — Determination of the Laws of Motion by Means of the Inclined Plane.

Apparatus:

(1) A smooth wooden plane not less than 4 m. long. (A stout board or plank planed on one side will do.)

(2) A wooden ball about 10 cm. in diameter, or a smaller ball of metal.

(3) A metronome set to beat seconds.

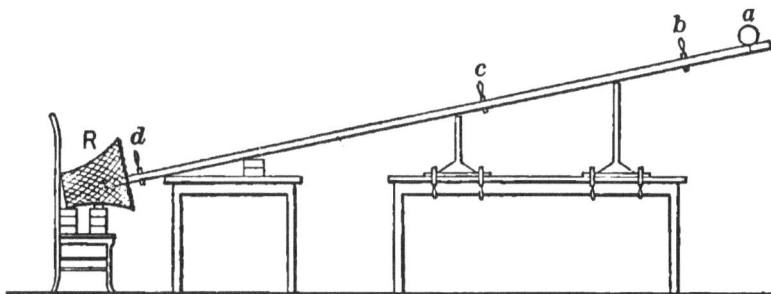

FIG. 18.

Procedure:

(*a*) Mount the plane upon two tables as shown in Fig. 18. It must be inclined at such an angle that the ball released at *a*, without initial impetus, will take rather more than three seconds to reach the lower end. Adjust a basket or other receptacle at *R* to catch the ball at the end of its trip. Provide three uprights or markers, capable of being fastened firmly to the edge of the board at any point, yet easily shifted. Small joiner's clamps (with which every laboratory and physics class-room should be amply supplied), screwed to the edge of the plank, as shown in the diagram, make convenient markers.

(b) Set up the metronome in a position where it can conveniently be started and stopped.

(c) Having set the metronome in motion, hold the ball opposite the line a, near the upper end of the plane.

(d) Release the ball with the hand, precisely at the stroke of the metronome, taking care to impart no initial impulse to it.

(e) Note the position of the ball at the third stroke after the release, and set a clamp or upright to mark the same. Repeat operations (d) and (e), readjusting the clamp until satisfied that further trials will not materially improve the result. (Trials in which the release is obviously too early or too late may be rejected.)

(f) Measure the distance from the clamp, the position of which we may designate as d, to the starting point (a). Lay off from a along the plane a distance $\overline{ab}$, equal to $\frac{1}{9}$ of $\overline{ad}$, and mark it by means of a clamp; also from a, a distance $\overline{ac}$ equal to $\frac{4}{9}$ of $\overline{ad}$, to be marked by a similar clamp.

(g) The positions (b) and (c), in case the previous operations have been carefully performed, will mark the passage of the ball at the first and second strokes of the metronome. *Verify the law by releasing the ball several times and noting its position at the end of each time interval.*

The above outline gives the experiment of the inclined plane in its simplest form. It admits of various refinements, of which the most important and most easily arranged is the automatic release of the ball at the stroke of the time-marker. Such a modification of the procedure is convenient and it adds to the accuracy of the results; but it is not essential to the success of the experiment.

The results of Experiment 2 may be most readily analyzed by arranging them in tabular form as below.

The inspection of these data brings out at once the following important features of the motion of a ball starting from a state of rest and rolling down an inclined plane. (Since, as has been pointed out, the law is the same in the case of a body falling freely through space, the conclusions are applicable to that case also.)

The relative distances apart of the uprights (column 4) give the distances traversed by the ball in successive seconds.

For the first second the previous speed was zero, the ball having been at rest. It would have remained at rest but for the action of the force of gravitation. The whole distance (a to b) is therefore due simply to the action of the force during the first second.

EXPERIMENT 2.

1	2	3	4	5
Marks.	Distance from (a).	Relative Distances from (a).	Relative Distances apart.	Times.
a	0·0 cm.	0		0 sec.
b	4·4+ cm. (calculated and verified)	d	(a to b) d	1 "
c	22·1− cm. (calculated and verified)	$4\,d$	(b to c) $3\,d$	2 "
d	397 cm. (observed)	$9\,d$	(c to d) $5\,d$	3 "

The distance $\overline{bc}$, traveled in the second time interval, is made up of two parts:

(1) The distance traversed by virtue of the initial speed of the ball, at the beginning of the interval.

(2) The distance traversed in consequence of the action of the force.

The whole distance $\overline{bc}$ is, however, 3 d, and, since the distance due to the continued action of the force is d (as in the first second), the distance ascribable to initial speed is equal to $2\,d$. This quantity ($2\,d$) is obviously the distance which the ball, if cut loose from all further action of force at the beginning of the second, would travel during that second. It is called the *velocity at the end of one second*.

From the path of the ball during the third time interval, *i.e.* $\overline{cd}$, we may, in the same manner, deduct d, the distance due to the continued action of the force. The distance $4\,d$, which remains, would be traversed without the continued action of the force; it is the *velocity* at the end of two seconds.

This doubled velocity is itself traceable to two sources:

(1) The velocity acquired during the first interval. This is equal to $2\,d$.

(2) The velocity due to the action of the force during the second interval. This is evidently equal to $2\,d$ also.

The velocity which a force is capable of imparting, in consequence of its action during one second, affords a means of measuring the force. It is called the *acceleration*.

The measure of the force is the product of the mass of the body acted upon and the acceleration.

The Unit of Force is called the *dyne.* *It is the force which, acting for one second upon a gram of matter, is capable of producing a velocity of one centimeter per second.*

The relations with which we have had to do in this discussion may be further studied by means of Fig. 19.

Conditions of the Supposed Motion. — A force f equal to 1 dyne, acting upon 1 gram of matter at O. (See Fig. 20.)

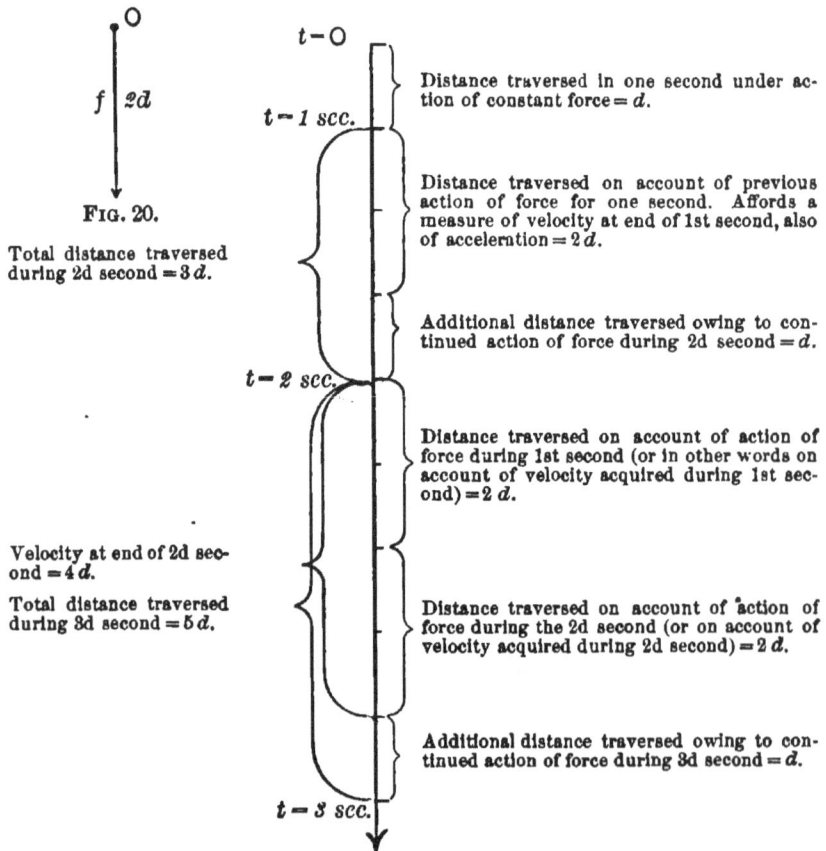

O

f | $2d$

FIG. 20.

Total distance traversed during 2d second $= 3\,d.$

Velocity at end of 2d second $= 4\,d.$

Total distance traversed during 3d second $= 5\,d.$

$t - O$

$t - 1\ scc.$

Distance traversed in one second under action of constant force $= d.$

Distance traversed on account of previous action of force for one second. Affords a measure of velocity at end of 1st second, also of acceleration $= 2\,d.$

Additional distance traversed owing to continued action of force during 2d second $= d.$

$t - 2\ scc.$

Distance traversed on account of action of force during 1st second (or in other words on account of velocity acquired during 1st second) $= 2\,d.$

Distance traversed on account of action of force during the 2d second (or on account of velocity acquired during 2d second) $= 2\,d.$

Additional distance traversed owing to continued action of force during 3d second $= d.$

$t - 3\ scc.$

FIG. 19.

We have thus far considered only the three seconds of time covered by Experiment 3. Had such measurements been extended over a longer interval, it would have been found that the ball traversed during the 4th second a distance equal to 7 d. Of this, 2 d is ascribable to the acceleration of the 1st second, 2 d to the acceleration of the 2d second, 2 d to the acceleration of the 3d second, while 1 d is due to the continued action of the force during the 4th second. The total distance traversed up to the end of the 4th second would be 16 d.

During the 5th second the ball would have traversed a distance 9 d made up in the same manner, and so on indefinitely. The total distance traversed during the five seconds would be 25 d.

26. Conclusions from Experiment 2. — (1) Comparing columns 3 and 5 of the table, we see that *the distance traversed is proportional to the square of the time occupied.* This relation may be expressed in the form of an equation, viz.

$$s = dt^2,$$

in which s is the total distance traversed in t seconds, and d is the distance traversed during the first second. Or in terms of the acceleration $(a = 2 d)$,

$$s = \tfrac{1}{2} at^2.$$

(2) The velocity is always proportional to the time spent in acquiring it and to the force acting. Expressed in the form of an equation, the relation between *velocity* (v), *time* (t), and acceleration (a), which serves as a measure of the force and is of course proportional to it, is as follows:

$$v = at.$$

(3) The acceleration due to a constant force is constant. It may be found by dividing the velocity at any instant by the time required to produce it.

Acceleration and *velocity at the end of the first second* are therefore represented by the same value, viz. 2 d.

(4) *Force acts upon moving bodies and upon bodies at
rest alike and in precisely the same way.* This is the
assumption upon which the experiment is based. There is
every evidence, from the character of the results, that the
assumption is warranted. Further direct evidence of its
truth will be found later. (See Chapter IV.)

As has been pointed out in Arts. 16 and 17, *velocity* and
acceleration, like *force*, are "directed quantities." They
may be resolved into components, or compounded to find
a resultant, like forces, by the use of the principle of the
parallelogram. It is in this way that the effect of a given
force upon a body already in motion is most conveniently
determined.

27. Acceleration due to Gravitation. — This important
quantity may be determined in a number of ways.

From the experiment of the inclined plane, for example,
we can determine the acceleration of the ball. If the pitch
of the plane be measured, the relation of the component
CE, which urges the ball down the plane (Fig. 18) to *BE*,
which represents gravitation, can be ascertained. The
acceleration due to the latter will bear the same relation
to the acceleration of the ball that *BE* bears to *CE*. The
result will be only a rough approximation to the true value,
however, unless we take account of the rotary motion of
the ball, and of the errors due to the resistance of the
air. The true value of the acceleration of gravitation
varies between 978·1 cm. at the earth's equator and
983·4 cm. at the poles. In latitude 45°, at the level of
the sea, it is 980·6 cm. (nearly). For the values, at
various places, determined with the utmost precision, see
Art. 49.

The above is the velocity which a body falling freely

through space for one second will acquire. The distance which the body will fall during the first second is one half the acceleration, or about 490 cm.

28. EXPERIMENT 3. — Verification of the Distance of Free Fall.

Apparatus:

(1) A hook or screw eye mounted about 500 cm. above the floor, viz. in the edge of a small bracket attached to the wall or to a suitable support (Fig. 21).

(2) An electromagnet; also a key and battery and conducting wires. A flexible, double lamp cord is to be preferred. The same should be about 600 cm. in length.

(3) A strong cord fastened to the yoke of the electromagnet and passing through the screw eye in the upper bracket. The other end of the cord should be attached to a clamp or cleat upon the edge of a conveniently placed table.

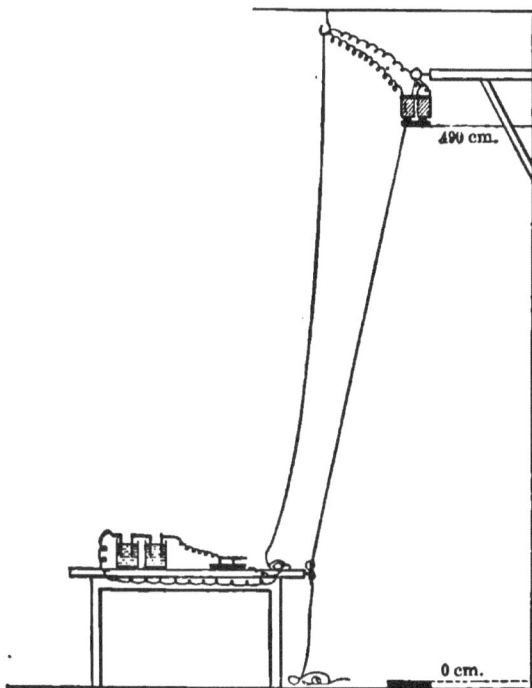

FIG. 21.

(4) The metronome described in Art. 6.

(5) A cast-iron weight. (An iron disk weighing 100 g. is suitable. It should have one face covered with paper to avoid its clinging to the magnet poles when the circuit is open.)

Procedure:

(*a*) The battery is connected in circuit with the key and electromagnet. The key is closed, and the iron weight is carefully adjusted upon the poles of the magnet.

(*b*) The magnet, with the weight still clinging to it, is hoisted by means of the cord to a position just below the bracket, such that the weight will be 490 cm. above the floor, or, better, above a block or pad placed vertically below the bracket. The magnet must be so attached to the cord as to swing freely, as shown in the figure, with the weight lowermost.

(*c*) The metronome is started, and the observer, with his finger upon the key, opens the circuit promptly at a stroke of the former. *Note how accurately the blow of the falling disk corresponds with the next stroke of the metronome.*

The observation is of the simplest character, but it affords evidence of the most direct nature concerning the distance through which a body falling freely passes in one second.

CHAPTER IV

GRAVITATION IN COMBINATION WITH OTHER FORCES

29. Projectiles. — A projectile is a body which has been given an initial velocity by being thrown from a gun or bow, or from the hand, or in any other manner. It follows a path which depends upon the size and direction of this initial velocity, and upon the action of gravitation, and of such other forces as may be brought to bear upon it.

Since, as we have seen (Art. 26), gravitation acts in precisely the same way upon bodies, whether they be at rest or in motion, we may apply to projectiles the laws of bodies falling freely from a state of rest.

30. EXAMPLE 1. (*Initial velocity horizontal.*) — Given a body, P, with an initial velocity, V_h (Fig. 22). The earth acts upon it with a vertical force, the acceleration due to which is represented by g. To find the position of the body at the end of 1, 2, 3, 4, etc., seconds.

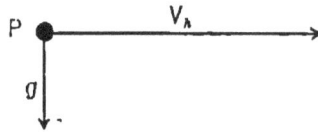

FIG. 22.

Were P at rest, it would fall in 1, 2, 3, 4, etc., seconds to positions y_1, y_2, y_3, y_4, etc., vertically below p (Fig. 23). Freed from the action of the earth, however, P, on account of its initial velocity, would move to x_1 in 1 sec., x_2 in 2 sec., x_3 in 3 sec., etc. Its positions under the combined actions of its own velocity and of gravitation will be respectively p_1, p_2, p_3, p_4, etc.

The curve drawn through these points is the path of the projectile P. It is a parabola, as is indeed the path of any projectile subjected to a constant force.

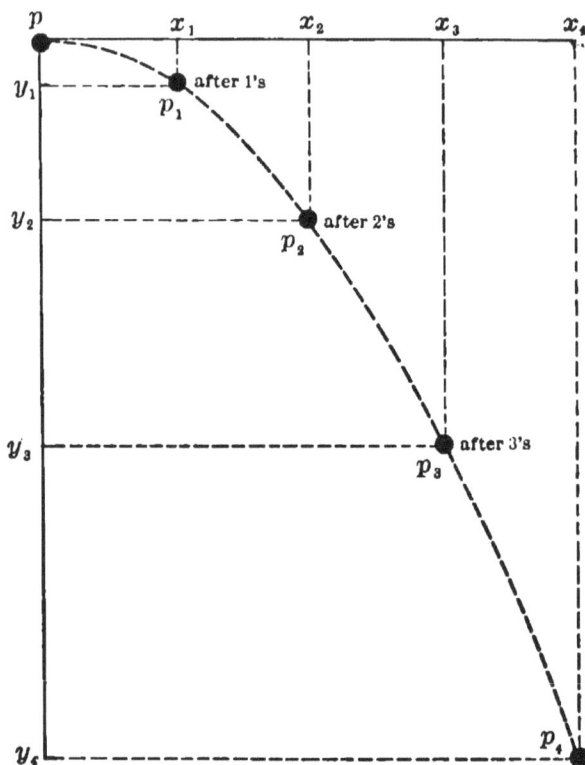

FIG. 23.

31. EXAMPLE 2. (*Initial velocity oblique.*) — The same construction is used as in Example 1, save that the line pv_4, along which the component of motion due to initial velocity is laid off, is oblique and has the same direction as the latter. (See Fig. 24.) The trace or path is parabolic.

As before, y_1, y_2, y_3, y_4 are the successive positions which would be reached in 1, 2, 3, and 4 seconds, under gravitation alone. The positions v_1, v_2, v_3, v_4 correspond to x_1, x_2, x_3, x_4 of Example 1. They are the positions which would be reached were no forces to modify the motion of p. The positions actually reached are p_1, p_2, p_3, p_4.

32. EXAMPLE 3. (*Initial velocity vertical.*) — There are two cases; of upward and of downward· initial velocity, respectively, but both are solved by the simple numerical process of computing the position which would have been occupied had gravitation not acted, and the position in case the initial velocity had been zero.

These are subtracted when the initial velocity is upwards, and added when it is downwards.

Given an upward initial velocity of 1000 cm. per second: to find the position of the projectile after 3 seconds ($g = 980$).

Without gravitation its position would be 3000 cm. above the starting point.

Gravitation, but for the

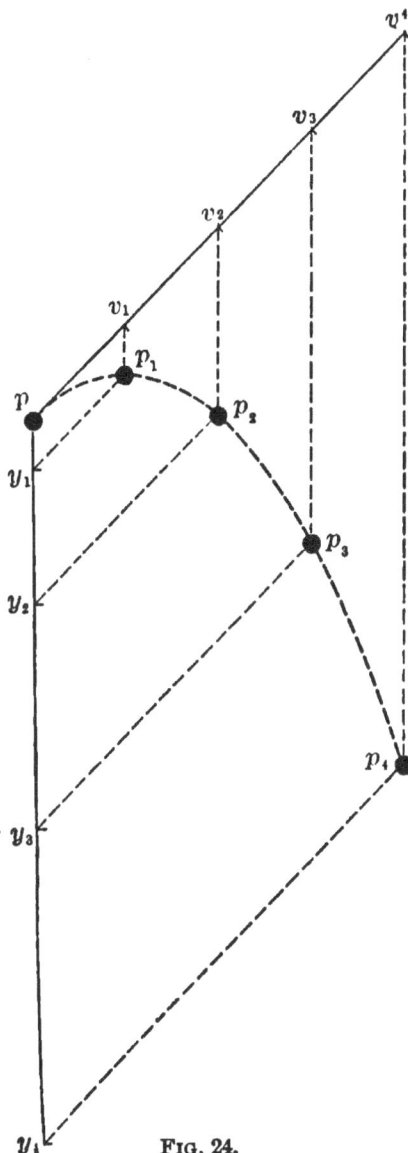

FIG. 24.

D

initial velocity, would carry it downwards over a path s, such that

$$s = \tfrac{1}{2} g t^2$$

or
$$s = \frac{980 \times 9}{2} = 4410.$$

P_3, the new position, is $4410 - 3000 = 1410$ cm. below the starting point.

Had the initial velocity been 1000 cm. downward the new position would have been found at $4410 + 3000 = 7410$ cm. below the starting point.

33. The Ballistic Curve. — Thus far we have considered projectiles moving under a single constant force. In reality, however, projectiles are subjected to other forces, such as the ordinary resistance of the atmosphere and the force of the wind. These modify the path of the projectile and cause it to deviate from the parabolic form. In the case of projectiles of high speed this divergence is considerable, and it has to be taken accurate account of in the science of gunnery. Along parabolic paths, for example, in order to carry a shot to the greatest distance, the elevation of the gun must be 45° above the horizon; whereas experiments with various weapons give for the maximum range very different angles, sometimes as low as 30°.

That for projectiles of low speed the ballistic curve is very nearly parabolic, the following experiment will serve to show:

34. EXPERIMENT 4. — **Plotting the Ballistic Curve of a Water Jet and Comparison of the Same with the Parabola.**

Apparatus:

(1) A blackboard.

(2) A long rubber tube of 0·5 cm. to 1·0 cm. bore leading from an

elevated tank to the neighborhood of the board. It is provided with a nozzle consisting of a glass tube drawn out at one end until the bore is reduced to about 0·2 cm. It is also provided with a pinchcock to stop the flow.

(3) A wooden scale divided to centimeters.

Procedure:

(*a*) Adjust the pressure so that the jet, when the nozzle is held horizontally near the top of the blackboard, will have a velocity sufficient to carry it a meter or more before it reaches the level of the base of the blackboard.

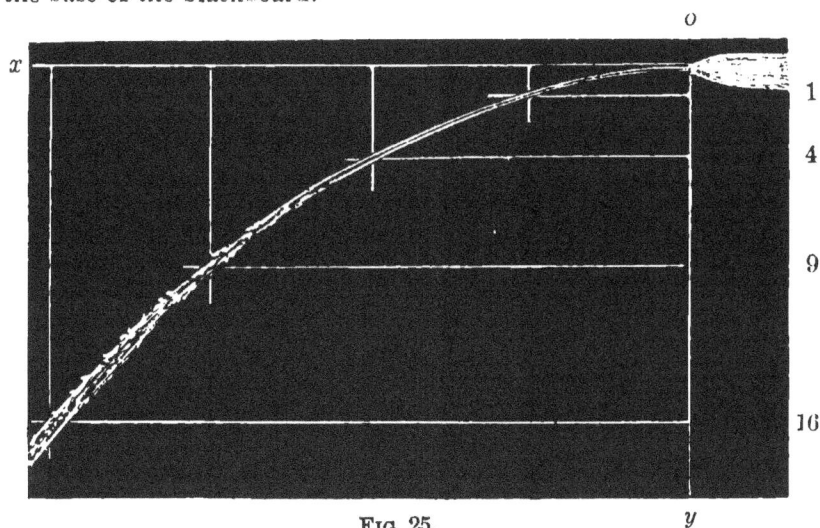

FIG. 25.

(*b*) Hold the nozzle in the position indicated above, the plane of the jet parallel to the blackboard. Then give the nozzle a single slight inward motion with the hand so as to wet the board. With a little practice it will be found possible to obtain a fairly good trace of the jet upon the board, which can then be perpetuated by outlining it with the chalk. The position of the nozzle should also be indicated.

(*c*) After the board has become dry, draw from the point where the jet may be considered to have its origin (point corresponding to the tip of the nozzle) a vertical line *oy* and a horizontal line *ox*. Extend the former to a point near the bottom of the board and divide it into 16 equal parts. From divisions 1, 4, 9, and 16, draw horizontal

lines to intersect the trace of the jet. From the intersections draw
vertical lines to cut the horizontal line *ox*.

Were the curve a parabola, these intersections of the verticals with
ox would be equidistant, as in Fig. 23, with which the diagram just
obtained should be compared. Note the amount and character of the
discrepancy. For such small velocities the deviation from the parabolic form should be insignificant.

Figure 25 shows the trace of a water jet obtained, not by the
method described above, but by photographing the jet itself against
a black background. The points of the corresponding parabola have
been drawn in. It will be seen that they lie very close indeed to the
path of the jet.

CHAPTER V

THE SIMPLE PENDULUM

35. The Pendulum. — Given a body B (Fig. 26), rigidly attached to a point O about which it is free to revolve. Such a body, when acted upon by a constant force as g, tends to come to rest in a position such that g is parallel to the line OB. This position, which is represented by the dotted lines in the figure, is called the *position of stable equilibrium*. When removed from this position and then released, the body goes through a series of oscillatory motions, the laws of which are to be experimentally determined. Such a body is called a *pendulum*. The simplest case is that of *a single material point attached to the center of oscillation O, by means of a rigid thread without mass.* Such a pendulum is called the *mathematical pendulum* (sometimes called the *simple pendulum*). Pendulums which do not fulfill the above condition are termed *physical pendulums.* The term *simple pendulum* will be used in this book, however, to designate any pendulum in which the mass is chiefly concentrated in a compact ball or bob, suspended by a wire or rod.

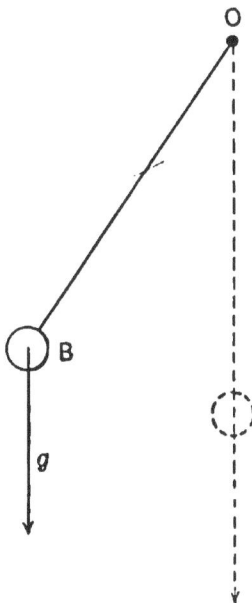

FIG. 26.

36. Definition of Terms referring to the Pendulum.

Center of suspension.—The point around which a simple pendulum oscillates. In the case of a physical pendulum, with freedom of motion in a single plane, there is a line or axis around which it swings. This is the *axis of suspension.*

Center of oscillation. — A point at which all the material of a physical pendulum may be considered as concentrated without affecting its rate.

Length. — The distance from center of suspension to center of oscillation.

Period. — The time in seconds between successive passages of the pendulum through its position of equilibrium. This is the *period of a single oscillation.*

Double period. — The time in seconds which elapses from the instant when a pendulum is at its turning point, on either side of its position of equilibrium, until it reaches the turning point upon the same side again.

Rate. — The number of single oscillations performed by a pendulum in one second.

Amplitude. — The path traversed by the pendulum in its motion. (Amplitude is measured in seconds, minutes, or degrees of arc.)

37. EXPERIMENT 5. — Law of the Simple Pendulum.

Apparatus:

(1) Two small leaden balls, also one of iron or brass, and one of wood. These are preferably all of a size and about 4 cm. in diameter. If they are bored through diametrically, with a hole large enough to admit a thread or small cord, it adds to the convenience of the operations.

(2) Several meters of strong thread or twine (the braided linen or silk cords, manufactured for fishing lines, are much the best).

(3) A stand consisting of a substantial base, an upright, and a

horizontal arm. The arm should be 40 cm. or more in length and must be situated at least 110 cm. above the base. It should contain

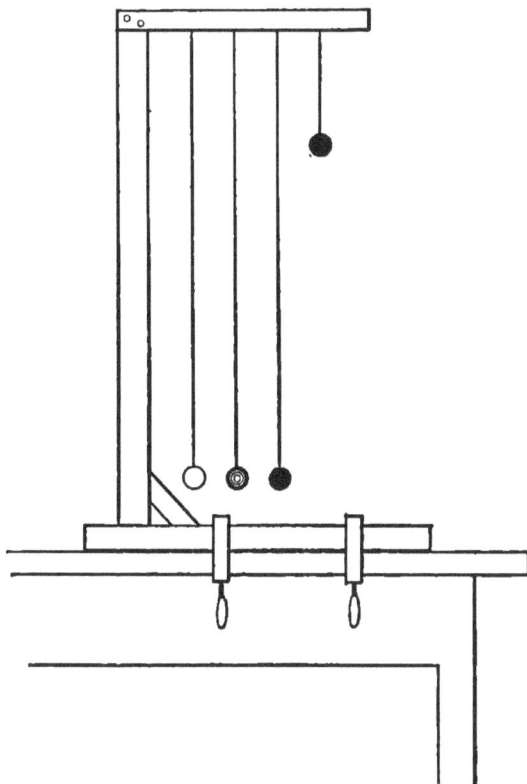

FIG. 27.

four vertical holes, running quite through, and distant about 10 cm. from one another and from the upright. The diameter of these holes should be about 0·1 cm. The horizontal arm should be fastened to the upright by mortising, in such a manner that one half of its thickness projects beyond the edge of the latter as in Fig. 28. The holes through the arm will then be in the same

FIG. 28.

vertical plane with the edge of the upright, an arrangement which greatly facilitates the measurements.

(4) A clock, the pendulum of which beats seconds, or a metronome.

(5) A meter scale.

Procedure:

(*a*) By means of small wooden plugs, fasten about 110 cm. of twine to each of the balls, passing the free ends up through the holes in the arm and securing them there by means of similar plugs. These last should be inserted from below. They should fit in such a way as to check the motion of the cord at the base of the opening, and yet permit it to be drawn through from above or below. After adjustment for length has been made, the cord may be further secured by means of a short peg inserted at the top of each hole.

Attach the wooden ball nearest the upright, then the iron or brass balls, then, outermost, the two leaden balls.

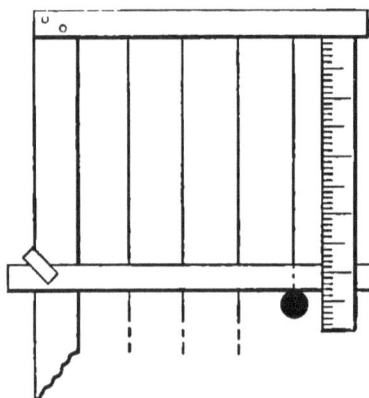

FIG. 29.

(*b*) Adjust the lengths of the suspension cords until the pendulums all swing in as nearly as possible the same time. Note that the lengths are then all the same. Note, further, that the wooden ball comes to rest much sooner than do the metal balls, and that the leaden balls, with their greater density, have the greatest persistence of motion.

(*c*) Shorten the suspension of one of the leaden balls until it swings twice as rapidly as does the other. Show by measurement that the lengths of the two pendulums are as 1 : 4. (In making the measurement of lengths, measure from a line tangent to the ball above [see Fig. 29] to the top of the suspension cord. Then add the radius of the ball.)[1]

[1] The true length of the pendulum is not the distance from the center of suspension to the middle of the ball, but that distance is sufficiently exact for the purposes of the present determination. The length l is

$$l = \frac{(\lambda + r)^2 + \frac{2}{5} r^2}{\lambda + r} = \lambda + r + \frac{\frac{2}{5} r^2}{\lambda + r},$$

where λ is the distance from the center of suspension to the top of the ball, and r is the radius of the ball. For a pendulum 100 cm. long, with a ball 2 cm. radius, $\frac{\frac{2}{5} r^2}{\lambda + r}$ is 0·016 cm., and for a pendulum 25 cm. long, with the same ball, it is ·064.

(*d*) From this relation of times to lengths determined in (*c*), viz.

$$l_1 : l_2 :: 1 : 4,$$
$$t_1 : t_2 :: 1 : 2,$$

it is reasonable to assume as the law of lengths, that the square of the time of vibration of a pendulum is proportional to its length. Verify this assumption by making the length of the shorter pendulum $\frac{4}{9}$ of that of the longer one, in which case the times will be found to have the ratio 2 : 3.

38. The Law of Equal Times.—Thus far we have assumed that the arc of vibration of a pendulum is without influence upon its rate. This assumption can be shown to be true with an exactitude sufficient for the purposes of these experiments. The law of equal times, or, as it is called, *the law of isochronism*, is in fact only an approximation; vibrations of large amplitude requiring a slightly longer time than those of small amplitude. The differences are such as require refined methods for their detection. For example, a vibration of infinitesimal amplitude, requiring 1 second, will have a period of 1·008 (nearly) when the amplitude is increased to 20°.

The approximate accuracy of the law of equal times may be verified as follows:

39. EXPERIMENT 6. — Isochronism of the Pendulum.

Apparatus:

(1) The pendulum stand described in Experiment 5, and the long pendulum with a leaden bob.

(2) The metronome.[1]

Procedure:

(*a*) The observer, who is seated opposite the pendulum, with the plane of vibration at right angles with his line of sight, directs his vision to a vertical mark previously made upon the upright of the pendulum stand. Past this swings the pendulum, which has been adjusted to a period which is very nearly an exact second.

---

[1] A loud-ticking clock which beats seconds will do as well.

(*b*) He watches until the pendulum passes the mark, at the beat of the metronome, and then begins to count seconds, continuing to count until the pendulum is on the mark again at the sound of that instrument. The time which has elapsed between these coincidences is that required by the pendulum to gain one beat over the metronome, or *vice versa*. It is only necessary to note further whether the pendulum gains or loses with reference to the metronome, in order to compute the relative period of the former.

Example: Suppose the pendulum to be losing, and that the observed interval between coincidences is 59 seconds. It makes 58 beats in 59 seconds (since it has made one less than the metronome), and its period is $\frac{59}{58}$ seconds $= 1\cdot0173$ seconds nearly.

(*c*) This observation of coincidences is to be repeated several times with amplitudes of about 20°, 5°, and less than 1° respectively.

40. The Method of Coincidences. — The method described above is much more accurate than the laborious method of counting the total number of vibrations in a given interval of time; and the likelihood of mistaking the count is much less than where a large number of vibrations has to be observed. It is known as the *method of coincidences*.

41. EXPERIMENT 7. — Relation of the Rate of a Pendulum to the Force under the Action of which it swings.

[For the purpose of this determination we use the same expedient as in a previous case (experiment of the inclined plane [Art. 25]); that of resolving gravitation into two forces, only one of which acts to produce motion.]

Apparatus:

(1) The pendulum stand described in Experiment 5.

(2) A protractor or other device for the rough measurement of angles.

(3) A watch or other timepiece with a second hand.

(4) A pendulum of the form shown in Fig. 30. It consists of a piece of stiff wire *a*, *b*, *c*, *d*, bent as shown in the figure. The ends *a*, *d* are soldered together and pass through the hole in one of the metal balls described in Experiment 5. The ball is fastened by

means of plugs. At *b*, *c*, is fastened, either by soldering or by lashing firmly with thread, a piece of steel wire (part of a knitting needle, ground to points at the ends) which serves as an axis of suspension.

The pendulum is mounted in a joiner's clamp, as shown in Fig. 31 ; small, smooth, conical holes being made in the end of the wooden screw, and in a corresponding posi-

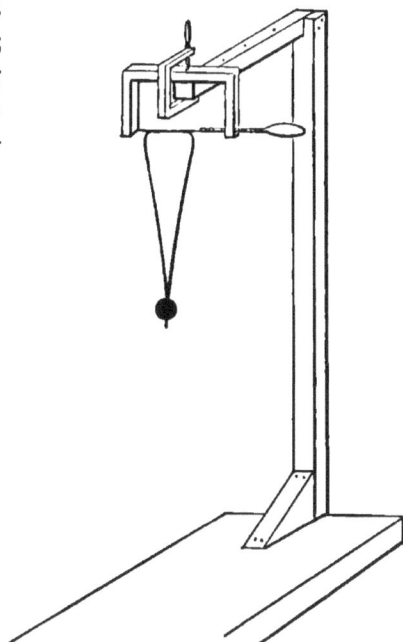

FIG. 30.

FIG. 31.

tion in the inner face of the jaw of the clamp (viz. in the line of the axis of the screw continued). The axis of suspension of the pendulum should bear only at the points and not on the periphery of the wire.

Procedure :

(*a*) Mount the pendulum vertically, clamping it to the pendulum stand, as shown in Fig. 31. Determine to the nearest second the time required for one hundred oscillations. Repeat the observation at least five times and average the results.

(*b*) Mount the pendulum at 45°, as shown in Fig. 32, and thus resolve *g* into two equal components, *a* and *b*, only one of which, the component *a*, tends to produce oscillation.

The component *a* is to *g* as 7 : 10, nearly (more accurately 7·07 to 10). It is the object of the observation to verify the fact that the square of the rate of the pendulum in this new position bears the above ratio to the square of its rate when mounted vertically.

The relation between force and period is as follows:

$$\frac{1}{t_g^2} : \frac{1}{t_a^2} :: g : a,$$

in which the period under action of the force corresponding to g is designated by t_g, and the period[1] under the action of a, by t_a. The law may be stated thus:

The square of the rate of a pendulum is proportional to the force to which its oscillations are due.

(c) Verify the law for an angle of 60° (at this inclination $a = \frac{1}{2}g$).

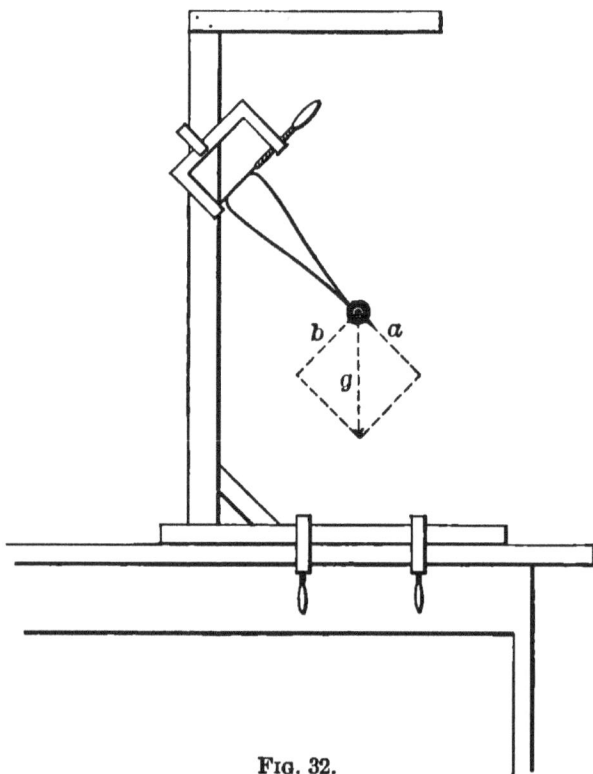

Fɪɢ. 32.

[1] The general relation of g to a is expressed in the formula

$$a = g \cos a,$$

where a is the angle which the line joining centers of suspension and of oscillation makes with the vertical line through the former.

42. Summary of the Results of Experiments 5, 6, and 7.

(1) The period of oscillation of a simple pendulum is independent of the material of which the bob is made, and also of the mass of the bob.

(2) The square of the period of oscillation is proportional to the length of the pendulum.

(3) The period is approximately independent of the amplitude of oscillation.

(4) The square of the period of oscillation is inversely proportional to the force to which the oscillations are due.

These results may be gathered into a formula which states what is generally called the *law of the simple pendulum*, viz.:

$$t = \pi \sqrt{\frac{l}{g}}.$$

In this expression t is the period in seconds; l is the length in centimeters; g is the acceleration due to the force of gravitation (or of whatever force produces the oscillations), also measured in centimeters; π is the ratio of the circumference of a circle to its diameter, viz. $3 \cdot 14159 +$. This formula applies to any pendulum whether simple or not, provided l is always taken to be the distance between the centers of suspension and oscillation.

By means of it we can compute :

(1) The length of a pendulum which will beat seconds, in any locality where g is known.

(2) The rate of any pendulum, of whatever length, where g is known.

(3) The value of g, in any locality where it is unknown, from the period of vibration of a pendulum of known length.

This last is an operation of the greatest importance in physics, since it is by means of it that our knowledge of variations in the force of gravitation in different parts of the world has been gained.

43. EXPERIMENT 8.— Testing a Metronome by Comparison with a Seconds Pendulum.

Apparatus :

(1) The pendulum stand already described, and the pendulum with a leaden ball.

(2) A meter scale.

(3) A metronome.

Procedure :

(*a*) Measure the diameter of the ball.

(*b*) Suspend the ball by a string or wire, making the length of the string, plus half the diameter of the ball, equal to 99·3 cm. (The pendulum thus constructed will have a period of almost precisely one second.)

(*c*) Mount just behind the suspension wire of the pendulum a card with a heavy vertical black line marked upon it.

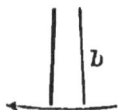

(*d*) Sit directly in front of the pendulum, the eye nearly on a level with the card, and the head in such a position that the suspension wire, when at rest, will coincide with the vertical line. Pull the ball to one side, about two or three degrees (*i.e.* about 5 cm.), parallel to the card, and release it. Listen to the metronome, previously set at 60 beats a minute, and at the same time watch the suspension wire. Note the successive positions which it occupies at each beat of the metronome, and thus determine whether the metronome is gaining or losing with reference to the pendulum. If the metronome is gaining, the position of the suspension wire at successive beats will be displaced in a direction opposite to the movement of the pendulum, as indicated in Fig. 33. If the metronome is losing, the reverse will be true.

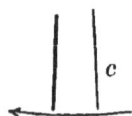

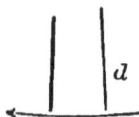

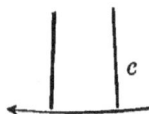

FIG. 33.

(*e*) Having determined whether the metronome is gaining or losing, watch the position of the suspension wire again, and when the wire is coincident with the mark behind the pendulum, at the tap of the metronome begin to count the beats of the latter. Continue to count until the wire coincides with the mark again, and then record the number of beats which have intervened. Compute from this interval the relative rates of the pendulum and the metronome. If, for example, the metronome is gaining, and it makes 65 beats from coincidence to coincidence, then its rate is $\frac{66}{64}$ beats per second, or its period (single) is $\frac{64}{66}$ seconds = 0·984+ seconds. This is obvious since the interval between coincidences is that required by the metronome *to gain one beat*.

If the metronome is losing, on the other hand, and the interval between coincidences is 65, its rate must be $\frac{64}{66}$ beats per second, and its period $\frac{66}{64}$ seconds = 1·016+ seconds.

(*f*) If a clock with a pendulum which beats seconds is available, find the rate of the metronome, as compared with the clock, by the method of coincidences. This part of the experiment is not a mere repetition of what has gone before, since it is to be performed entirely by the ear. The process consists in listening simultaneously to the ticking of the clock and the metronome until they beat in unison, and then in counting the beats of the latter until unison is reached again.

Students who have the time and patience to cope with somewhat greater difficulties than are demanded by this determination, may try the following very interesting and instructive experiment:

44. EXPERIMENT 9. — **Approximate Length of a Pendulum which beats Seconds.**

Apparatus:
(1) A pendulum with leaden ball. (Experiment 5.)
(2) A meter scale.
(3) A metronome. (Experiment 5 [*e*].)
To obtain the best results in this determination, a fine wire of iron or brass should be used instead of the suspension cord. This, at the upper end, should be held within the jaws of a hand vise, which, in turn, is clamped to the horizontal arm of the pendulum stand. Thus a much more definite point from which to measure lengths is obtained.

Procedure:
(*a*) Measure the diameter of the ball.
(*b*) Adjust the pendulum to a length of about 105 cm. Deter-

mine its rate by the *method of coincidences*. Measure the length to the upper periphery of the ball.

(*c*) Repeat the above observations of rate and length with the pendulum set approximately at 103 cm., 101 cm., 99 cm., and 97 cm., etc.

(*d*) Tabulate the results as shown below.

TABLE: LENGTH OF A SECONDS PENDULUM.

Obs.	Time in beats of the metronome from coincidence to coincidence.	Rate.	Period from each set.	Lengths (observed).	Lengths[1] (corrected).
a	22	$\frac{21}{22}$	$\frac{22}{21} = 1\cdot045$	103·4 cm.	105·8 cm.
b	33	$\frac{32}{33}$	$\frac{33}{32} = 1\cdot0312$	101·7 "	104·1 "
c	53	$\frac{52}{53}$	$\frac{53}{52} = 1\cdot0192$	99·5 "	101·9 "
d	400	$\frac{401}{400}$	$\frac{400}{401} = 0\cdot9975$	95·0 "	97·4 "
e	91	$\frac{92}{91}$	$\frac{91}{92} = 0\cdot989$	93·4 "	95·8 "

(*e*) To determine from these measurements the length of a pendulum, the period of vibration of which would be the same as that of the metronome, take a piece of cross-section paper and mark along the base (as abscissas) times in seconds, and along the vertical edges (as ordinates) lengths as shown in Fig. 34. (See Appendix II.)

Find the points on this sheet corresponding to corrected lengths and rates, for sets *a*, *b*, *c*, *d*, *e*, etc., respectively.

These will lie approximately in an oblique line (strictly speaking along a curve; the curvature of which is so slight, within the limits of this experiment, as to be scarcely perceptible). Wide divergence of *one* of the values from the line joining the others indicates a blunder. If the values do not lie so that a nearly straight line can be drawn through them, the experiment has been poorly performed.

A certain amount of divergence is to be expected among experimental results, however painstaking the observer may be. Thus it will be seen in Fig. 34 that the points *a*, *b*, *c*, *d*, and *e* do not lie upon

---

[1] These lengths are obtained by adding to the observed length the radius of the ball, which, in the case to which these data refer, was 2·4 cm.

a common line. They do, however, lie near a line *mn*, drawn in such a way as to equalize the divergence as far as possible.

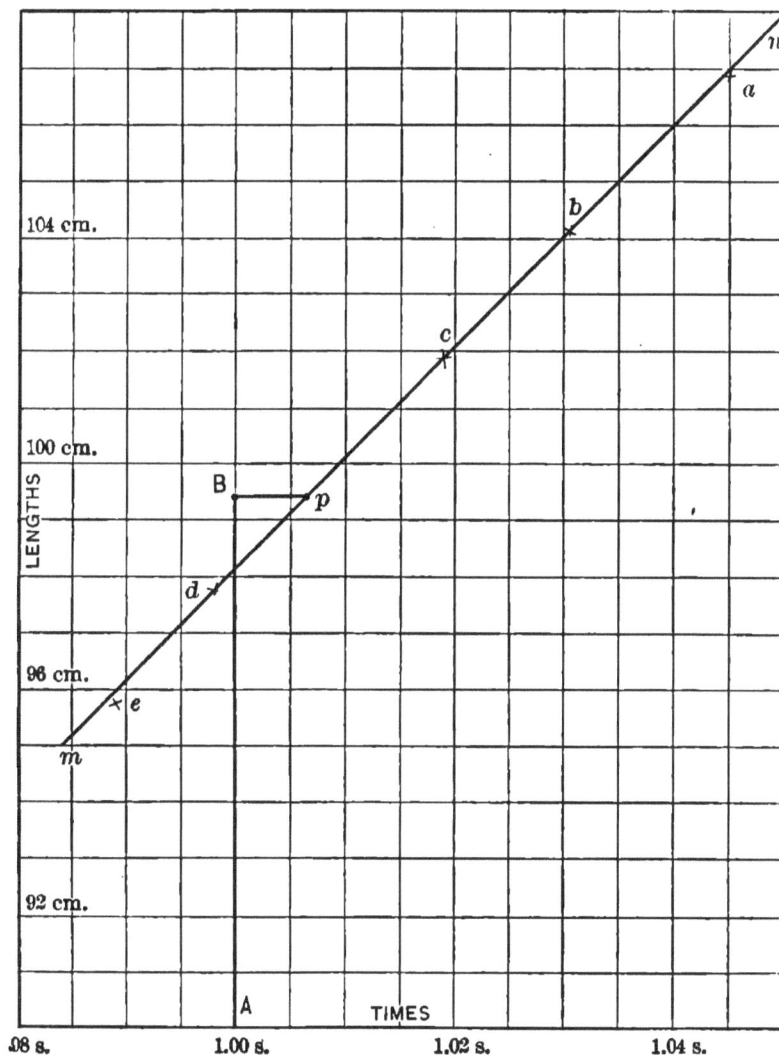

Fɪɢ. 34.

Draw such a line through the group of observation points, after the latter have been marked upon the paper. Note the reading of the intersection of this line with the ordinate corresponding to 1 second.

E

This reading is the length of the pendulum beating in unison with the metronome, as determined by this experiment.

Systematic Errors. — The divergence of the observation points a, b, c, d, etc., from line mn is called in each case the *error of observation* or the *accidental error.* There may be sources of error, however, which would cause the entire group to have too large or too small values. Such errors are called *systematic errors.* In this experiment, for example, the metronome might be running very fast or very slow, so that it did not beat exact seconds; or the scale might not be a true meter, or the observer might measure from the wrong level at the top of the pendulum. Or he might use a wrong value for the radius of the ball in correcting the lengths. All of the above would affect the entire series of observations. In the present case we can detect the size and direction of the first of these systematic errors, that due to the rate of the metronome, by computing the length of a pendulum which would beat seconds, and representing the result graphically upon the same sheet with the experimental curve. For this purpose we use the formula

$$t = \pi \sqrt{\frac{l}{g}},$$

which solved for l gives us

$$l = \frac{t^2 g}{\pi^2}.$$

A value of g nearly enough correct, in most localities, for the purpose of this computation is $g = 980.5$ cm., which is the acceleration due to gravitation in latitude 45° at the level of the sea.

This gives for l, since $t^2 = 1$, and $\pi^2 = 9.870$,

$$l = \frac{980.5}{9.870} = 99.34 \text{ cm.}$$

Represent this result upon the cross-section paper by means of a vertical line AB, extending from the point (A) upon the base line which corresponds to 1 second, to (B) at the height corresponding to 99.34 cm.

Draw the horizontal line Bp. This, expressed in *seconds*, will give the error of the metronome. For example, in the case given for illustration in Fig. 34. One true second $= 1.007$ seconds by the metronome, whence we conclude that the period of the latter is $\frac{1.000}{1.007} = 0.9993$.

CHAPTER VI

THE PHYSICAL PENDULUM

45. EXPERIMENT 10. — To find by Experiment the Center of Oscillation.

Apparatus:

(1) The pendulum stand and one of the pendulums with leaden ball described in Experiment 5.

(2) A physical pendulum consisting of a straight cylindrical rod or bar, preferably of iron or brass, about one meter long. It should be of the same material throughout. The axis of suspension of this pendulum consists of a piece of steel wire, pointed, as described in Experiment 7. It should be soldered to one end of the cylinder, especial pains being taken that the axis of suspension passes through the axis of the cylinder and is perpendicular to the same (Fig. 35).

The cylindrical pendulum may be suspended from a joiner's clamp, prepared as in Experiment 7.

Procedure:

(*a*) Hang the simple pendulum and the physical pendulum side by side (Fig. 36), and adjust the length of the former until the two swing as nearly as possible in the same time.

(*b*) Measure the length of the simple pendulum as prescribed in Experiment 5, and make the correction for the true value of *l*. That length, measured from the axis of suspension of the physical pendulum along the axis of the cylinder, determines the position of its center of oscillation.

Measure the length of the cylinder and state the distance F<small>IG</small>. 35. (*l*) of the center of oscillation from the axis of support in terms of the total length, *L*.

Compare your result with the position which would be given by

computation under the assumption that the cylinder is homogeneous.[1]

The above experimental method is applicable to any physical pendulum, whatever its form. The cylindrical bar is selected because
the result can be checked by computation.

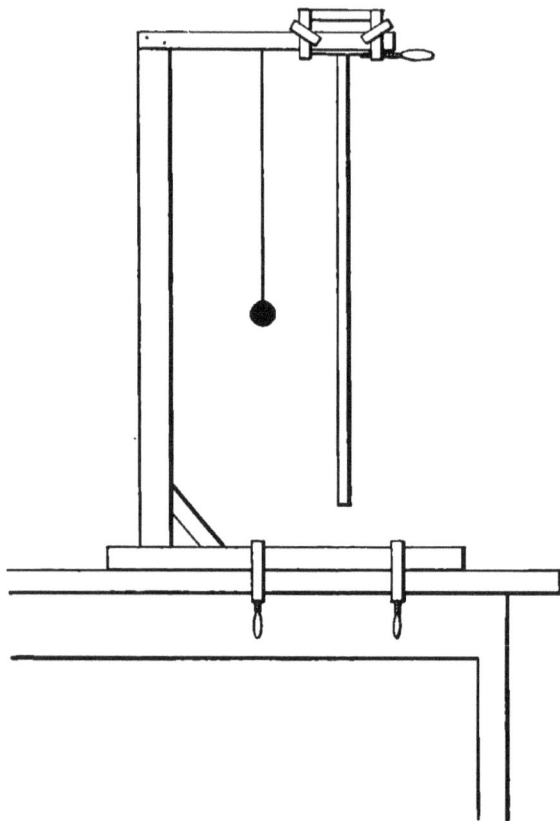

FIG. 36.

[1] The distance between axis of suspension and center of oscillation of
a thin bar swinging from one end is

$$l = \tfrac{2}{3} L.$$

The formula is $l = \dfrac{K}{mx}$, where K is the moment of inertia of the bar,
m is its mass, and x is the distance of the center of mass from the axis of
suspension. K, however, is $m\dfrac{l}{3}$, and x is $\tfrac{1}{2} L$. (See Nichols and Franklin,
Elements of Physics, Vol. I, pp. 66 and 75.)

**46. Further Analysis of the Motion of the Pendulum:
Simple Harmonic Motion.** — The motion of the pendulum
follows very closely indeed the law of what is known as
simple harmonic motion.

This is the motion which a point a, traveling back and
forth along the diameter of a circle (Fig. 37), would
make, were it to follow the movement of a point b upon
the periphery in such a manner that while the latter
travels on the circumference with a uniform speed, the
line which connects it with a
is always perpendicular to the
diameter cd.

Evidently the speed of a will
be variable, rising from zero at
c, and at d, when b, which con-
trols its motion, is traveling
at right angles to cd, and reaching a maximum speed when-
ever it passes the center o, at
which times b moves parallel
to the diameter cd.[1]

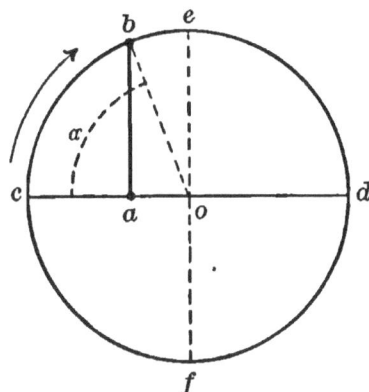

FIG. 37.

The harmonic motion of a is indeed of the nature of an
oscillation or vibration about the position of equilibrium o.
Its characteristics are those of the motions of the pendulum, the tuning fork, etc.

[1] Simple harmonic motion may be defined as "the *projection* upon a
fixed line of uniform motion in a circle" (*Elements of Physics*, Vol. I,
p. 46). The speed of a in terms of b may be expressed by means of the
equation

$$v_a = v_b \sin a,$$

where v_a is the speed of a, v_b the linear speed of b, and a is the angle
which the radius bo makes with cd. (Fig. 37.)

The position of a with reference to o is given by the equation

$$x = r \cos a,$$

where ao is the distance required, and r is the radius of the circle.

47. The Curve of Sines. — If we imagine the paper in Fig. 37 moved uniformly along under the circle and at right angles to the diameter $c\,d$, and the point a capable of tracing its course upon the paper, the result will be a curve characteristic of simple harmonic motion. This curve, which is called the *curve of sines*, is given in Fig. 38. It describes very completely such a motion as that of a pendulum. Times, for example, are measured along the line OY. The intervals y_1y_3, y_3y_5, y_5y_7, etc., correspond to the single period of oscillation, while the intervals $y_2\ y_6$, etc., mark the double or complete period.

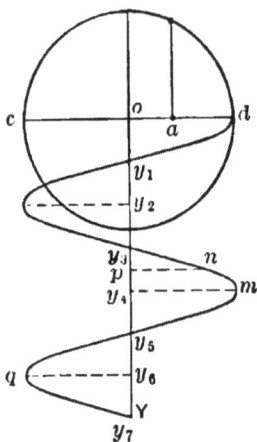

FIG. 38.

Lines like $y_4\ m$, $y_6\ q$, etc., drawn at right angles from OY to the crest of the curve, measure the amplitude, while a line $p\ n$, drawn from a point p to the curve, gives the position of the pendulum with reference to its position of equilibrium at any time t, which the point p represents.

48. The Tracing of a Sine Curve by Means of a Pendulum. — That the motion of a pendulum is in fact of the character just described, may be shown as follows:

A pendulum is mounted so as to swing transversely above a plate of smoked glass (Fig. 39). The bob of the pendulum must have a vertical hole, coincident with its axis, within which plays a stylus. •(See Fig. 40.)

The experiment consists in placing the plate beneath the pendulum in such a position that the stylus, when at rest, touches the smoked surface near the middle of the shorter edge. The longer diameter of the plate must lie

at right angles to the plane of vibration of the pendulum.
The pendulum is now given an amplitude of about 5 cm.,

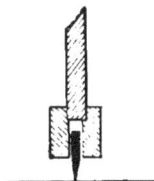

FIG. 40.

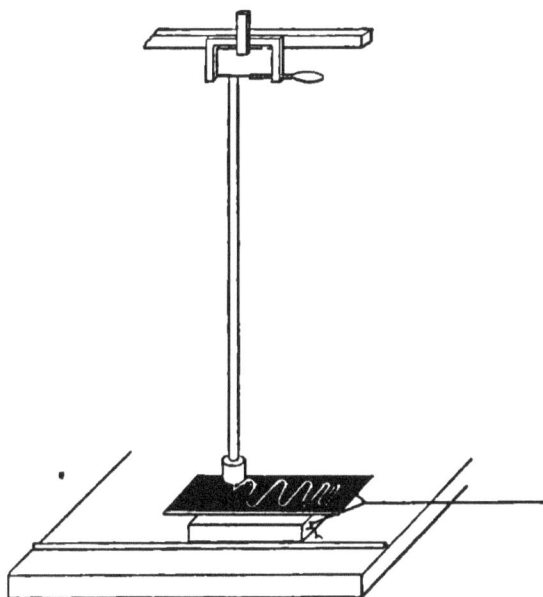

FIG. 39.

and after it has completed a few excursions, the plate is
drawn under it with as steady a motion as possible. The
path of the stylus will be found to consist of a curve varying from the curve of sines only in so far as the motion of
the plate varies from uniformity of direction or speed.

49. The Measurement of Gravitation. — The first direct
experimental evidence that the earth's attractive force
differs on different parts of the surface of the planet,
appears to be due to the astronomer Richer, who in 1672
took a clock from Paris to Cayenne for the purpose of
certain observations. He found that the timepiece lost
about $2\frac{1}{2}$ minutes daily in its new locality. The pendulum was shortened sufficiently to restore it to its normal

rate, and when later it was brought back to Paris it was found to be gaining $2\frac{1}{2}$ minutes.[1]

This result, which is quite in accordance with our conception of the nature of gravitation and our knowledge of the shape of the earth, has since been verified by determinations of the value of g in various parts of the world.

For the measurement of gravitation the pendulum is by far the most accurate instrument. By the method of coincidences the period of oscillation can be found with extraordinary precision. The determination of the corrected length (l) with a corresponding degree of accuracy is a more difficult matter. Originally a simple pendulum with a heavy spherical bob was used. Borda, with such an instrument, found the length of the seconds pendulum in Paris to be 99·332 cm., and the acceleration due to gravitation in that place, 980·96 cm. Later, Captain Kater of the British Navy (1818) introduced a pendulum based upon the principle of the interchangeability of the centers of suspension and oscillation.[2] This instrument, which is called the *reversion pendulum*, has two sets of adjustable knife-edges. These are so placed that when the pendulum swings from one of them, aa (Fig. 41), the other, bb, is as nearly as may be at the center of oscillation. The rate in this position having been found, the instrument is suspended from bb. Any change in rate indicates that bb requires adjustment. When, finally, the pendulum swings with the same

FIG. 41.

[1] Lommel, *Experimental Physik*, p. 71.

[2] For the demonstration of this principle, which was discovered by the Dutch physicist Huygens (1673), see *Elements of Physics*, p. 75.

period in both positions, the distance *aa* to *bb* is the length (*l*). All the quantities in the formula

$$t = \pi \sqrt{\frac{l}{g}}$$

are then known, excepting *g*.

The figure shows in cross-section a standard form of the reversible pendulum. It is designed to give complete symmetry of outward form, with the concentration of mass chiefly in one end.

By means of such pendulums, precise determinations of gravity have been made in many parts of the world. In the following table are given some of the values thus obtained.

TABLE I.

VALUES OF THE ACCELERATION DUE TO GRAVITATION.

Locality.	Latitude.	Elevation above the sea.	Value of G. (not reduced to sea-level).
Boston, Mass.. . .	42° 21' 33"	22 meters	980·382 cm.
Philadelphia, Pa. .	39° 57' 06"	16 "	980·182 "
Washington, D.C. .	38° 53' 20"	10 "	980·100 "
Cleveland, O.. . .	41° 30' 22"	210 "	980·227 "
Cincinnati, O.. . .	39° 08' 20"	245 "	979·990 "
Chicago, Ill. . . .	41° 47' 25"	182 "	980·264 "
St. Louis, Mo. . .	38° 38' 03"	154 "	979·987 "
Kansas City, Mo. .	39° 05' 50"	278 "	979·976 "
Denver, Col. . . .	39° 40' 36"	1638 "	979·595 "
San Francisco, Cal..	37° 47' 00"	114 "	979·951 "
Greenwich	51° 29' 00"	47 "	981·170 "
Paris	48° 50' 11"	72 "	980·960 "
Berlin	52° 30' 16"	35 "	981·240 "
Vienna	48° 12' 35"	150 "	980·852 "
Rome.	41° 53' 53"	53 "	980·312 "
Hammerfest . . .	70° 40' 00"	——	982·580 "

CHAPTER VII

KINETIC ENERGY; POTENTIAL ENERGY; WORK

50. Work. — The study of the motion of the pendulum
leads us to a very important topic :
that of *energy*. When we wish to
start a pendulum into oscillation,
we first lift the bob from its position of rest *a* (Fig. 42), into a
new position *b*, causing it to follow
the circular path in which the
pendulum is constrained to move.
In the language of physics, when
a body is moved from one position
to another against any force, *work*
is done upon it.

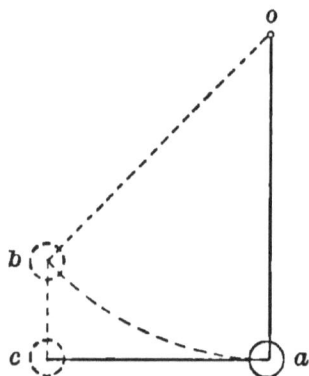

FIG. 42.

51. Work; how measured. — Work is measured by the
product of the force against which the motion takes place
and the distance through which the body moves against it.

In this case the force is gravitation, and if the bob of
the pendulum have a mass m, the force of gravitation acting upon it is mg.

It cannot be said, however, that the bob has been moved
along the whole path ab in opposition to gravitation. We
must resolve the path ab into two components, ac (horizontal), and bc (vertical). To the movement along ac,
gravitation offers no opposition, and no work against that

force need be done; along the vertical component *bc*, however, work must be done, and it is this vertical component only which we have to consider. The work in the case of the pendulum may, therefore, be expressed by the equation

$$W = mgs,$$

where *m* is the mass of the bob, *g* the acceleration due to gravitation, and *s* the vertical component (*bc*) of the distance through which it has been moved.

52. Energy; Kinetic and Potential. — Suppose the pendulum bob, at *b*, to be released. It will soon regain its former level, but in so doing it will acquire *velocity*. Because of its velocity it will possess what is called *energy of motion* or *kinetic energy*.

In the case of the pendulum, the bob is not free to fall vertically along the path *bc*, but in following the circular path from *b* to *a* it acquires precisely the same velocity and the same amount of kinetic energy.

This energy it is which enables the bob to rise against gravitation on the other side of its position of equilibrium. As it rises towards the turning point it loses velocity and, consequently, kinetic energy. What is lost in kinetic energy is, however, gained in energy of another kind: a sort of latent or stored energy due to the position of the bob at higher and higher levels.

When the turning point is reached, the kinetic energy has disappeared. The pendulum is at rest. It is not in the condition in which it was at the beginning of the experiment, however; it is in the condition which it was after work had been done upon it to raise it to the position *b*.[1]

---

[1] Losses due to atmospheric resistance and to the existence of other forces than that of gravitation are not taken into account.

The energy of position thus gained is called *potential energy.* Potential energy may be stored in bodies in a variety of ways other than that of lifting them from a lower to a higher level. Whenever a body is forced into a strained position, as when a spiral spring is stretched or compressed, or a bow is bent, or even when the parts of a body are placed in unstable atomic or molecular relations to each other, as in the case of gunpowder and other explosives, the body is imbued with potential energy. Such forms of potential energy, to distinguish them from the energy of mere position, are called *potential energy of configuration.*

Kinetic energy is conveniently measured by means of the mass and the velocity which it possesses. It is, however, as has been demonstrated by countless careful experiments, the precise equivalent of the work previously done in raising the bob to its position *b.*

This work is *mgs.* The quantity *s* is, however, $\frac{1}{2} v$, where *v* is the velocity after the bob has fallen to *c.* (See Arts. 25 and 26.) The quantity *g*, moreover, is itself equivalent to *v*, so that the kinetic energy which is equal to the work *W* may be expressed thus:

$$E = \tfrac{1}{2} mv^2 = mgs = W.$$

53. The Erg. — *The unit of energy and of work is the erg. It is the work done by a force of one dyne acting through a distance of one centimeter.* This is a very small quantity of work indeed. The earth exerts a force of from 980 to 981 dynes upon each grain of matter at its surface; and the definition of the erg refers, therefore, to the work of lifting a gram one centimeter against a force scarcely more than a thousandth as great as gravitation.

The Foot-Pound.—Engineers who use British units employ, as the unit of work, the work necessary to raise a pound weight avoirdupois one foot. This is called the *foot-pound.* Where metric systems prevail, the corresponding practical unit is the *kilogram-meter.*

In localities where $g = 980.5$, the foot-pound is 13,560,-000 ergs and the kilogram-meter is 98,050,000 ergs.

54. Energy of the Pendulum.—In attempting to express the energy of a physical pendulum, in terms of the mass and the velocity, the difficulty arises that the different particles of which the pendulum is composed are situated at various distances from the axis of suspension, and consequently possess different linear velocities.

We may regard the pendulum as made up of an infinite number of simple pendulums, all bound rigidly together and oscillating with a common·period (that, namely, of the simple pendulum, the mass of which lies in the center of oscillation).

While these simple pendulums vary as to linear velocity, they all swing through a given angle in the same time. The motion of a rotating body, or of a body oscillating about a fixed axis, can be expressed more conveniently by means of its rate of angular motion than by the various linear motions of its parts.

55. Angular Velocity.—The rate of rotary motion is expressed by means of the term *angular velocity*, which is the *rate of rotation.* For the uniform motion of rotation the angular velocity is simply the angle (in radians)[1]

---

[1] Radian: the angle which incloses an arc, the length of which is equal to the radius of the circle to which the arc belongs. Since there are in an entire circumference 2π radians, a body revolving uniformly at the rate of n revolutions per second possesses an angular velocity $\omega = 2\pi n$.

passed through in one second. In the case of variable rotary motion or of oscillations, we may conceive the body, at the instant for which the velocity is required, to be released from the action of all forces except those which compel its particles to follow their circular paths. The angle in radians through which it would then move (uniformly) in one second measures its velocity at the instant in question.

56. Moment of Inertia. — The energy of a rotating or oscillating body is proportional to the square of its angular velocity (ω), just as that of a body possessing motion of translation is proportional to the square of its linear velocity. To obtain the amount of energy in the former case, however, we must not multiply ω^2 by $\frac{1}{2}m$, but by another quantity, $\frac{1}{2}K$, which depends not only upon the mass, but likewise upon the distribution of the mass around the axis of rotation or of suspension. The quantity K in the equation

$$W = \tfrac{1}{2} K \omega^2,$$

which gives the kinetic energy of a rotating body, is called the *moment of inertia of the body about its axis of revolution.*

That K must depend upon the distance of the mass from the axis of rotation, as well as upon the mass itself, is obvious. Consider, for example, the case of two wheels, each of which weighs 10 kilograms. One of them has its mass chiefly in a rim, at a radius of 60 cm. from the axis, the other an equally large amount of matter in a rim with a radius of 20 c. To put these two wheels into revolution at the rate of 1 turn per second ($\omega = 2\pi = 6.283$), will take very different amounts of work, although the masses are the same and the angular velocities are the same.

In the large wheel the matter in the rim will be traveling about 360 cms. per second, in the small wheel only
about 120 cm. Their kinetic energies, if we may neglect
the masses of hubs and spokes, will be about as 9:1.
The large wheel in coming to rest will do about nine
times as much work.

57. EXPERIMENT 11. — Energy stored in a Fly Wheel.

Apparatus:

(1) Thirty or forty kilograms of the iron disk weights described in
Appendix III, one of them provided with a hook. Weights of 10 kg.
each are to be preferred.

(2) An iron wheel and axle. The wheel should be at least 25 cm.
radius, and rather light. The driving-wheel of an ordinary sewing
machine is a good pattern.

The wheel should be mounted upon an axle not less than 50 cm. in
length. A keyed shaft of steel is the best. The wheel should be near
one end of the shaft. The shaft must have substantial bearings of
wood or metal in which it runs smoothly. It is important that the
wheel be well balanced and the shaft straight.

(3) Five or six meters of stout cotton rope from 0·5 cm. to 0·8 cm.
in diameter. This should be soft, pliable, and capable of much
stretching without rupture. It must be strong enough to sustain
several times the weight employed in the experiment.

Procedure:

(*a*) The wheel is mounted between two strong tables to which the
bearings are firmly clamped (see Fig. 43). The weights are placed
upon the floor directly below the longer arm of the shaft. One end
of the rope is tied around a spoke of the wheel, the other to the hook
of the weights.

(*b*) The operator turns the wheel from him with one hand, guiding
the rope so as to wind it snugly upon the shaft until all the slack is
taken up. He then draws upon the rope with a steady but increasing
motion, thus setting the wheel into briskly accelerated revolution.
When the rope is all unwound, the wheel, now at high speed, begins
to rewind it in the opposite sense. The operator still holds the slack
of the rope, giving sufficient tension to secure a snug, smooth wind-

ing. Just before the last of the slack is taken up he releases the rope. The wheel, still running with a considerable velocity, will be seen to raise the heavy weights, perhaps fifteen or twenty times its own mass, bodily from the floor and to a height of many centimeters.

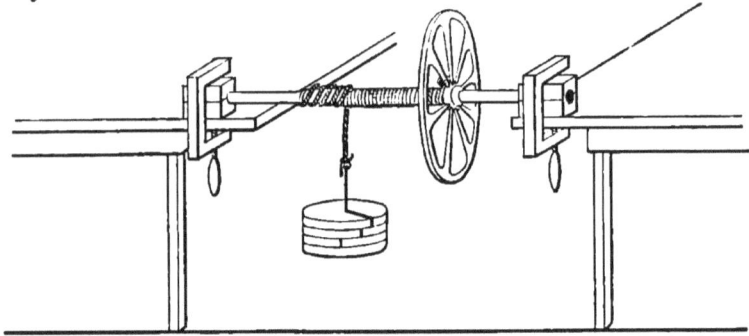

FIG. 43.

58. The Conservation of Energy. — In these changes, as regards energy, which the bob of a pendulum undergoes in the course of its oscillation, we have one of the simplest examples of what is called the *transformation of energy*. Energy changes continually from the kinetic to the potential form, and *vice versa*, and it is *transferred* from one body to another; but it can neither be created nor destroyed. This fact is expressed in what is known as the *law of the conservation of energy*, which may be stated thus:

The sum of the potential and kinetic energy of a body always remains constant excepting as the body may receive energy by the action upon it of forces from without (viz. forces due to the action of other bodies upon it), or as it may give up energy to other bodies by doing work upon them.

Left to itself, in other words, a body or system of bodies never gains or loses energy. If its amount of energy is increased, it is at the expense of the energy of other neighboring bodies. If it loses energy, neighboring bodies gain a like amount. The sum total of energy in the universe is constant, but it is transferred from body to body and is transformed continually.

CHAPTER VIII

MACHINES

59. A machine, in mechanics, is a device for the advantageous application of force.

Simple machines are those in which a single device is used. The most important of these are:

The pulley.

The lever in its various forms.

The inclined plane and its modifications.

60. The Pulley. — A single pulley (Fig. 44) is simply a device for changing the direction of a force. It is a form of the lever with equal arms, as will be seen when that device is considered (Art. 61). By the combination of pulleys, however, the principle of work is introduced, and large forces are obtained by the intelligent application of a small one. Figure 45 shows the simplest case; that of one fixed pulley *A*, and a movable one *B*.

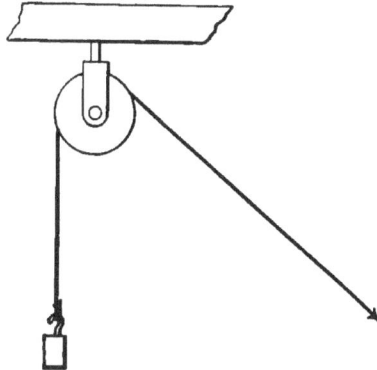

FIG. 44.

The effect of this combination may be gathered from the following considerations:

F

(1) The stress upon the rope must be everywhere the same; consequently the downward force at P_2, necessary to produce equilibrium, must be twice as great as that at P_1. (The latter force is balanced by the stress upon the rope; the former, by twice that stress, *i.e.* by the stress in the branch b + the stress in the branch c.)

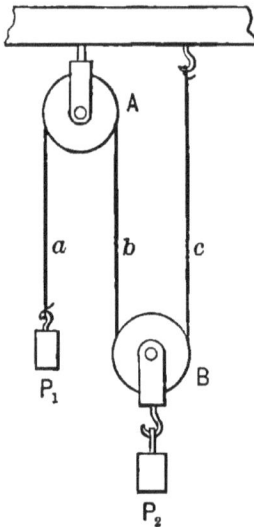

(2) Since the rope remains taut, every vertical movement of P_1 will be accompanied by one half as great a movement of P_2.

(3) The force of the earth upon P_1 is mg where m is the mass of P_1, so that the body at P_2 (including the pulley) must have a mass $2\,m$, which will exert under the action of the earth a force $2\,mg$, or twice that at P_1.

Fig. 45.

(4) The work done by P_1 in falling n centimeters is mgn. The work done upon P_2 to raise it $\dfrac{n}{2}$ centimeters is $2\,mg \times \dfrac{n}{2} = mgn$.

This simple method of reasoning will serve in the analysis of any problem dealing with systems of pulleys.

It will be seen from the above that there is *no saving of work by this device, a statement which is true of all machines, however complicated they may be.* On the other hand, there are losses due to friction, etc., of which no account has been taken, and these losses sometimes consume a considerable proportion of the work done.

61. The Lever. — In its simplest form the lever is a rigid bar supported at a point (Figs. 46 and 47) called the *fulcrum.* At two points, *a* and *b*, upon the bar, forces acting at right angles to its axis are applied. These are opposed to each other; that is to say, they tend to produce

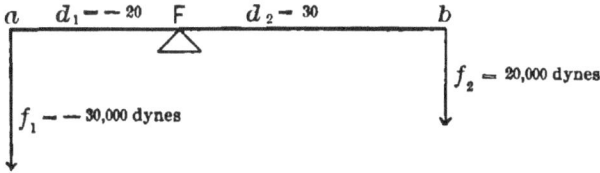

FIG. 46.

rotation around the fulcrum in opposite directions. The distances d_1 and d_2 from the points of application, *a* and *b* respectively, to the fulcrum, are called the *lever arms.*

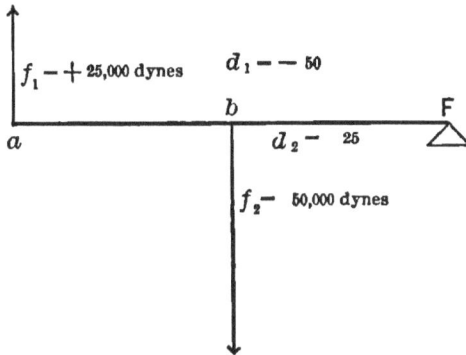

FIG. 47.

It is convenient to distinguish two classes of levers.

(1) Those in which the fulcrum lies between the points of application.

(2) Those in which the points of application are both on the same side of the fulcrum.

62. The Moment of a Force *is the product of the force into its distance from the fulcrum.*

The moment of the force f_1, for example, the point of application of which is at *a*, is $f_1 d_1$; that of f_2 is $f_2 d_2$.

63. Law of the Lever (applicable to all cases of both classes and to all combinations of levers).

The sum of the moments of the forces must be equal to zero.

(In the above statement it is the algebraic sum which is referred to; lever arms and forces being counted positive or negative according to their direction.)

64. EXAMPLE. — In Fig. 46 let lever arms to the right of the fulcrum and forces acting upwards be positive. Let f_1 be 30,000 dynes, f_2 20,000 dynes, d_1 20 cm., and d_2 30 cm.

The moment of f_1 is

$$f_1 d_1 = -30,000 \times -20 = +600,000,$$

and the moment of f_2 is

$$f_2 d_2 = -20,000 \times +30 = -600,000.$$

Their sum is

$$f_1 d_1 + f_2 d_2 = 0,$$

which is the equation of equilibrium.

In Fig. 47 we may assume

$$d_1 = -50 \text{ cm.,}$$
$$d_2 = -25 \text{ cm.;}$$
$$f_1 = +25,000 \text{ dynes,}$$
$$f_2 = -50,000 \text{ dynes.}$$

The same condition will be fulfilled.

65. Case of a Body Free to revolve upon a Fixed Axis. — The principle of the lever may be extended to the case of a body of any shape to which forces are applied, provided the body is free to revolve around a fixed axis.

The straight line perpendicular to the axis, and connecting the point of application of a force with the latter, is the lever arm of that force. If the force does not act at right angles to the lever arm, only its component perpendicular to the arm is to be taken in finding its moment.

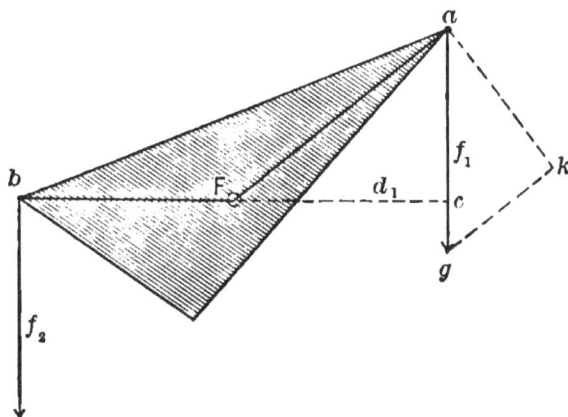

FIG. 48.

In Fig. 48, F is the fulcrum, or axis of revolution. At a and b the forces of f_1 and f_2 are applied. The former is not perpendicular to its lever arm $\overline{Fa}$.

To find the moment of f_1, we find its component ak, perpendicular to the arm, and take the product

$$\overline{Fa} \times \overline{ak} = \text{moment}.$$

Another and simpler method is to drop a perpendicular d_1 to the force f_1 and take the product

$$d_1 f_1 = \text{moment of } f_1.$$

The result is the same.[1]

---

[1] The identity of the two methods is readily shown.
Since $\overline{ak} = f_1 \cos \beta$, we have

$$\overline{Fa} \times \overline{ak} = \overline{Fa} f_1 \cos \beta.$$

Since $d_1 = Fa \cos a$, we have likewise

$$d_1 f_1 = \overline{Fa} f_1 \cos a.$$

The triangles Fac and akg, however, are similar and a is equal to β;

$$\therefore d_1 f_1 = \overline{Fa} \times \overline{ak}.$$

66. The Principle of Work applied to the Lever. — The truth of the statement already made with reference to all machines, that no work is gained, is obvious in the case of the lever.

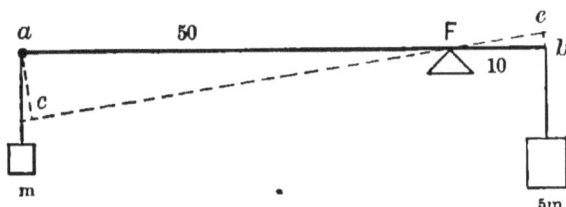

FIG. 49.

Let the weights m and $5\,m$ be balanced upon a bar ab (Fig. 49). The lever arms are therefore as $5:1$.

If sufficient additional weight be added at a to overcome the friction, and m move down through a distance equal to ac, the weight at b will be raised vertically through the distance bc. The triangles Fac and Fbe, however, are similar, and

$$\overline{ac} : \overline{bc} :: 5 : 1.$$

Let the distance be be 1 cm.

The force at a is mg dynes, the distance traversed is 5 cm., and the work

$$W = 5\,mg \text{ ergs.}$$

The work done upon the weight at b is the product of 1 cm. and $5\,mg$ dynes, which is the precise equivalent of the above.

67. The Pulley and the Wheel and Axle considered as Levers. — The ordinary lever is of advantage where forces are to be balanced against one another, or for small movements, rather than for the movement of masses through considerable distances. As the lever arms are turned the moments of the forces change. Thus in Fig. 50 the moments of f_1 and f_2 are $\overline{aF} \cdot f_1$ and $\overline{bF} \cdot f_2$ when the arms are horizontal. When turned into the position a_1b_2, however, these are reduced to d_1f_1 and d_2f_2.

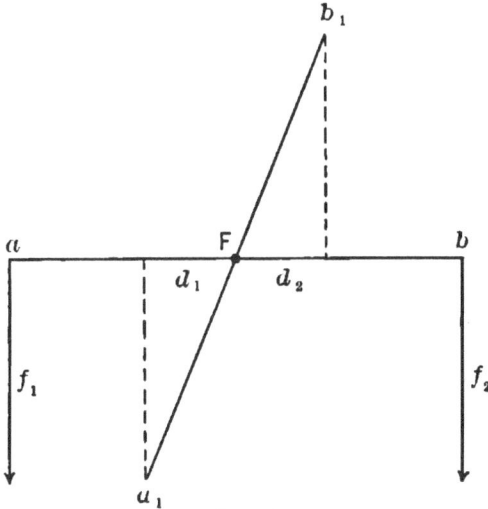

Fig. 50.

The pulley is a form of lever in which this defect is remedied. The effective lever arms d_1 and d_2 (Figs. 51 and 52) are radii of the wheel drawn to the points upon

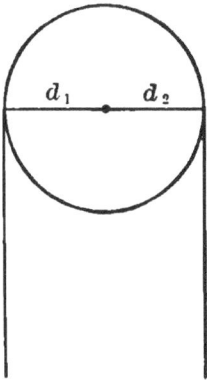

Fig. 51.

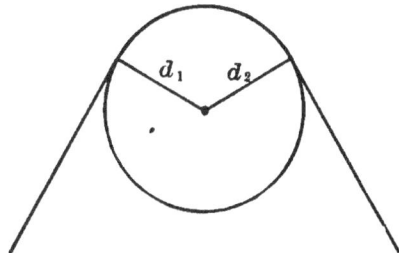

Fig. 52.

the periphery at which the rope leaves the pulley. These arms are of unvarying length, however great the movement may be.

The common pulley is a lever with equal arms; the *wheel and axle* (Fig. 53), however, combines the advantages of *constant moments* and of unequal arms.

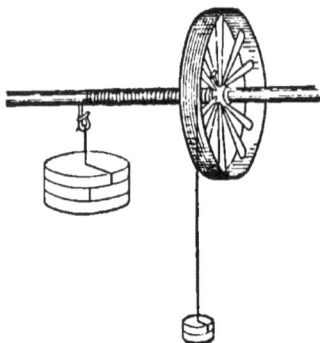

68. The Inclined Plane. — The inclined plane and its modifications, the screw and the wedge, present a slightly different method of overcoming a large force by means of a smaller one which acts through a correspondingly greater distance.

FIG. 53.

Suppose it to be required to raise the ball B (Fig. 54), which rests upon a horizontal surface, through a vertical distance n.

The work required is $n\,mg$, where m is the mass of B. Along the vertical path the force would be mg, but by roll-

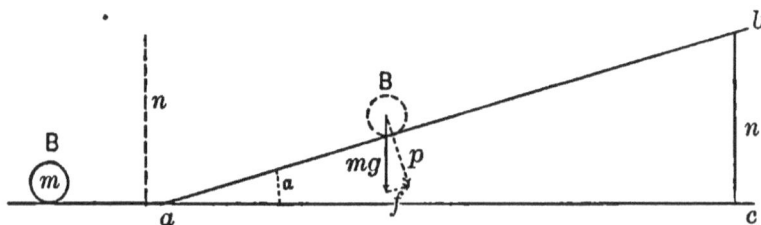

FIG. 54.

ing the ball along an inclined plane of proper pitch, any given smaller force f may be made to carry it to the required height. The work expended is the same as before, plus that necessary to overcome friction.

To find the force necessary to carry the ball up a given plane ab, suppose it placed upon the plane, and the force mg with which the earth attracts it resolved into two components. Of these p is perpendicular to the plane and produces pressure, but no motion, while f, parallel to the plane, is the required force.[1]

The work done by f is

$$f \times \overline{ab} = n\,mg.$$

It is the precise equivalent of that required to lift the ball vertically.

69. The Relation of the Screw to the Inclined Plane may be shown by rolling a triangular strip of paper aof (Fig. 55) about a cylinder. The diagram needs no elucidation.

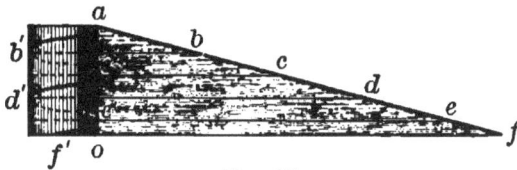

FIG. 55.

Figure 56 shows two well-known forms of screw threads and half of a nut or screw-bearing in which the screw plays; the latter, when fixed, takes up the thrust or reaction of the axial motion of the screw.

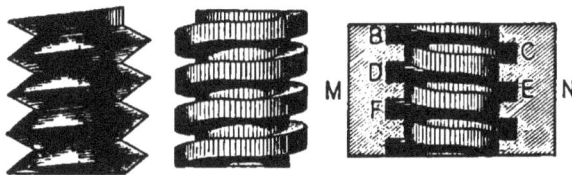

FIG. 56.

[1] Since the triangle of forces is similar to abc (Fig. 54), the value of f is
$$f = mg \sin a$$
where a is the angle which measures the pitch of the plane.

The following are some of the chief advantages of the screw:

(1) It may be given a motion of translation, in the direction of its axis, by the application of a rotary motion. The action of the screw may, therefore, readily be made *continuous*, which is a great advantage in any mechanical device.

(2) It is a form to which great strength is easily given, so that by the use of a long lever arm an enormous force can be exerted in the direction of the screw axis.[1]

---

[1] The screw affords a good example of the law of reaction (third law of motion) and of the nature of force considered as a stress. The backward thrust upon the nut of a screw, which is exerting pressure in the direction of its axis, is precisely equal in amount to the pressure exerted and opposite in direction.

CHAPTER IX

THE BALANCE

70. Forces in Equilibrium. — Bodies near the surface of the earth, or indeed in proximity to matter of any kind, are continually acted upon by forces. When a body is at rest, it is because the forces acting upon it are balanced against each other.

71. Conditions of Equilibrium. — (1) Case of a body free to revolve about a point or an axis of rotation. The condition of equilibrium is that stated in a previous section, viz. the sum of the moments of the forces must be zero. (2) Case of a body sustained in any manner against the attraction of the earth (or against any parallel forces). The forces may be regarded as supplanted by a single force (*mg*) acting at the *center of mass* (sometimes called the *center of gravity*).

In regular homogeneous solids, the center of mass is the geometrical center of the body. (The method of determining the position of the center of mass of any irregular body is given in Art. 74.)

72. Stability. — It is usual to describe equilibrium as STABLE when any motion to which the body can be subjected, without change of the level of the support, will *raise the center of mass;* as INDIFFERENT when such motions will *neither raise nor lower the center of mass;* and as UNSTABLE when such motions will *lower the center of mass.*

73. Examples. — (*a*) *Point of support above the center of mass.* Equilibrium will be had only when the center of mass is vertically below the point or axis of support. The only motion of which the body is capable is that of revolution, and the center of mass will rise (see Fig. 57). If the body be turned from its position of equilibrium, as shown in the figure, the force *mg* acting upon the center of mass will always have a component tending to restore it to that position.

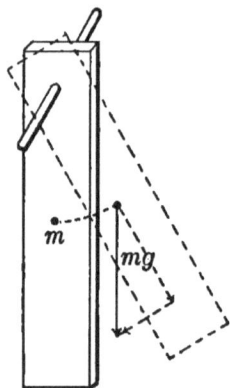 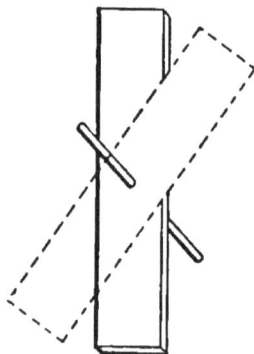 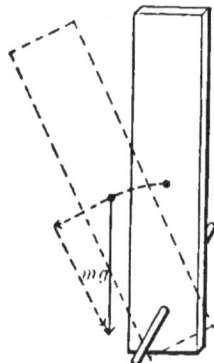

FIG. 57.   FIG. 58.   FIG. 59.

(*b*) *Center of support at the center of mass* (axis of support passes through center of mass) (Fig. 58). The body will be in indifferent equilibrium. It may be turned about the axis at will without raising or lowering the center of mass. It has no tendency to return to any given position.

(*c*) *Support below the center of mass.* If the support be at a point or along an axis, as in Fig. 59, the body will be in *unstable* equilibrium when the center of mass is vertically above the support. Any movement will *lower* the center of mass, and if the body be in the slightest degree displaced from the position of equilibrium, the force *mg* will have an active component which tends to increase the displacement.

Stable equilibrium with support below the center of mass can be obtained by the use of three points of support. A cylinder lying upon its side (Fig. 60, *a*) is in stable equilibrium as regards motion in the direction of its axis, and in indifferent equilibrium to motions at right angles to its axis. If it be flattened, and lie with its

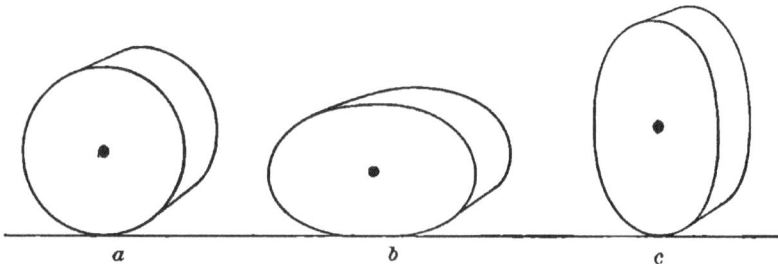

Fig. 60.

major diameter horizontal (*b*), it is in stable equilibrium. If the major axis is vertical (*c*), there will be stability in the axial direction and instability at right angles to the same.

74. EXPERIMENT 12. — To find the center of mass of an irregular solid.

This experiment depends upon the principle stated under Example (*a*), (Art. 73), viz. that a suspended body comes into stable equilibrium with its center of mass vertically below the center of support.

Apparatus:

(1) A large open basket.

(2) A plumb-line, with pointed bob.

(3) The pendulum stand described in Experiment 5.

(4) A kilogram weight, some strong twine, and some double hooks of wire; a thread and needle.

Procedure:

(*a*) Suspend the basket from the arm of the pendulum stand by means of the hook and twine, in any position such that the line of suspension, continued, would pass through some portion of the basket mesh. By means of the plumb-line, which is to be suspended from the lower branch of the hook which holds the basket (Fig. 61), care being taken that it is so placed as to form the continuation of the line of suspension, determine the point where the latter penetrates the mesh. (This can be quite accurately done by lowering the bob until its point is nearly in contact with the wall of the basket.)

(*b*) Mark the above determined point, also that in which the plumb-line, continued, would pierce the handle of the basket behind

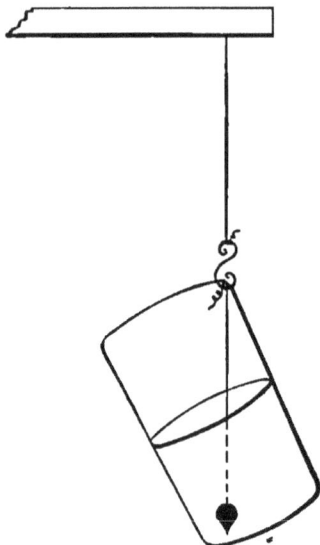

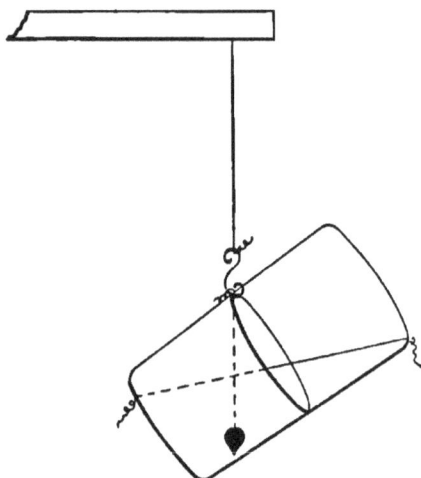

FIG. 61.   FIG. 62.

the hook. Take the basket down, and, by means of the needle, stretch a thread so as to mark the position which the vertical line from the center of support had occupied, *i.e.* the position of the plumb-line.

(*c*) Repeat the above operation with the basket suspended in quite a different position (Fig. 62). If the two trials have been carefully made, the two threads will cut one another, or very nearly so. *Since the center of mass lies upon both threads, it must be located at their intersection.* Note that the center of mass lies at a point not occupied by any of the material of which the basket is composed.

(*d*) Fasten the weight in one corner of the basket and repeat the experiment. Note that the center of mass has shifted.

75. The Balance. — The balance is a lever with equal arms, which is used for the comparison of masses.

Few instruments have been brought to a higher degree of accuracy and delicacy. With a good balance, for example, it is possible to detect with certainty differences between two masses amounting to less than one part in a million.

Balances of precision owe their delicacy chiefly to the fact that the center of mass is very near the axis of support. If m_1 (Fig. 63) be the center of mass of a balance which oscillates about the center o, and if a small excess of weight upon the left-hand pan turn the beam through an angle a, the entire mass, which we may consider as concentrated at the center of mass, is moved to n_1. The vertical distance through which it is lifted is $\overline{n_1 p_1}$, with an expenditure of work

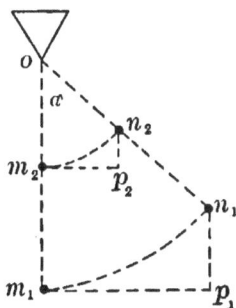

FIG. 63.

$$W_1 = \overline{n_1 p_1} \times mg.$$

If the center of mass is at m_2, nearer to o, the same angle corresponds to a displacement of m_2 to n_2, with a vertical rise $\overline{n_2 p_2}$ and work

$$W_2 = \overline{n_2 p_2} \times mg.$$

The work in the two cases, to which the sensitiveness of the balance is inversely proportional, is in the ratio

$$W_1 : W_2 :: om_1 : om_2.$$

The important features in the construction of the balance are indicated in Fig. 64.

The *beam* is made of a single piece of metal, designed with a view to great stiffness and to minimum of mass.

There are three wedges, technically called *knife-edges:* a
central one which forms the axis of support, and one at
each end, upon which the scale pans hang freely by means
of stirrups. The knife-edges are accurately ground, and
are constructed of hardened steel or sometimes of agate.
They rest against plane bearings of agate. The three
knife-edges are placed in the same horizontal line, and

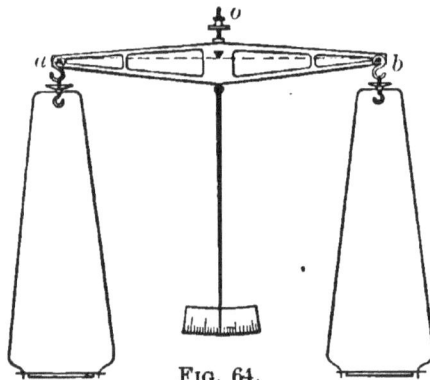

Fig. 64.

this line is located just above the center of mass. A long
pointer, which swings in front of a scale, indicates the position of the beam. A short upright rod, vertically above
the knife-edges, is threaded and carries a milled nut. By
raising and lowering this nut, the center of mass can be
raised and lowered, and the sensitiveness of the balance
changed.

The advantages gained by this method of construction
are as follows:

(1) Since the scale pans, with their contents, hang
freely from the knife-edges, we may consider their masses
as concentrated at *a* and *b*, and since these points are at
the same level with *O* and approximately with *m* the
center of mass, *the balance may be loaded without sensibly
shifting the center of mass.*

(2) By the use of knife-edges two points are gained:
(*a*) The movements of the beam are rendered almost frictionless.

(*b*) The lengths of the arms are accurately fixed and are rendered unchangeable.

The latter point (*b*) is of especial importance, because it is the *moments of the forces* acting at *a* and *b* that are compared. These are the products

$$\overline{Oa} \times m_a = \overline{Ob} \times m_b,$$

where m_a and m_b are the masses balanced against each other. Now Oa and Ob must be strictly unchangeable and equal (or of known ratio), in order that we may deduce the relation of the masses with certainty.

So accurate is the workmanship of the best balances, that it is not uncommon to find that the arms differ in length by only a few millionths.

(3) Sensitiveness of action is obtained by bringing m and O as near together as it is possible to do without losing the stability of equilibrium, and by reducing the mass of the beam to a minimum. The use of a long pointer makes it possible to detect very small movements of the beam.

76. EXPERIMENT 13. — **The Balance.** **Relation between Length of Arm and Sensitiveness.**

Apparatus :

(1) A wooden strip or beam 60 cm. long and 3 cm. × 1 cm. in cross-section (Fig. 65). Through a transverse hole in the same, midway between the ends, is inserted a three-cornered file or other triangular bar of steel, one edge of which is to serve as a central knife-edge for support. This edge is perpendicular to the principal axis of the beam, and so situated that when the latter is swinging, its center of mass will be just below the support. Along the axis of the beam are bored

G

a set of transverse holes, a_1 a_2, b_1 b_2, c_1 c_2, d_1 d_2, e_1 e_2, each about 2 mm. in diameter. These are situated to the right and left of the center of mass, at distances 5 cm., 10 cm., 15 cm., 20 cm., and 25 cm., respectively.

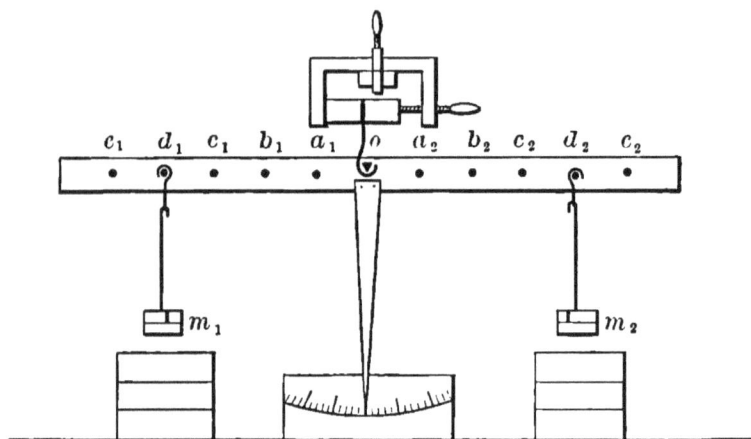

FIG. 65.

A set of short, straight, metallic rods or pins are fitted within these holes, their centers lying in the axis of the beam.

The beam hangs from a stirrup of stiff iron wire, of the form shown in Fig. 66. Two similar stirrups support the weights m_1 m_2 as in that figure. These smaller stirrups must be adjusted to equal one another in mass by filing.

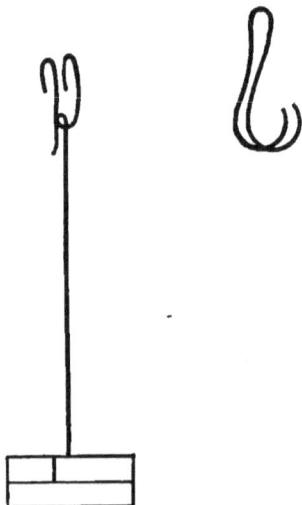

FIG. 66.

The beam is provided with a light stiff pointer, the tip of which is vertically below the center of mass when the beam is horizontal. The pointer moves in front of a circular scale with equal divisions. (Pointer and scale may be constructed of cardboard. The former, if of metal, must be as light as is consistent with stiffness.)

Under each of the weights blocks should be piled to such a height as to limit the oscillations of the beam to a small arc.

It will be seen that the above apparatus approaches in a simple manner the conditions to be fulfilled in a good balance.

The various supports for weights a_1, a_2, b_1, b_2, etc., are in a straight line passing through axis of support of the beam, and the latter point is near the center of mass.

The length of arm can, however, be varied stepwise from Oa_1 to Oe_1, a thing not readily accomplished with ordinary balances.

Procedure:

(*a*) The weights m_1, m_2 (200 g. each, or less if that weight produces appreciable flexure of the beam), are suspended from e_1 and e_2. Shot are added to the lighter side until the pointer oscillates through *nearly* the same range on both sides of the zero of the scale.

(*b*) The "position of rest" of the pointer is found by the *method of oscillations*; that is to say, seven successive turning points are noted, and these are averaged.

For example, when the observations begin, the pointer is swinging to the left. We follow it with the eye until it reaches its greatest elongation and starts on its return swing. The position of greatest elongation is noted; also the six following ones to right and left alternately, estimating tenths of the small divisions. The readings are to be tabulated as below.

LEFT.	RIGHT.
8·8	
7·9	9·9
7·0	9·0
6·2	8·1
4)29·8	3)27·0
7·475 av.	9·000 av.

Treating the quantities on the left side as negative and averaging, we find the position of the zero:

$$\frac{9 \cdot 000 - 7 \cdot 475}{2} = 0 \cdot 762 + \text{ to the right.}$$

(*c*) Taking the utmost care not to alter in any way the relation of the parts of the balance to one another by touching the beam or even by checking the oscillations, as any such disarrangement tends to shift the zero point, add to the mass on one side a small weight from the box of weights (1, 2, or 5 cg., according to the sensitiveness of the balance), sufficient to shift the zero point through several scale divisions. Redetermine the zero by the method of oscillations, and divide the number of scale divisions through which it has been displaced by the number of centigrams added. *The result is the sensitiveness per centigram.*[1]

(*d*) Determine the sensitiveness, per centigram, of the balance with the weights at d_1 d_2, c_1 c_2, b_1 b_2, and a_1 a_2, *redetermining the zero before adding the deflecting weight each time.* Vary the deflecting weight in inverse proportion to the lever arm.

Express the results obtained graphically, using lengths of lever arm as abscissas, and sensitiveness as ordinates. With what degree of approximation does your curve agree with the relation

$$a : l,$$

where *a* is the angle of displacement due to one centigram and *l* the length of the balance arm?

77. EXPERIMENT 14. — **Weighing by Substitution.**

Apparatus:

(1) A well constructed balance, with a pointer and scale, and a set of weights.

(2) Two evaporating dishes or other flat open vessels.

(3) About 500 g. of fine shot.

Procedure:

(*a*) Place a 200 g. weight upon the right-hand pan of the balance, and in the other pan one of the dishes.

(*b*) Pour shot into the dish until the pointer remains upon the scale; then read seven oscillations, as in Experiment 13, and compute the position of rest. (Call this dish of shot *A*.)

(*c*) Remove the dish with its load, and place the other dish upon the pan. Add shot until the point reaches the same position of rest, as nearly as you can determine it without taking another set of readings. Whatever the value of the weight upon the other pan, and *whatever the ratio of the arms of the balance*, the second dish of shot, which we may call *B*, is of the same weight as *A*, within the limits of accuracy of your estimation of the agreement of the two positions of rest. This is called *weighing by substitution*.

[1] In the case of delicate balances the sensitiveness is expressed by the *displacement per milligram.*

(*d*) To estimate the accuracy of the weighing, obtain the position of rest with dish *B* upon the pan, employing the method of oscillations. Subtract the same from the position of rest for *A*. Add to one pan a centigram weight (or more in the case of a rough balance), and redetermine the position of rest by oscillations.

The difference between the positions of rest for *A* and *B*, divided by the change in the position of rest produced by the centigram weight, gives the difference in weight between *A* and *B* in centigrams.

Example:

With dish *A*:

Seven oscillations gave as position of rest,   1·19 (left)

With dish *B*:

Seven oscillations gave,   0·37 (left)

                     Difference =   0·82 s. d.

One centigram added to right-hand pan gave, from seven oscillations, 3·12 (right).

The influence of 1 cg. was therefore

$$3·12 + ·82 = 3·94 \text{ s. d.,}$$

and the difference of weight between *A* and *B* is

$$\frac{0·82}{3·94} \text{ cg.} = 0·0021 - \text{grams.}$$

CHAPTER X

COHESION, ADHESION, AND FRICTION

78. The Molecular Forces. — Friction, adhesion, cohesion, elasticity, capillarity, and many kindred phenomena are due to forces, the sphere of action of which is very small, comparable indeed in size to the distances which separate the molecules of a solid or liquid.

Such forces are called the *molecular forces*. Their existence, upon which the stability of all material structures depends, may be demonstrated in a variety of ways.

If a piece of clean dry plate glass be laid upon a similar piece, for example, we find upon picking it up again that the lower piece tends to follow. The nearer we succeed in bringing the two surfaces to one another the stronger the tendency will be. Commonly there will remain between the upper and lower plates a considerable layer or cushion of air. To demonstrate its presence we have only to lift the upper plate a few centimeters, and bring it down forcibly upon the lower. If we do so, keeping the surfaces parallel, we shall be made aware by the dull sound uttered at impact, — altogether different from the semi-metallic ring of glass against glass, — also by the yielding character of the material against which the upper plate impinges, that we do not bring the two glass surfaces into contact at all.

It is an air cushion which receives the blow, and the effect, both as to the sound and the character of the impact, is very like that which would be obtained were a thin film of soft rubber interposed.

This film of air can be in great measure removed by pressing the plates together, and by sliding them back and forth upon one another. Then, if the surfaces be clean, so that no grains of dirt keep them apart, and, if the surfaces be parallel (which is by no means always the case with chance specimens of glass), they will come well into the range of the action of the molecular forces, and the lower plate, if the upper one be lifted, will cling to it permanently.

That this phenomenon is indeed chiefly due to the molecular forces, and not to atmospheric pressure, may be shown by means of the following experiment:

79. EXPERIMENT 15. — Adhesion of Glass Plates in Vacuo.

Apparatus:
(1) Two glass plates with plane faces.
(2) An air pump.

Procedure:
(a) The plates are thoroughly cleaned and one is laid upon the other. If the surfaces match well enough to enable one to lift the lower by its adhesion to the upper they will serve; if not, other plates must be tried.

(b) If a bell jar of the form shown in Fig. 67 is available, it should be selected for this experiment. In this receiver, which is a well-known form, a brass rod passes through the metal cap (k) with freedom of vertical motion.

To the upper face of the upper glass plate glue a disk of bristol-board, and to the center of this fasten the end of the brass rod by means of sealing-wax or other cement, in such position that the plate will be horizontal when the bell jar is in place.

FIG. 67.

(c) Bring the lower glass plate into its place, working it as well

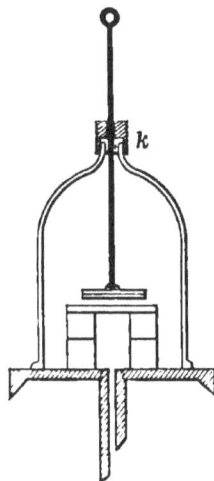

into close contact with the upper as is possible without endangering the fastening of the latter. Invert the bell jar over the plate of the air pump, giving the glass plates support until they are in a horizontal position. Then the bell jar being in place, lower the plates by means of the rod until they rest upon a block previously adjusted to receive them. After exhausting the air, the rod may be raised again, carrying both plates with it. It will be found that their attraction for one another is decidedly greater after the exhaustion, on account of the absence of the air film which had previously intervened.

80. Experiment 16. — Cohesive Forces in the Case of Water.

Apparatus:

(1) A simple balance and a set of weights.

(2) A glass disk about 10 cm. in diameter; a battery jar or other glass dish of somewhat greater diameter than the disk.

Procedure:

(a) Remove one of the scale pans from the balance and hang in its place the glass disk, its surface carefully leveled (Fig. 68). The height

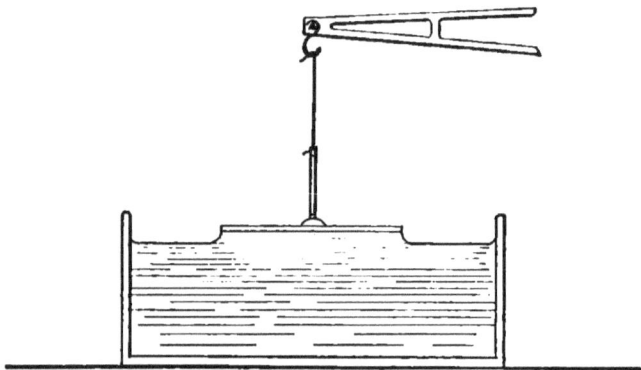

Fig. 68.

of the lower surface of the disk should be such that when the balance arm is level it will be 1 or 2 cm. above the rim of the jar. Balance the disk and place directly under it the jar, which is then to be filled to the brim with water.

(b) Cause the disk to fall gently upon the surface of the water by removing a weight from the scale pan, and take pains that no considerable air-bubbles are entrapped beneath the glass. Now add

weights gradually to the scale pan and note the following phenomena:

(1) A very considerable excess of weight over that necessary to balance the disk in air can be placed upon the scale pan at the other end
of the balance beam without tearing the disk loose from the water
upon which it rests.

(2) As weights are added, the disk rises above the general level of
the liquid, lifting the underlying portions with it. The surface of the
water becomes curved, as shown in the figure, rising on every side to
meet the periphery of the disk.

(3) When the disk finally leaves the water *it comes away with wetted
surface.*

This indicates that of the two sets of molecular forces brought into
play by the experiment, the attraction of the glass for the contiguous
water particles was greater than that of those particles for the liquid
next to which they were situated. *This is the case whenever a solid
dipped into a liquid has its surface wetted by the same.*

If mercury be substituted for water in this experiment, the same
phenomena are observed, excepting that the plate, after a considerably
greater force has been applied, is detached with its surface dry. We
conclude from this form of the experiment *that the attraction between
glass and mercury is less than between mercury and mercury,* and, *a forti-
ori,* that the latter is *much greater than the cohesive force of water.*

By means of the weights necessary to detach the disk, in the case
of various liquids, these adhesive and cohesive forces may be compared and measured. The difficulties of controlling the conditions
are so great, however, that the determination is one not to be recom-
mended to the beginner.

81. Friction.—The resistance which the molecular forces
offer to the motion of a body that slides along a surface
with which it is in contact, is termed *sliding friction.*
It is measured by means of the force necessary to maintain the sliding body in uniform motion along a horizontal
surface of the character in question.

82. Coefficient of Sliding Friction.— The coefficient of
sliding friction is the ratio of the force necessary to main-

tain a sliding body in uniform motion to the force with which the latter presses against the surface upon which it slides. Thus in Fig. 69, 50 kg. are. kept in uniform

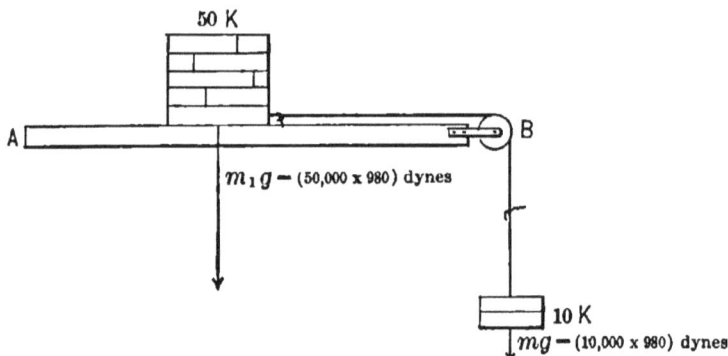

50 K

A

$m_1 g - (50,000 \times 980)$ dynes

B

10 K
$mg - (10,000 \times 980)$ dynes

FIG. 69.

motion along the horizontal plane AB by the force due to the attraction of the earth upon 10 kg. The forces in question are

$$50,000 \times 980 \text{ dynes}$$

and $20,000 \times 980$ dynes.

The ratio of the two is

$$\frac{m_2 g}{m_1 g} = \frac{1}{5} = \cdot 20,$$

which is the coefficient of sliding friction.

83. EXPERIMENT 17. — **Laws of Sliding Friction.**

Apparatus:

(1) An inclined plane of soft wood, about 100 cm. long. At one end this plane is hinged to a substantial base. The other end plays between two stiff uprights to which it can readily be clamped at any desired angle. The upper surface of the plane is to be freshly sandpapered before the beginning of the measurements.

(2) A wooden block 20 cm. × 10 cm. × 5 cm., also freshly sandpapered. In the middle of one of the broader sides is bored a hole,

into which fits a tall wooden peg. The same peg may also be used in a similar hole in the middle of one of the narrower sides.

(3) Several 5 kg. iron disk weights, a meter scale, some joiner's clamps, and a wedge.

Procedure:

(*a*) The inclined plane is clamped to a table, as in Fig. 70. The wooden block is placed upon the plane and the free end is lifted until the block, when tapped into motion with the finger, shows a tendency to continue sliding.

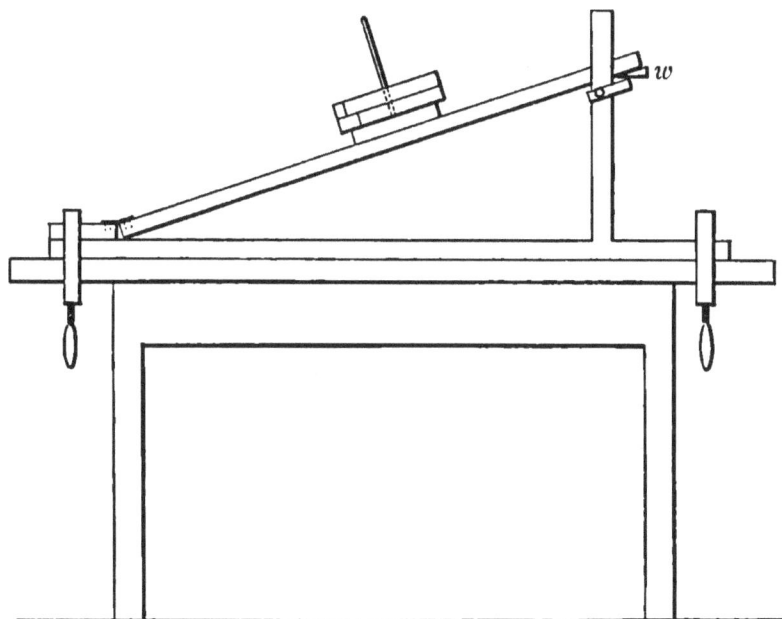

Fig. 70.

(*b*) Fasten two joiner's clamps to the uprights just below the plane, to afford a support for the latter, insert the wedge (*w*), which must be broad enough to reach across between the clamps.

By means of the wedge adjust the pitch further, until the *critical angle* is found, *i.e.* the angle at which the block, *once started*, will continue to slide with a uniform motion.

That this angle measures the coefficient of friction is evident from a consideration of forces acting upon the sliding body. The force *mg* between the earth and a body on an inclined plane (as in some previous examples) is resolved into p and f (Fig. 71), the functions of which are to produce pressure against the plane and motion along the same respectively. The ratio $\dfrac{f}{p}$ is therefore the coefficient of friction. Since the triangle of forces in this case, however, is similar to ABC, the triangle made by the plane, its base, and its upright, the ratio $\dfrac{BC}{AC}$ likewise measures the coefficient.

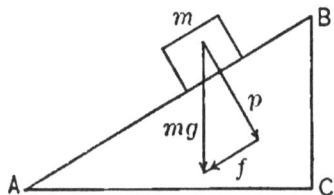

FIG. 71.

(c) By means of the present experiment the following laws of friction are to be verified :

(1) *The friction is proportional to the pressure;* i.e. to the mass moved. Otherwise stated, this means that *the coefficient is a constant* and is independent of the mass employed.

(2) *The friction is independent of the area of surfaces in contact.*

To demonstrate the first law, insert the peg in the block, and load the latter successively with 5 kg., 10 kg., 15 kg., and 20 kg. It will be found that the critical angle is the same for all loads.

To demonstrate the second law, turn the block upon its edge and repeat the determination. It will be found that the critical angle is still unchanged.

(d) Determine the coefficient for the following substances.

(1) A piece of plate glass sliding upon wood.

(2) A piece of plate glass sliding upon glass.

(3) A piece of polished metal (not lacquered) sliding upon wood.

(4) Paper upon wood. (For this purpose, paste smoothly a piece of paper to one face of a wooden block; sand paper the edges and test it upon the wooden plane.)

84. Starting Friction. — This term is used for the force necessary to start a body into motion upon a horizontal surface. The apparatus just described will serve for the demonstration of the following points, with reference to starting friction :

(1) Starting friction always exceeds sliding friction. (It will be found that, when the plane is at the critical

angle, the block may be placed upon it and will remain at rest.)

(2) To start the block the angle must be considerably increased. The increase is not constant, however. Starting friction is variable, and depends upon the extent to which the molecular forces are brought to bear upon the resting mass. Once started, the sliding body comes into a constant relation to these forces.

The following experiment upon starting friction is instructive:

85. EXPERIMENT 18. — Starting Friction of a Lubricated Surface.

Apparatus:
The inclined plane described in Experiment 17; also two glass plates. (One longer than the other.)

Procedure:
(a) Secure the longer plate to the plane by means of rubber bands or twine.

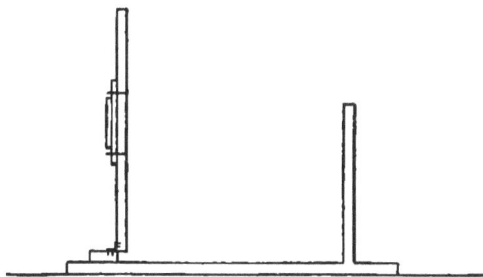

FIG. 72.

(b) Flood the upper surface of the plate with water and lay the other lightly upon it. It will be found that the critical angle is exceedingly small. Indeed, it is difficult to level the apparatus accurately enough to prevent the upper plate from sliding.

(c) Press the plates together for an instant, thus reducing the intervening layer of water to a thin fiber. The plane may now be raised to 90°, as in Fig. 72, without causing the upper plate to slip.

(*The coefficient of starting friction has risen from an infinitesimal value to more than unity.*)

If the experiment be repeated with oil the result will be the same.

86. Rolling Friction. — The term *friction* has been applied, although inaptly, also to the resistance which a rolling body experiences. The French physicist, Coulomb, who investigated the subject with great care, found the resistance to rolling to be *directly proportional to the pressure* and *inversely proportional to the radius of the wheel.*

87. Friction of a Shaft in its Bearings. — This is an interesting and important case of *sliding friction.* It may be briefly stated that the shaft rolls in its bearing until the line of contact is upon a portion of the surface of the latter,

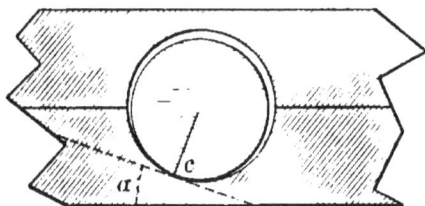

FIG. 73.

c (Fig. 73), the tangent to which makes the critical angle with the horizon. Then it begins to slide. The materials of the shaft and bearing are selected with a view to the reduction of the angle *α* to a minimum, and lubricants are applied for the same purpose.

CHAPTER XI

ELASTICITY

88. Stress and Strain. — All material bodies owe their structure, volume, and form to the interaction of the molecular forces between the particles of which they are composed. These forces are in equilibrium. When forces from without are brought to bear, this equilibrium is disturbed, and there are movements of all the particles with reference to each other. These movements continue until equilibrium is re-established. The body is then said to be under *stress*, and the distortion which results is called *strain*.

The changes which are thus brought about are of two kinds, viz. :

(1) Changes of form.

(2) Changes of volume.

When the body is again released from the action of the outer forces, it tends to return to its former volume and shape.

A body which returns completely to its former shape is said to have *perfect elasticity of form*; a body which resumes exactly its former volume has *perfect elasticity of volume*.

The term *elasticity* is also used with reference to the power of resisting change of form or volume.

Fluids, on account of the greater freedom or mobility of their molecules, do not ordinarily possess elasticity of form, but their elasticity of volume is perfect.

89. Limit of Elasticity. — Solids possess elasticity both of form and volume. When the stress passes a certain value, however, the solid no longer returns to its original condition; it is then permanently distorted. We say in such a case that *the limit of elasticity has been passed.*

It will be possible to consider in this book only a few of the simplest phenomena in the domain of elasticity.[1]

90. EXPERIMENT 19. — **Relation of Stretching Force to Elongation of a Wire.**

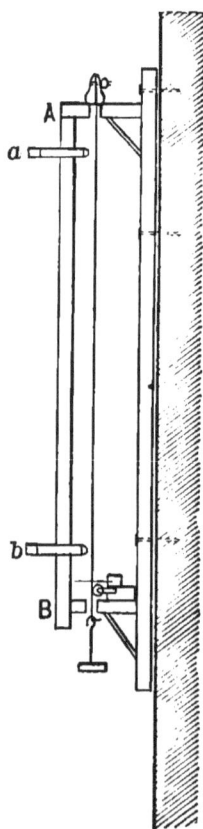

FIG. 74.

It is the object of this experiment to study the behavior of a wire which is subjected to homogeneous longitudinal stress.

Apparatus:

(1) A fine wire of brass or steel, suspended vertically. (Pianoforte wire, about ·04 cm. in diameter is to be preferred.)

(2) A set of iron disk weights with hook; a meter scale.

(3) A pair of microscopes of low power, provided with eyepiece micrometers. (See Appendix IV.) (If due care be taken with reference to the rigidness of support of the wire, one of these may be dispensed with without seriously vitiating the result.)

One of the simplest forms which the apparatus for this experiment may be given is that shown in Fig. 74.

It consists essentially of a pair of substantial wall-brackets, placed about 150 cm. apart, with two connecting vertical wooden strips. One of these is fastened to the wall of the laboratory, while the other, AB, supports the microscopes.

The wire is fastened above in a clamp with steel jaws. (An ordinary hand-vise makes an excellent clamp for this purpose.) It then passes

[1] For an elementary discussion of the theory, see *Elements of Physics,* p. 83.

through a hole in the upper bracket. The lower end is attached to the hook of a 1 kg. weight as shown in the figure. The weight, the hook of which passes through a hole in the lower bracket, keeps the wire tense.

To mark upon the wire two points, the movements of which are to be observed, take a hair, and loop it around the wire opposite each microscope. Draw it taut, using a single knot, then shift it along the wire gently with a match end or splinter of wood, until it is within the field of the microscope and near the (apparently) lower end of the scale. Dip the splinter into a shellac solution and touch the knot, which should be on the side of the wire away from the microscope. As seen in the field of the microscope, the wire and marker will present the appearance shown in Fig. 75.

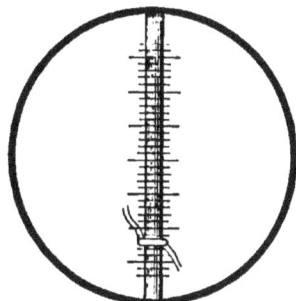

Fig. 75.

To prevent troublesome oscillation and the turning of the wire, the following device may be employed. It consists of a pulley fastened to a wooden block. Upon this block is placed another smaller one. Around the wire, about 10 cm. above the lower bracket, is twisted a smaller wire, the free ends of which project horizontally. One of these, coming into contact with the smaller block, prevents the stretched wire from turning. The pulley, which is pushed forward until its face is in light contact with the latter, checks its oscillations. This

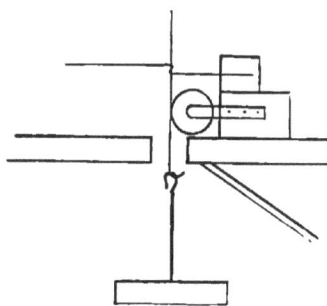

Fig. 76.

little device is not fastened to the bracket, but is adjusted as may be necessary, with the hand.

The microscopes are inserted through holes, about 1 m. apart, which are bored in the upright piece, *A B.* They should fit the holes snugly, but should be adjustable for the purpose of rough focussing.

Procedure:

(*a*) Place the suspended wire under the tension of a kilogram weight.

(*b*) Focus the microscopes (*a*) and (*b*) upon their respective markers,

H

and note the positions of the latter by means of the micrometer scales.

(c) Add carefully four kilogram weights, and observe the positions of the markers after the addition.

(d) Remove the weights stepwise, redetermining the position of the markers after each change; then restore the weights, one at a time, continuing to read the positions, until the full load has been restored.

The results of these observations should afford a verification of the law that the elongation is proportional to the stretching force.

(e) Tabulate the results obtained, as follows, and from them plot a curve for increasing and decreasing weights.

TABLE.

Observations upon the Stretching of a Wire.

Stretching weight m.[1]	Position of the marks.	
	Lower.	Upper.
4000 g.	34·7	21·6
3000 g.	26·4	20·4
2000 g.	18·9	19·6
1000 g.	10·4	18·6
0	1·8	17·2
1000 g.	10·2	18·9
2000 g.	18·7	19·8
3000 g.	26·3	20·7
4000 g.	34·6	21·5

From these readings we can easily compute the movement of the upper and lower marks, by subtracting each from 1·8 for the latter and from 17·2 for the former. The movement of the lower mark

---

[1] The weight previously applied, *i.e.* 1 kg., to keep the wire taut is not included, nor need it be since the object of the measurements is to determine the influence of *changes* in the stretching weight.

minus that of the upper gives the total elongation, and this divided by the corresponding stretching weight (m) gives the elongation per gram.

M.	Movement of the marks.		Total elongation (Δl) expressed in scale divisions.	Elongation per gram of stretching weight.
	Lower.	Upper.		
4000 g.	32·9	4·4	28·5	0·07125
3000 g.	24·6	3·2	21·4	0·07133
2000 g.	17·1	2·4	14·7	0·07175
1000 g.	8·6	1·4	7·2	0·07200
0	0	0	0	0
1000 g.	8·4	1·7	6·7	0·06700
2000 g.	16·9	2·6	14·3	0·07150
3000 g.	24·5	3·5	21·0	0·70000
4000 g.	32·8	4·3	28·5	0·71250
20000 g.			142·3	

The average elongation per gram is obtained by dividing the sum of the total elongations by the sum of the stretching weights. This gives us

142·3 s. d. ÷ 20000 = 0·07115 s. d. per gram.

For the method by which scale divisions may be reduced to centimeters, and the stretch modulus of the wire obtained from these readings, see Experiment 20.

The data in the foregoing table are intended to indicate the character of the results which may be obtained with the apparatus just described. They are given graphically in Fig. 77, for the purpose of showing the method of treating such data. Ordinates are total elongations (Δl) expressed in scale divisions, and abscissas are weights in grams (multiplied by $g = 980$ they become dynes).

The fact that the observations lie along a straight line verifies the above-mentioned relation, *i.e. proportionality of stretching force and elongation.* Their close coincidence with one another pair-wise indicates that within the range of this experiment steel is very nearly perfectly elastic.

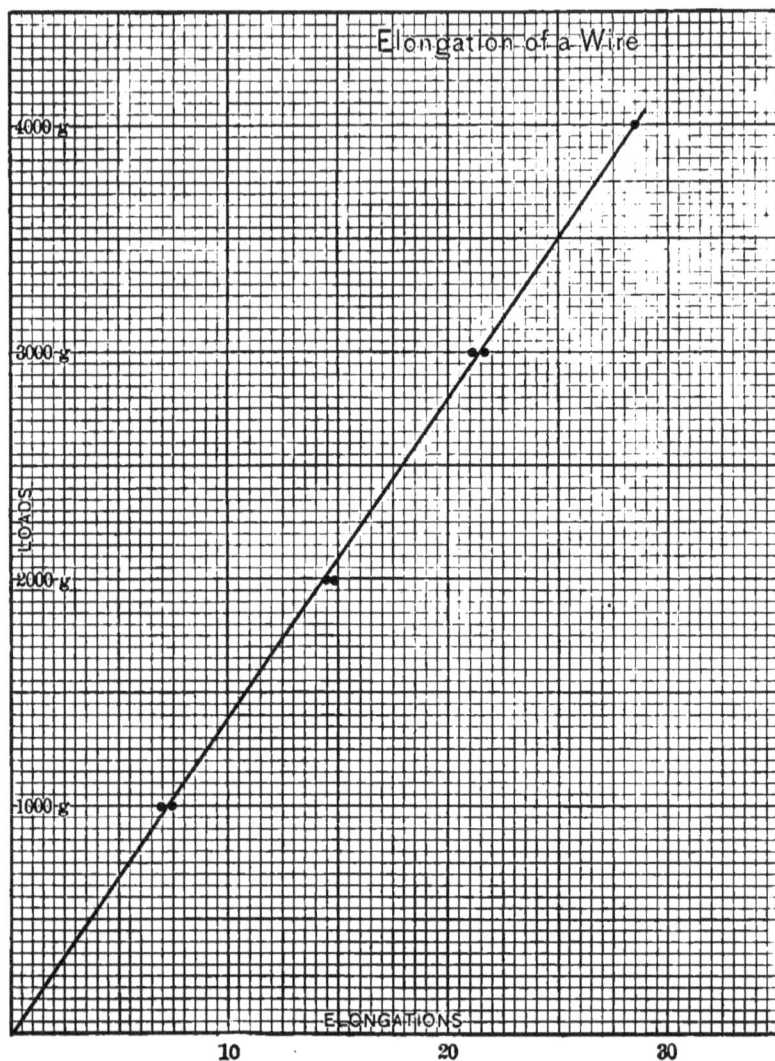

FIG. 77.

91. EXPERIMENT 20. — The Stretch Modulus (sometimes called Young's Modulus) of a Steel Wire. — Were observations like those of the foregoing experiment extended to wires of various diameters, it would be found that *the elongation for a given force* is inversely proportional to the cross-section (q) of the wire.

Assuming this relation, which indeed is too evident to require verification, we may use the data obtained in that experiment for the quantitative determination of the *modulus of the wire*. This quantity, which, to distinguish it from other coefficients in use in the study of elasticity, is called the *stretch modulus*, or sometimes *Young's modulus*, is a constant by means of which the power of a substance such as steel to resist elongation is quantitatively expressed. If we call the elongation Δl, we shall have:

$\Delta l : F$, where F is the stretching force.

$\Delta l : l$, where l is the length of the wire.

$\Delta l : \dfrac{1}{q}$, where q is the cross-section.

These relations may be embodied in an equation,

$$\Delta l = \frac{1}{E} \frac{Fl}{q},$$

where $\dfrac{1}{E}$ is a constant depending upon the nature of the material of which the wire is made. The quantity E is called the stretch modulus; its value is evidently

$$E = \frac{Fl}{q\Delta l}.$$

It expresses numerically the resistance to stretching, and consequently is chosen in such a way as to be *inversely* proportional to the elongation.

To find the value of E in the case of the wire under observation we must know:

(1) The force in dynes. This will be $F = m \cdot g$ dynes.

(2) The length l, in centimeters, between the markers. This measurement need not be more exact than that of the small quantities q and Δl. The apparently rough method of measuring from the middle of the eyepiece of (a) to the middle of the eyepiece of (b), with an ordinary meter scale, is abundantly accurate.

(3) The cross-section (q). This may be measured by means of a micrometer gauge, or a piece of the same wire 100 cm. in length may be weighed, and its cross-section computed upon the assumption that the density is 7·85 (for steel). (We have for this computation $q = \dfrac{m}{d \cdot l}$, where m is the mass of the weighed piece, l its length, and d its density.)

(4) The elongation, Δl, in centimeters.

For this purpose the eyepiece micrometers must be calibrated. (Without changing the distance between eyepiece and objective, turn the eyepieces so that the micrometer scale in each has its lines parallel to the wire, and note to tenths of a scale division the space which the wire occupies upon the micrometer scale. From this, and value of the diameter, compute *the value of one scale division in centimeters*.)

From the curve of elongations read the ordinate corresponding to a given force (mg). This value is in scale divisions of the micrometer: reduce it to centimeters.

Having thus found l, Δl, and g, in centimeters, and $mg = F$ in dynes, E is determined. It is a very large number indeed, being, in fact, the force in dynes necessary to stretch a rod of steel 1 cm. in cross-section to double its length (were that possible without passing the limit of elasticity).

The following is the computation, for example, of the modulus of the steel wire, data for the stretching of which were given in Experiment 19 (Art. 90).

Length of wire between marks $= l = 100$ cm.
Sum of stretching forces applied $= 20,000$ g.
$\qquad\qquad\qquad\qquad\qquad\qquad\quad = 980 \times 20,000$ dynes
$\qquad\qquad\qquad\qquad\qquad\qquad\quad = 19,600,000$ dynes $= F$.
Diameter as measured with micrometer gauge $= \cdot 039$ cm.
radius $= r = \frac{1}{2}$ diam. $= \cdot 0195$ cm.
cross-section $= q = \pi r^2 = \cdot 00119 \overline{\text{ cm.}}^2$
Sum of elongations expressed in scale divisions $= 142\cdot 3$ s. d.
Diameter of the wire in scale divisions $7\cdot 9$ s. d.

Value of 1 s. d. in centimeters $= \dfrac{\cdot 039}{7\cdot 9} = \cdot 0049 +$ cm.

Sum of elongations in centimeters $= 142\cdot 3 \times \cdot 0049 = \cdot 697$ cm.
Substituting these values in the formula,

$$E = \frac{F \cdot l}{q \cdot \Delta l},$$

we have $\dfrac{19,600,000 \times 100}{\cdot 00119 \times \cdot 697} = 2,360,000,000,000 = 2\cdot 36 \times 10^{12}$

$\qquad\qquad\qquad\qquad\qquad\qquad\quad = $ stretch modulus for steel.

92. Other Phenomena resulting from Longitudinal Stress. —

That the diameter of a tube diminishes when the tube is

under longitudinal stress can be easily shown by stretching an ordinary rubber tube between the hands. What is true of caoutchouc can be shown by refined methods to be true even in such rigid materials as glass and steel, and what is true of a tube will be found true in the case of rods and wires.

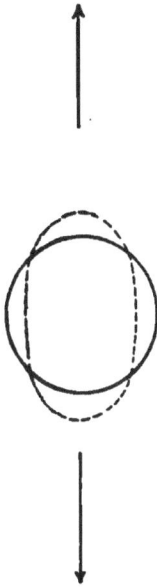

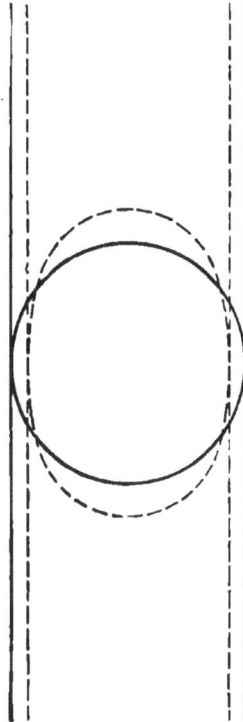

FIG. 78.   FIG. 79.

If we could select a small spherical element or portion in the interior of a rod, and observe its change of form when the rod is stretched, we should find the sphere converted into an ellipsoid with the major axis in the direction of the pull (Fig. 78).

What happens to this particular particle is, however, happening to all its neighbors, and we may extend our consideration to a large sphere. Suppose the sphere to have a diameter equal to that of the rod. When the rod is stretched its surface will be tangent to the ellipsoid into which the strained sphere has been converted. (See Fig. 79.)

In spite of this diminution of diameter the volume of the rod or tube is increased by stretching, as may be shown by means of the following simple experiment:

93. Experiment 21.—Influence of stretching upon the Volume of a Tube.

Apparatus:

A rubber tube; a piece of barometer tubing which fits the bore of the former; disk weights with hook.

About 10 cm. of the barometer tubing (any thick-walled glass tube will do) is cut off, and one end is closed in the flame of a blast-lamp and blown into a small strong bulb (Fig. 80).

In the middle of a leather strap, 30 cm. long, a hole is cut of such size as to allow the tube to be passed through, but not large enough to admit the bulb. The free ends are fastened so as to form a strong loop.

The closed glass tube is now inserted into one end of the rubber tube and the remaining piece of glass tubing into the other, and both are securely wired (Fig. 81).

Procedure:

(*a*) Clamp the rubber tube in a vertical position to a substantial support, as shown in Fig. 81, and fill it with water until the surface of the liquid is in the upper part of the open glass tube at *w*.

(*b*) Hang weights (5 to 10 kg.) to the strap, and note the marked fall in the level of the column of water.

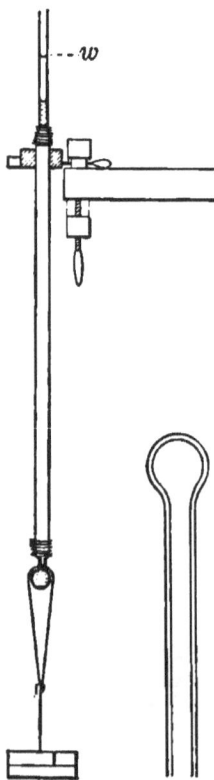

Fig. 81.   Fig. 80.

The same effect can be demonstrated in the case of glass by means of the apparatus depicted in Fig. 82.

This consists simply of a strong glass tube, closed at one end and enlarged sufficiently to facilitate the application of weights by the method just described. The tube is nearly filled with water, and the end is then drawn out into a capillary neck and bent twice through 90°; after which the filling is completed by successive heatings and coolings of the tube.

FIG. 82.

The effect in glass is exceedingly small on account of the high-stretch modulus of this substance. It is easily observable, however, if the bore of the neck is very narrow. The apparatus is a sensitive thermometer, and on that and other accounts the experiment is a more difficult one than the foregoing.

94. Elasticity of Torsion.—When outer forces producing stress are so applied as to twist a body about a given axis, the tendency of the body to resist such twisting, and when released, to return to its original position, is termed *elasticity of torsion*.

The relation between the force applied to produce torsion and its effects, in the case of a rod one end of which is fixed, may be stated as follows:

Let θ be the angle through which the free end of the rod is turned: then

$\theta : F$, where F is the torsional force;

$\theta : l$, where l is the length of the twisted rod;

$\theta : \dfrac{1}{r^4}$, where r is the radius of the twisted rod.

In the case of torsional forces, it is really with the moment of the force (or the torque) that we have to do.

It can be shown [1] that the laws of torsion are fully expressed by means of an equation

$$T = n \cdot \frac{\pi r^4 \theta}{2 l},$$

in which T is the torque (or moment of the twisting force), and n is a constant called the *constant of torsion*, which depends upon, and indicates the power of the material to resist torsion.

The relations between T, θ, and l are easily demonstrated by means of the following experiment:

95. EXPERIMENT 22. **Torsion of a Rod.** — It is the object of this experiment to verify the fact that the angle through which a rod or wire, one end of which is fixed, will be twisted by a torsional force applied at the free end *is proportional to the length of the rod and to the moment of the force.*

FIG. 83.

Apparatus:

(1) A straight brass rod rather more than 100 cm. long and about 0.5 cm. in diameter. This must be rigidly fastened to a block at one end; 10 cm. from the other end a wheel is mounted. (See Fig. 83.)

[1] See *Elements of Physics*, p. 102.

(2) A pointer mounted upon a collar which fits the rod and can be set and loosened at will; also a divided semicircle capable of being set up in a vertical plane at right angles to and with its center in the axis of the rod (Fig. 84).

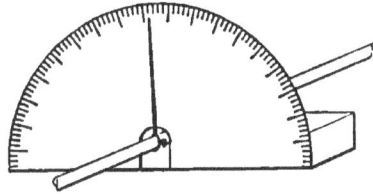

(3) A meter scale, a set of 1 kg. iron disk weights with hook.

Procedure:

(*a*) The rod is mounted upon a table, one end fixed, the wheel just

FIG. 84.

beyond the edge of the same, and a second table or stand affording support for a wooden bearing (b_1), in which the free end of the rod rests. The pressure upon this bearing is just sufficient to relieve the rod from the weight of the wheel (Fig. 83). A second similar bearing (b_2) should be placed just behind the wheel.

(*b*) By means of a cord running over the periphery of the wheel and to which the disk weights may be attached, apply successively 1 kg., 2 kg., 3 kg., 4 kg. The torque is rmg in each case, where r is the radius of the wheel, and m the mass of the weights applied.

Note that wherever the pointer may be attached to the rod, with circular scale behind it, it will be turned through an angle

$$\theta : rmg.$$

Note further, that using a given force and placing the pointer at different points upon the rod,

$$\cdot\ \ \theta : l,$$

where l is the distance from fixed end of the rod to the pointer.

CHAPTER XII

THE PROPERTIES OF LIQUIDS

96. Hydrostatic Pressure. — When a fluid, either a liquid or a gas, previously at rest, is subjected to any force, motions will be set up; and these will continue until an arrangement of the parts of the fluid has been attained, such that at any given point within the same the forces at work counterbalance each other and thus have zero for their resultant. The fluid will then be under stress of the only kind to which it is possible for it to be subject while at rest. This type of stress is called *hydrostatic pressure.* The nature of it is such that if we select any imaginary area whatever the pressure upon that area will be normal to the same.[1]

The fact of the balance of forces within a liquid at rest may be verified by the observation of the form which small globules of any liquid assume when suspended in a liquid mass with which they do not tend to mix. Drops of oil falling through water afford the most familiar example. The spherical form of such drops is due to the fact that the forces acting upon their surfaces are everywhere normal to the surface and equal to one another.

The form of soap bubbles floating in still air affords an equally complete verification of the principles in the case of gases, while the distortions which such bubbles undergo

---

[1] For a proof of this, which is known as *Pascal's principle*, see *Elements of Physics*, p. 87.

when subjected to a draught indicate that the principle is to be applied only to fluids at rest.

97. EXPERIMENT 23. — **The Transmission of Pressure in Liquids.**
Apparatus:
A glass cylinder about 30 cm. in height (Fig. 85), a glass tube or series of tubes coupled together and fitted so as to enter the cylinder

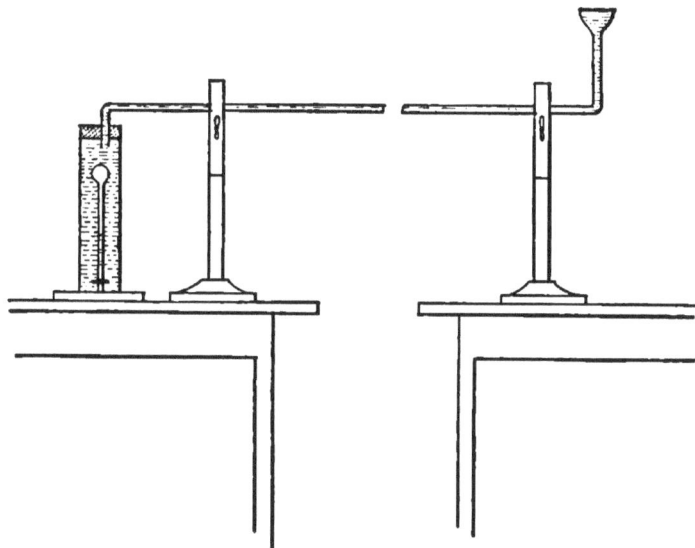

FIG. 85.

through a water-tight cork as shown in the figure. These may extend horizontally to any convenient distance.

Procedure:
(*a*) The cylinder is filled to the brim with water, and a glass tube *t*, the closed end of which is blown out to form a bulb, is immersed therein with the mouth downwards. This tube, the air within which is somewhat compressed, acts as a pressure gauge.

(*b*) The cork is now inserted into the mouth of the cylinder, the excess of liquid flowing into the long, horizontal tube. The latter is completely filled, and its somewhat enlarged aperture is closed by means of a membrane of soft rubber.

This arrangement of the apparatus completed, it will be found that the water column within the gauge will respond sharply to every change of pressure produced at the diaphragm. However great the distance of the latter from the cylinder may be, it is only necessary to tap it with the finger tip to throw the water column into oscillation.

The observer should note the promptness with which the indication at the gauge follows the movement of the diaphragm. In point of fact the pressure is transmitted by means of wave motion in the liquid. This motion is much too rapid to admit of direct observation with the unaided eye. Its velocity in water is about 1435 meters per second. (See further the chapters on sound.)

98. The property of transmitting pressure, described in the foregoing experiment, is made use of in the apparatus known as the hydraulic press. It consists essentially of a strong, closed reservoir containing water or other liquid. Through the walls of this reservoir play two pistons. One of them, A (Fig. 86), is of large, the other, a, is of small area.

The reservoir is always full, and the pressure upon its walls, including the areas of the two pistons, is everywhere the same, and is normal to the surface. Since by pressure we mean the force per unit of area, it follows that loads placed upon the two pistons, as in the diagram, will always balance one another, provided their masses are proportional to the areas of the pistons. Moreover, if the smaller piston be forced downwards through its cylinder, the larger piston will rise and carry its load with it. If the ratio of the areas be 1,000,000 : 1, for example, the force of gravity upon somewhat more than one kilogram will serve to lift one million kilograms

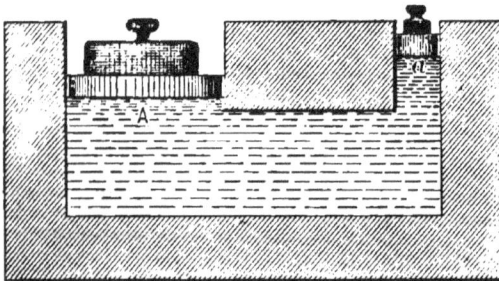

Fig. 86.

against the earth's attractive force. Note that the law of machines holds in this case. The work done by the one kilogram in falling one centimeter, for instance, would be 980,000 ergs. The large piston with its load of 1,000,000 kilograms would be lifted far enough to afford room for the water displaced by the small one, *i.e.* 0·000001 centimeters, and this movement corresponds likewise to 980,000 ergs. As in the case of nearly all machines for the movement of very large masses by the application of small forces, there is a considerable waste of energy in overcoming friction.

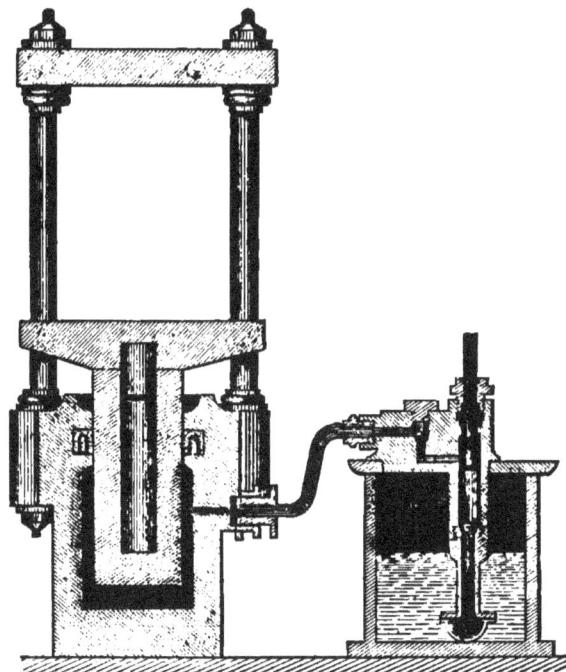

FIG. 87.

In order to make the action of the hydraulic press continuous, the smaller piston is driven like the plunger of a pump and operates a valve by means of which water is introduced into the reservoir at each stroke. A common form of press is shown in Fig. 87.

99. The Distribution of Pressures within a Liquid contained in an Open Vessel. — The stress to which such a fluid is subject is that due to the action of gravity upon each particle, and to the pressure of the superincumbent atmosphere. For the purpose of the present discussion the latter may be disregarded. The following relations are of importance:

(1) *The pressure within a liquid mass increases from the surface downwards in direct proportion to the depth.*

(2) *The pressure at a given depth below the surface is proportional to the density of the liquid.*

(3) *The pressure in a given liquid is dependent only upon the depth. It is independent of the form of the vessel and of the amount of liquid which it contains.*

Statements (1) and (2) may be verified as follows:

100. EXPERIMENT 24. — **Relation between Pressure and Depth of a Liquid.**

Apparatus:

(1) A tall cylindrical glass jar.

(2) A glass funnel and tubing; a burette tube with two arms; blocks, supports, etc.; a wooden scale.

Procedure:

(*a*) Bend a piece of glass tubing (*jj*) about 80 cm. long as shown in Fig. 88. Attach one end of it, with an air-tight joint of rubber tubing, to a glass funnel, the other end similarly to one arm of a U-shaped tube (preferably a double burette tube with stopcock). Support the whole at such a height above a laboratory table that the cylindrical jar, when standing upon the table, will slide under the mouth of the funnel.

(*b*) Warm the lip of the funnel and coat it with beeswax-resin cement, and while the latter is still soft lay over the mouth of the funnel a thin diaphragm of pure soft rubber. The diaphragm should be tense enough to form a plane septum, but should not be tightly stretched.

(*c*) Fill the jar with water, place it under the funnel, and raise it by the insertion of blocks until the mouth of the submerged funnel is near the bottom.

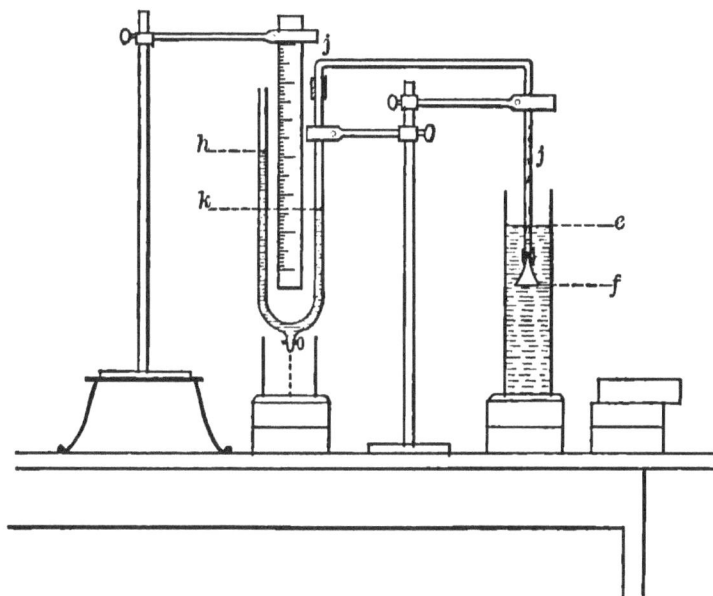

Fig. 88.

(*d*) The upward pressure against the rubber diaphragm, being unbalanced, will distend the latter, rendering it concave, as seen from below (Fig. 89), and forcing it into the funnel. This force is now to be balanced by pouring water into the open arm of the U-tube, which serves as a pressure gauge, until the diaphragm has resumed its original plane contour.

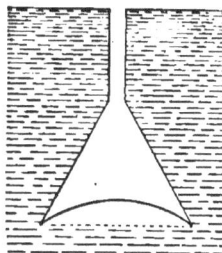

Fig. 89.

(*e*) Measure the difference of level *h*, *k* of the two columns in the U-shaped tube, also the distance *e*, *f* from the surface of the water in the jar to the plane of the diaphragm. Remove the blocks from beneath the jar, one at a time, and repeat the above measurements for the various depths of submergence thus obtained. The data should accord with statement (1).

I

(*f*) Fill the jar with oil, or with strong brine, and repeat the determination for a single depth. Water should still be used in the U-tube. Note that the height *hk* is no longer equal to *ef*. Find the density of the liquid in the jar, using one of the methods described in Chap. XIII. The result should verify the relation $\overline{hk} : ef :: d : 1$, or

$$\frac{\overline{hk}}{ef} = d.$$

In this formula *d* is the density of the liquid in the jar as compared with the density of water, which is taken equal to unity.

The height, *hk*, of the liquid in the U-tube is a measure of the pressure upon the diaphragm, and the relation expressed in the equation therefore verifies statement (2).

101. EXPERIMENT 25. — **Pascal's Vases.**

The statement that pressure within a given liquid depends upon the depth alone, and is independent of the form and volume of the containing vessel, may be verified by means of the following device, which is a modification of the classical apparatus known as *Pascal's vases*:

FIG. 90.

Apparatus:

(1) A strong brass tube, about 7 cm. in diameter and 10 cm. long. This carries a broad flange at its middle, as shown in Fig. 90. One end, which is threaded, fits any one of three similar tubes, also of brass, which form the lower ends of three glass vessels. These vessels are given widely different forms, as shown in Fig. 91.

(2) A rubber diaphragm, which is to be fastened to the bottom of the flanged tube.

FIG. 91.

(3) A long pointer, pivoted as shown in Fig. 92.

(4) A scale along which the pointer plays.

Procedure:

(*a*) The rubber diaphragm is fastened to the unthreaded end of the flanged tube, which may conveniently be constructed with a sliding collar for securing the same.

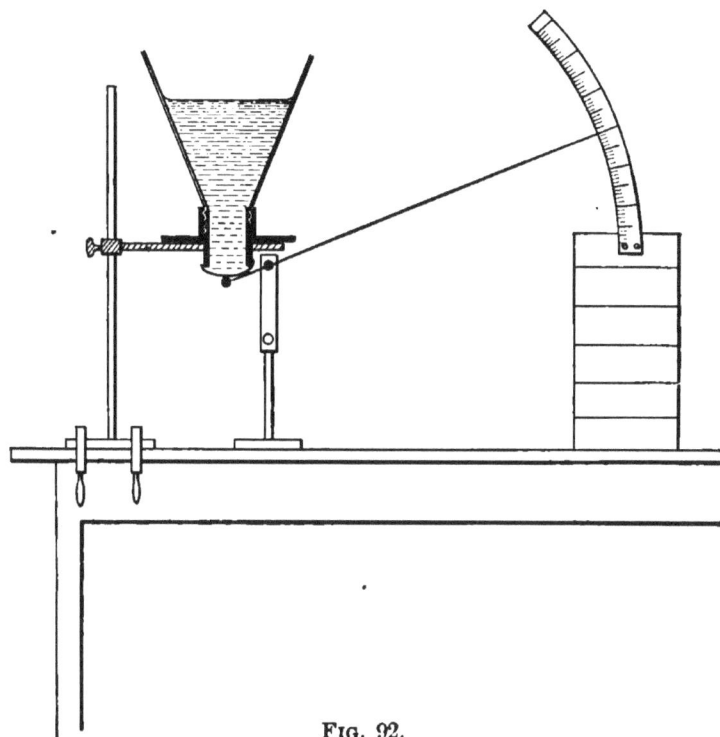

FIG. 92.

(*b*) The tube is mounted with the flange horizontal and diaphragm below, as shown in Fig. 92, and one of the glass receptacles with its brass base is screwed into place above. We thus have a vessel with a flexible bottom and capable of being filled with water.

(*c*) The pointer is adjusted with its short lever arm in contact with the middle of the diaphragm and the long arm indicating the zero of the scale.

(*d*) Water is poured into the vessel. The rubber diaphragm being distended downward moves the pointer along the scale. When the surface of the water reaches a level previously marked near the lip of the vessel, the scale reading is noted.

(*e*) The vessel is emptied with a siphon, and one of the other receptacles is substituted for the one just used. The lower portion should remain undisturbed. The pointer being readjusted to zero, the filling with water is repeated, the level being brought to the same height above the diaphragm.

(*f*) The operation is repeated with the remaining receptacle in place.

The object of the experiment is to show that the pressure upon the flexible bottom is the same for a given height of liquid, whichever receptacle be used. This fact is indicated by the uniformity of the scale readings.

102. The Principle of Archimedes. — It is obvious from the preceding experiments that the pressure upon any plane area within a liquid at rest is that due to the weight of the column of superincumbent liquid.

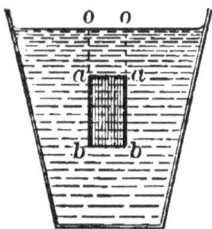

Fig. 93.

Consider the case of a submerged solid cylinder (Fig. 93). The pressure is everywhere normal to its surface, and the lateral forces, which at each point are proportional to the depth of the point below the surface of the liquid, precisely counterbalance one another. Were this not the case there would be lateral movement of the cylinder, which does not occur. Upon the upper face of the cylinder the force is *downwards* and is that due to the weight of the liquid column *aaoo*. The force acting upon the lower face is *upwards*, and is that due to the weight of a liquid column equal to *bboo*. The resultant of these forces is a force equivalent to that due to the weight of a mass of liquid of the same size as the cylinder and acting vertically upwards. If we consider this force, which is called the *buoyant force of the liquid*, in connection with the attractive force of the earth upon the cylinder, we see that these being opposite in direction the result is a loss of weight. The statement of this fact is known as the *principle of Archimedes.* It is as follows :

A body, when submerged in any liquid, loses weight by an amount equal to the weight of the liquid which it displaces.[1]

. **103.** The principle of Archimedes is usually verified in the case of liquids by means of what is known as the *cylinder and bucket apparatus.*

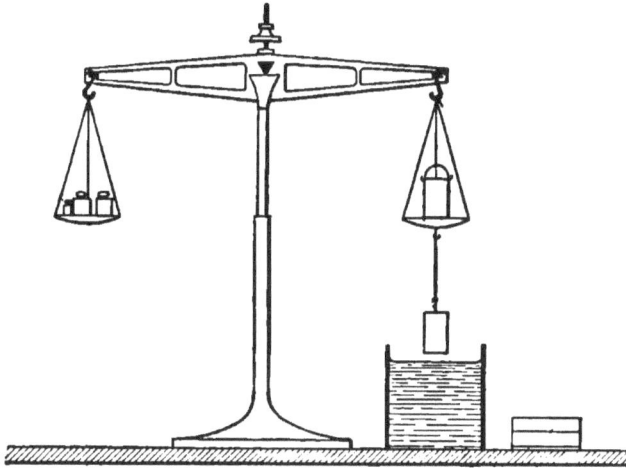

FIG. 94.

A small, cylindrical, brass bucket is placed upon one scale pan of a balance (Fig. 94), beneath which hangs a cylinder which exactly fits the interior of the bucket. Its volume, therefore, is that of the contents of the bucket.

---

[1] The discussion is readily extended to the case of bodies of irregular form. We may regard such a body as made up of vertical rods, or ele-
ments of infinitesimal cross-section. For each of these taken separately the considerations given above suffice, and what is true of each of these elements will be true of the whole body, whatever may be its form.

If the balance be brought into equilibrium and, the arms being still free, a beaker of water be brought up from below, it will be found impossible to submerge the cylinder on account of the buoyant force of the fluid. (See preceding article.)

If, however, water be poured into the bucket by degrees, partial submergence of the cylinder will take place, and equilibrium will be restored when the portion submerged equals the portion of the bucket which has been filled (Fig. 95). When the bucket is full, the upper surface of

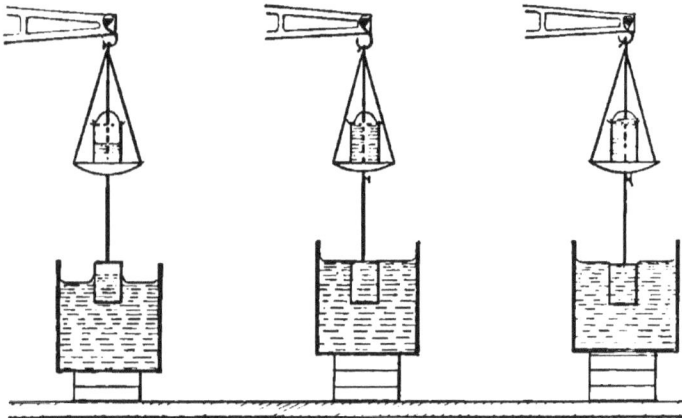

Fig. 95.

the cylinder will be at the level of the water in the beaker. If, up to this point, the top of the cylinder has been kept dry, it will be found possible to carry the experiment a step further. It is possible by taking advantage of capillarity, to add a considerable quantity of water to the bucket, sinking the cylinder considerably below the general surface of the liquid, before the bucket overflows or the water in the beaker encroaches upon the top of the cylinder. This final stage is, likewise, indicated in the figure.

104. The Beaker gains in Weight what the Submerged Cylinder loses. — By balancing the scales with the beaker of water upon one scale pan and the bucket upon the other (Fig. 96), and attempting to submerge the cylinder

FIG. 96.

sustained from without in the beaker, the *reaction* corresponding to the buoyant force of the liquid may be made evident by apparent increase of weight of the beaker. That this reaction is precisely equal and opposite to the buoyant force (third law of motion) may be shown by filling the bucket with water, by which means equilibrium is restored.

105. EXPERIMENT 26. — **The Influence of Partial and Total Submergence upon the Weight of a Body.**

Apparatus:

(1) A balance and weights. The former should be provided with a mechanical device for the arrest of the beam.

(2) A cylindrical glass vessel containing water. The depth must exceed 20 cm.

(3) A scale divided to millimeters and a micrometer gauge.

(4) A cylindrical rod of brass or copper about 20 cm. long, also a similar rod of wood, or other light material, loaded at one end with a cylindrical sinker of the same diameter and of such mass that the rod will float in a vertical position and more than half submerged. To take the place of the latter a floating tube may readily be constructed out of any tube of thin metal closed at one end. One of the cases in which thermometers are packed will serve the purpose. It may be ballasted by the introduction of shot, and then capped at the upper end.

Procedure:

(*a*) Measure the diameter of the heavier rod with the gauge, and compute its volume per centimeter of length in cubic centimeters.

(*b*) Beginning at one end of the rod, which end should be plane and normal to the axis of the rod, graduate the latter in centimeters, and mark the divisions by means of a triangular file ground to a smooth edge, or with any suitable tool. These divisions may be painted with a very narrow brush.

(*c*) Weigh the rod.

(*d*) By means of a thread fastened to a loop of fine wire previously soldered to the end which was not used as the zero of graduation, suspend the rod from the hook in the bottom of one of the scale pans. It should hang with axis accurately vertical.

(*e*) Bring the balance to equilibrium, and arrest the beam. Then bring from below the cylinder of water, raising it slowly until about 2 cm. of the rod are submerged (Fig. 97). Support the cylinder in this position by blocks.

(*f*) Free the beam of the balance slightly, and readjust by adding weights to the scale pan from which the rod hangs. *Note the amount thus added and the depth of submergence* of the rod when the beam is free and the pointer of the balance is at zero.

(*g*) Raise the cylinder by steps of about 2 cm. each, repeating operation (*f*) for each position, until complete submergence is reached. Compute the mass of water displaced at each stage of the determination,[1] and compare it with the corresponding masses of the weights added to balance the buoyant force of the liquid. Your result should accord with the principle of Archimedes.

(*h*) Repeat the above-mentioned measurements, using the lighter

---

[1] One cubic centimeter of water weighs one gram.

rod. From the final step of this experiment, which is reached when the buoyant force entirely supports the rod, you should be able to deduce, from the weights added and the weight of the rod in air, the following **law of floating bodies,** viz.

A floating body displaces its own weight of liquid.

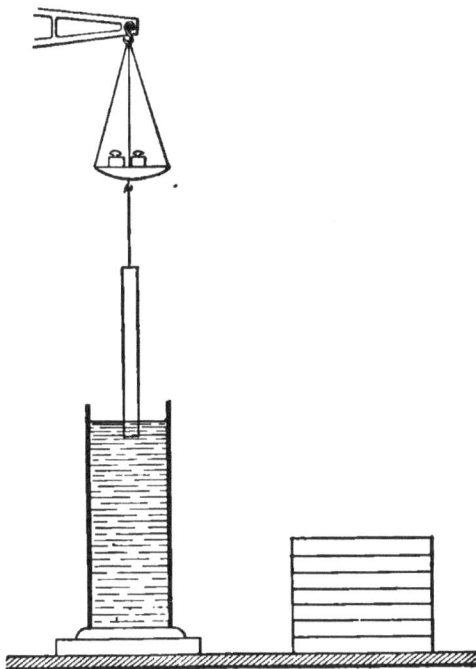

FIG. 97.

CHAPTER XIII

DENSITY

106. Application of the Principle of Archimedes to the Measurement of the Densities of Solids and Liquids. — The buoyant force upon submerged bodies affords some of the most convenient and accurate methods of comparing the densities of both solids and liquids. The basis of the comparison, in the case of solids and liquids, is the density of water; in the case of gases, air is sometimes taken as the standard of reference, while sometimes hydrogen is used and sometimes water.

107. Density. — *Density is defined as the mass per unit volume.* The formula which gives the relation between these three quantities is accordingly

$$D = \frac{M}{V},$$

in which D is the density, M the mass, and V the volume.

In the comparison of the densities of other substances with that of water, the term *specific gravity* is commonly used to express the result. The specific gravity S is

$$S = \frac{D}{d},$$

in which d is the density of water at the temperature of the experiment.

108. EXPERIMENT 27. — The **Measurement of Density by weighing in Air and in Water.**

Apparatus:

(1) A balance and weights.

(2) A beaker of water.

(3) A block of metal weighing about 200 grams.[1]

Procedure:

(*a*) Suspend the body, the density of which is to be determined, from the hook of the right-hand scale pan and find its weight. (*b*) Bring the beaker of water into the position shown in Fig. 98, so as to submerge the body, and weigh again. The loss of weight in water affords a measure of the buoyant force of the latter, and is equivalent, as we have already seen, to the weight of the water displaced by the submerged body. The ratio of the weight in air to the loss of weight in water,

$$\frac{\text{weight in air}}{\text{loss of weight in water}} = S = \frac{D}{d};$$

for the mass of the body is $M = DV$, and that of the displaced water is $m = dV$, so that

$$\frac{M}{m} = \frac{D}{d} = S.$$

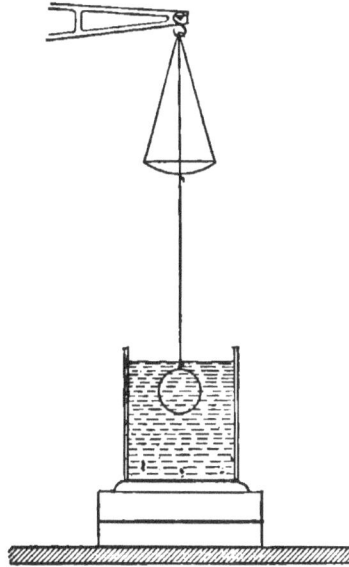

FIG. 98.

The method just described may be used to determine the density of liquids also. In that case a body of *known mass and density* is submerged in the liquid the density of which is desired, and is weighed. A comparison of its loss of weight in that liquid and in water gives the specific gravity of the former.

---

[1] It is desirable to use for this experiment a solid with smooth surface and free from cavities or indentations. When it is necessary to find the density of materials of rough exterior, or porous, they should be submerged, and the vessel of liquid containing them should be placed under the exhausted receiver of an air pump, until the air entrapped within the interstices of the material is set free.

109. Method of the Hydrometer. — The fact that a floating body always displaces its own mass of the liquid which sustains it, is made use of in the hydrometer, which is an instrument designed for the ready approximate indication of the densities of liquids. The hydrometer consists of a glass float of the form shown in Fig. 99. It is ballasted by means of mercury or shot in the bulb b, so that it will rest in an upright position, all submerged excepting a portion of the narrow cylindrical stem. The extent to which the stem itself will be submerged depends obviously upon the volume of liquid necessary to equal the hydrometer in mass. Hydrometers are usually constructed with a scale which gives the densities directly in terms of that of water. If the instrument is to be used in liquids denser than water it is ballasted so as to float almost submerged in the latter. (See (1), Fig. 99.) With increasing density of the surrounding liquid the stem emerges above the surface more and more. For use in liquids lighter than water the ballast is adjusted so that the hydrometer will float in that liquid with nearly the whole of the stem exposed. As the density of the liquid diminishes, the hydrometer sinks, and new parts of the scale upon the stem come to the surface.

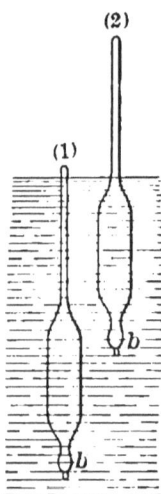

FIG. 99.

Many special forms of hydrometer are used in the industrial arts, and these possess scales convenient to their various purposes. Such instruments are the alcoholometer, which indicates directly, by the depth to which it sinks, the percentage of alcohol which the sample of spirits in which it floats contains; the salinimeter and acidimeter, which give in direct readings the degree of concentration of brine or the strength of an acid.

110. Hydrometers of Constant Immersion. — All of the hydrometers mentioned in Art. 109 have a common principle of action. They are called *hydrometers of variable immersion.* Fahrenheit, the originator of the type of thermometer which goes by his name, devised an hydrometer, the slender neck of which carries but one mark. The instrument is constructed with a tiny scale pan above (Fig. 100),

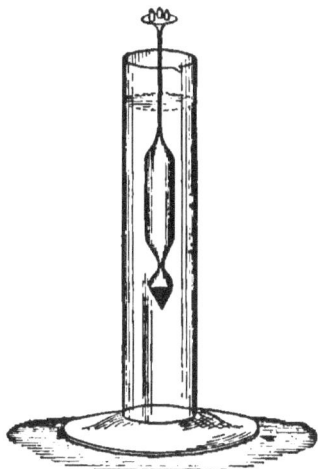
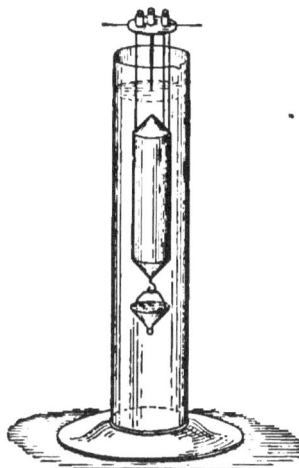

FIG. 100.   FIG. 101.

and the method of using it consists in placing weights upon the pan until the mark coincides with the level of the liquid. The mass thus added, plus the mass of the hydrometer itself, measures the mass of the displaced liquid. Nicholson extended the usefulness of the hydrometer of constant immersion to the measurement of the density of solids. The Nicholson hydrometer (Fig. 101) is constructed of metal instead of glass. The ballast is placed in a conical vessel below the float, and the cover of this vessel serves as a submerged scale pan. The method of using this type of hydrometer is indicated in the following experiment:

111. EXPERIMENT 28. — The Measurement of Densities with the Nicholson Hydrometer.

Apparatus:

(1) The hydrometer and a set of weights from 1 mg. to 20 g.

(2) A cylindrical vessel of water; a piece of metal, or glass, or a crystal insoluble in water. The weight of the specimen should be about 10 g.

Procedure:

(a) Float the hydrometer, and add weights to the upper scale pan until the point below the pan touches the surface of the water; or if the instrument be one of the form with a single upright, until the mark coincides with the surface.[1]

(b) Note the weights upon the pan, and remove them, repeating the count. Place the specimen upon the upper pan, and add weight sufficient to bring the point again into contact with the surface film. Count and remove the weights as before. *The difference between this weight and the previous one is the weight of the specimen in air.*

(c) Place the specimen upon the submerged pan, and add weights to the upper pan until the adjustment is restored. The difference between this weight and the weight used in operation (a) is the weight of the specimen in water.

112. Method of the Specific Gravity Flask. — In measuring the density of fluids it is not necessary to have recourse to the indirect methods based upon the principle of Archimedes. Indeed the customary method, when the

---

[1] The instrument here described is a modification of the hydrometer of Nicholson, in which the adjustment is made by means of a pointer projecting downwards from the middle of the scale pan. This device affords a much more delicate and definite indication of the depth of submergence than it is possible to get with a mark upon the neck of the hydrometer as found in the original form.

density of a liquid is required with greater accuracy than that easily obtainable with the hydrometer, consists simply in weighing a flask of known volume (and of known mass when empty) which has been filled with the liquid in question. Such a vessel is called a *specific gravity flask.* It is commonly given the form shown in Fig. 102, and consists of a bottle of thin glass, the accurately ground stopper of which contains a capillary opening, through which at the time of filling the excess of liquid may escape. In the case of gases the procedure is analogous, but the

Fig. 102.

flask becomes a metal globe containing several liters of gas, and provided with a stopcock.

113. EXPERIMENT 29. — Density of Liquids by the Method of Liquid Columns (Hare's Method).

Apparatus:
(1) The device shown in Fig. 103, which consists of two long straight tubes of glass, terminating beneath the surface of two beakers, which contain the liquids to be compared.

The upper ends of the tubes are connected together and to the air pump by means of the " *T* " tube (*t*) and connecting tubes of soft rubber.

(2) Some water, alcohol, solution of brine, oil, mercury, etc.

(3) An air pump.

(4) A scale divided to centimeters.

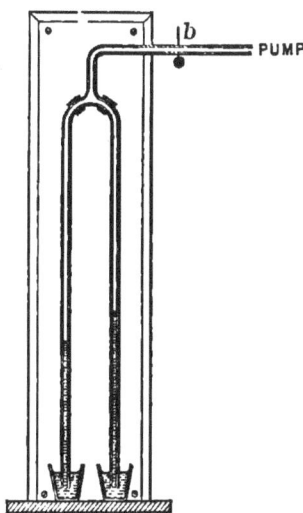

Fig. 103.

Procedure:
(*a*) One of the liquids to be tested is poured into one of the beakers. The other beaker is nearly filled with water.

(*b*) The pressure within the tubes is reduced (necessarily by like amount) by the action of the air pump, whereupon the atmospheric pressure upon the surface of the liquids within the beakers is no longer balanced by the pressure within the tubes, and drives the liquid upwards in each of these to a height inversely proportional to its density. (See further Chapter XV, *Properties of Gases.*)

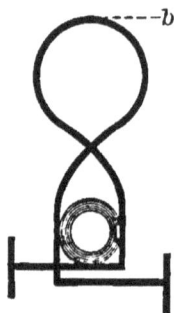

The liquid columns are sustained by closing *b*, which is a pinchcock of the well-known form shown in Fig. 104. The experiment consists in measuring the heights of the liquid columns thus formed, using substances of known density, which should have been previously determined by one of the methods already described, and verifying the statement just made.

FIG. 104.

114. The Density of Certain Substances.

TABLE.

I. THE COMMON ELEMENTS.

Substance	Density	Substance	Density
Aluminium	2·56 to 2·80	Lead	11·20 to 11·45
Antimony	6·70 to 6·72	Magnesium	1·69 to 1·75
Bismuth	9·76 to 9·93	Manganese	7·10 to 8·03
Bromine	3·15	Mercury	13·5958
Carbon { (graphite)	2·17 to 2·32	Nickel	8·57 to 8·93
Carbon { (diamond)	3·49 to 3·53	Nitrogen	0·001254
Cobalt	8·30 to 8.70	Oxygen	0·001429
Copper	8·88 to 8·95	Phosphorus { (common) 1·836	
Gold	19·30 to 19·34	Phosphorus { (red) 2·15 to 2·34	
Hydrogen		Platinum	21·2 to 21·7
Iodine	4·948	Potassium	0·875
Iron { (cast)	7·03 to 7·73	Silver	10·42 to 10·57
Iron { (wrought)	7·79 to 7·85	Sodium	0·978
Iron { (steel)	7·60 to 7·80	Sulphur	1·97 to 2·13
Iron { (pure)	7·85 to 7·88	Tin	6·97 to 7·37
		Zinc	6·87 to 7·24

II. Various Materials.

Anthracite	1·40 to	1·80
Asbestos	2·05 to	2·80
Clay	1·80 to	2·60
Glass	2·50 to	3·90
Granite	2·50 to	2·90
Ivory	1·80 to	1·90
Marble	2·65 to	2·80
Porcelain	2·24 to	2·49

Beeswax	0·960 to	0·965
Butter	0·865 to	0·868
Gasoline	0·660 to	0·690
Linseed oil	0·930 to	0·935
Milk (cow's)	1·028 to	1·035
Palm oil	0·905	
Paraffin	0·880 to	0·930
Petroleum (refined) 250°	0·800 to	0·830

K

CHAPTER XIV

PROPERTIES OF THE SURFACE FILM OF LIQUIDS

115. The Surface of a Liquid is Normal to the Force acting upon it. — The surface of a liquid at rest is always *perpendicular at each point to the resultant force acting at that point.* If gravity be the only force, the surface will be level, *i.e.* at right angles to a line drawn from the surface to the center of the earth. If other forces act, the surface takes other forms, but always in accordance with the above principles. When solids and liquids come into contact, the molecular forces, the existence of which between solids and liquids has been demonstrated in Chapter X, produce easily observable modifications of the form of the surface.

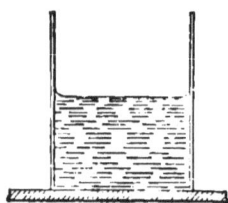
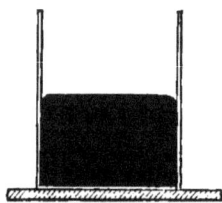

FIG. 105.   FIG. 106.

A glass vessel of water partly filled presents the appearance of Fig. 105, the liquid rising on every side to meet the glass. If water be added to a level slightly above the lip of the glass, the molecular forces prevent outflow. The excess, which may be plainly seen in a flat dome above the vessel, is held in position by the surface layer of the liquid, the particles of which are locked together so as to form a film or skin.

Mercury in a glass vessel shows analogous phenomena save that the surface is convex instead of concave

(Fig. 106). The film in the latter case does not make molecular contact with the glass, but shrinks away from it.

The surface of a liquid contained in a vessel will be concave whenever the liquid wets the containing vessel; that is to say, whenever the molecular forces between the solid and liquid exceed those between neighboring particles of the liquid. It will be convex whenever the molecular forces between the liquid particles are in excess.

. Analogous deformations of the surface occur wherever solids and liquids are brought into contact. If a glass rod be brought into contact with water from above, the liquid film rises along the surface of the glass, and the rod may be raised above the general level, lifting with it, by virtue of the

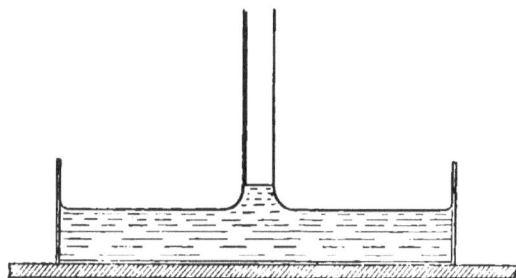

FIG. 107.

strength of the surface film, a considerable volume of liquid (see Fig. 107).

When such a point impinges upon mercury, the film bends under it so that a cavity is formed within which the end of the rod lies below the general level of the mercury, but not within the mass of the liquid itself (Fig. 108).

Even if the rod be plunged to great

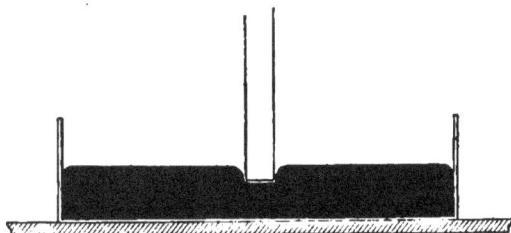

FIG. 108.

depth below the level of the mercury, there will be no true junction between the glass and the liquid like that which

exists between glass and the water in which it is submerged.
The surface film will still separate the body of the liquid
from the glass, and a thin layer of air will remain between
the two substances.

If open tubes be plunged into any liquid, this same action
of the surface film produces either a rise in the level of the
liquid within the tube (Fig. 109) where the surface is concave, or a corresponding depression (Fig. 110) where the
surface is convex.

In tubes of small bore the rise is very noticeable; whence
the name **capillarity**, which has been extended to all the
phenomena depending upon the action of the surface film.

116. The Law of Diameters. — The rise or depression of
liquid in tubes, measured from the general level of the
surrounding liquid, is *inversely proportional to the diameter
of the tube.*

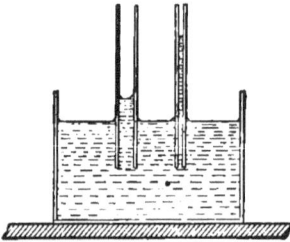
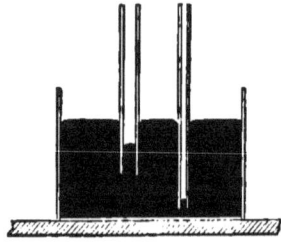

FIG. 109.   FIG. 110.

117. Theory of the Surface Film. — The reason for as-
cribing these phenomena to action of a surface film is as
follows: Consider any particle within the body of a liquid.
If it is far below the surface, the molecular forces due
to the surrounding particles will be the same in all directions. If, however, the particles lie close to the surface,
as in Fig. 111, where m is the particle, and r is the radius
of the sphere within which the action of the molecular

forces is appreciable, there will be forces due to particles immediately below *m*, which are not counterbalanced by forces from above, owing to the proximity of the surface. The resultant of all these unbalanced forces is inwards at right angles to the surface, and the resultants which exist in

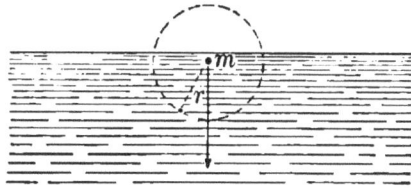

Fig. 111.

the case of all the liquid particles near the surface, constitute a *pressure* normal to the surface exerted by the particles which form the surface layer (or film) upon the body of the liquid within.

It is evident that this pressure will be greater in the case of a convex film (Fig. 112), and less in the case of a concave film (Fig. 113), than in that of plane film.

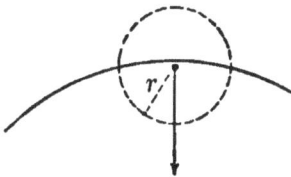

Fig. 112.

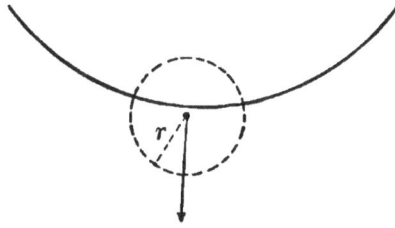

Fig. 113.

The cause of the rise or depression of liquids in capillary tubes obviously lies in the curvature of the surface film. The pressure exerted by a plane film is capable of driving liquid above the general level in tubes which the liquid wets, because the film which caps the liquid in such tubes is concave. When, on the other hand, the film is convex, which is the case when the solid is not wet, the film is stronger, and it depresses the liquid below the general level, as already described.

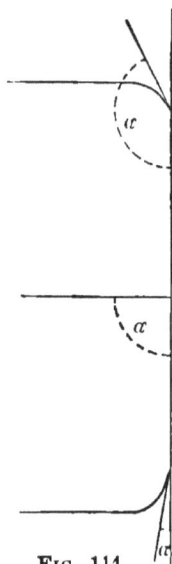

118. Angle of Contact. — The curvature of the liquid film in the neighborhood of solids depends, as has already been shown (Art. 115), upon the relative strength of the molecular forces within the liquid and between liquid and solid. This relation determines the *angle of contact*, by which the capillary behavior of liquids in various containing vessels may be defined. This is the angle *α*, Fig. 114, which the film makes with the solid wall at the point of contact. The angle may be greater, equal to, or less than 90°

In the first case there will be *depression*, in the last *elevation*, in capillary tubes.

Fig. 114.

119. EXPERIMENT 30. — **Van der Mensbrugghe's Experiment showing that a Liquid Film is always under Tension.**

Apparatus:

(1) A wire ring about 10 cm. in diameter, within which hangs a loop of thread as shown in Fig. 115.

(2) A dish containing a soap solution.

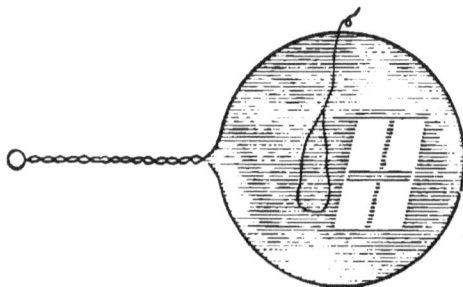

FIG. 115.

Procedure:

(*a*) Dip the ring into the solution, whence it emerges carrying a plane soap film bounded by the ring and completely filling it. The loop of thread may be seen floating in the liquid layer within which it moves freely, excepting as it is constrained by the strand which connects it with the ring.

(*b*) The film consists of two distinct regions: that outside of the loop of thread and that within the loop.

Puncture the latter region with any convenient tool, such as a pin or needle or a bit of wire. The result, which follows instantly, is most beautiful and striking. The thread, hitherto lax, is drawn into circular form by the tensile forces of the outer region of the film which are no longer balanced by those of the region within the loop. It is now held tensely to its position of equilibrium within the film

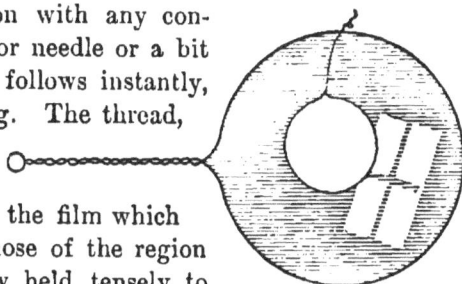

FIG. 116.

(Fig. 116), and its laxity and freedom of motion have disappeared.

If the ring used in this experiment be mounted in the field of a lantern, this phenomenon can be exhibited to many observers simultaneously.

120. EXPERIMENT 31. — **Contractile Power of a Water Film.**

Ernest Nichols has extended the principle of Van der Mensbrugghe's experiment to the study of the surface film of a body of water. In

FIG. 117.

this form of the experiment an ordinary rubber band, of the oblong, slender form used to hold packages of postal cards, etc., together, is made to float upon the surface of water (Fig. 117).

The inner region of the film, that, namely, which is bounded by the loop of caoutchouc, is then touched with a wire or splinter of wood which has been dipped into alcohol or oil. The result is to alter the angle of contact and with it the tension of the film. The outer region, still uncontaminated, exerts in full its previous pull upon the loop, and these forces, no longer balanced by those from within, distend the band into the form shown in the figure. The application of a trace of the alcohol or oil to the outer region will instantly restore the band to its original lax position and shape.

121. EXPERIMENT 32. — Contractile Tendency of the Soap Bubble.

Apparatus:

A glass funnel about 8 cm. in diameter at the mouth, some soap solution, a lighted candle.

The soap bubble consists of a layer of liquid the outer film of which, being convex, must always exert a greater pressure than does the inner concave film. There is therefore a resultant pressure tending to expel the contained air from the bubble and to reduce the size of the latter.

Procedure:

Blow a bubble with the above apparatus, and hold the open tip of the funnel horizontally near the candle flame, as in Fig. 118. The outflow will be found sufficient to distort and almost to put out the flame.

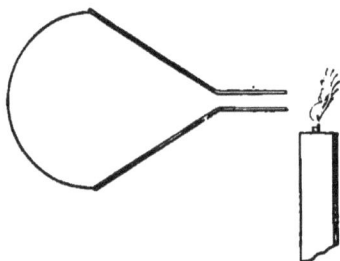

FIG. 118.

Note that this effect does not cease when the soap film reaches the mouth of the funnel, but that the film continues to move toward the apex, driving the air before it. This tendency to recede into the part of the funnel which has the least diameter is due to the difference in curvature of the two films which inclose the liquid layer (Fig. 119). The resultant pressure is always inward. Where the angle between opposite walls is considerable, the force is sufficient to lift the layer bodily against gravity.

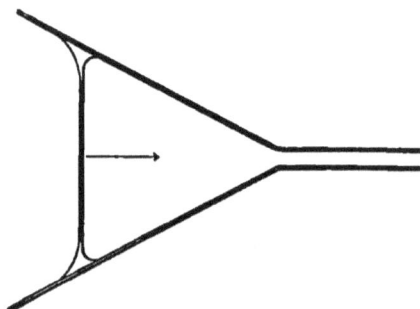

FIG. 119.

This statement should be verified by turning the tip of the funnel upwards and observing the behavior of the soap film.

CHAPTER XV

PROPERTIES OF GASES

122. Definition of a Gas.—Gases differ from liquids in not having a surface film. There is much greater freedom of motion between the particles, and the molecular forces do not produce the same degrees of constraint as in the case of solids and liquids. When subjected to pressure, liquids change volume but slightly (water under 265 atmospheres of pressure loses only about $\frac{1}{10000}$ of its original volume). Gases, on the other hand, are freely compressible.

123. Boyle's Law (Mariotte's Law).—The relation between volume and pressure in the case of gases is a perfectly definite and very simple one, and it is the same for all gases. It was announced by Boyle (1662) and independently by Mariotte (1679), and is as follows:

The volume of a gas varies inversely as the pressure to which it is subjected.

The equation $$\frac{V_2}{V_1} = \frac{P_1}{P_2},$$

in which V_1 and V_2 are the volumes of a given body of gas at pressures P_1 and P_2 respectively, is a statement of Boyle's law.

124. EXPERIMENT 33.—**Measurement of the Mass of a Gas.**
Apparatus:
(1) A thin-walled brass sphere containing about 1000 cc., with stopcock threaded to fit the air pump (such spheres are usually fur-

nished as accessories to that instrument, and afford the best apparatus for this experiment).

(2) An air pump.

(3) A balance and weights.

Procedure:

(a) Weigh the receptacle filled with air; then exhaust upon the air pump, close the stopcock, and weigh again. Note the marked diminution of mass due to the absence of gas which the receptacle contained when first weighed. (This would amount to 1·3 grams per liter in case the exhaustion were complete.)

(b) After weighing again, attach the receptacle to a gas jet by means of a rubber tube, open the stopcock, and turn on the gas. (The gas should have been on previously for a second or two and allowed to escape, so as to free the pipes from air.) Turn off the gas, close the stopcock, detach the receptacle, and weigh again. Note that there is increase over the mass of the empty receiver, but that the weight is decidedly less than when filled with air. If carbon dioxide, nitrous oxide, or hydrogen, in cylinders, are accessible, the receptacle may be exhausted and filled with these also and weighed.

Accurate measurements of the density of gases cannot be expected with the apparatus described above. The various weighings should, however, indicate densities relatively of the same order as those given in the following table.

TABLE.

DENSITY OF CERTAIN GASES AT 76 CM. PRESSURE AND 0° C. (EXPRESSED IN GRAMS PER LITER).

Gas.	Density compared with air.	Mass in grams per liter.
Air	1·0000	1·2934
Carbon monoxide (CO)	0·9671	1·2505
Carbon dioxide (CO_2)	1·5198	1·9651
Chlorine (Cl_2)	2·4495	3·1674
Hydrogen (H_2)	0·0692	0·0895
Methane (marsh gas) (CH_4)	0·5530	0·7150
Nitrogen (N_2)	0·9702	1·2546
Nitrous oxide (N_2O)	1·5229	1·9692
Oxygen (O_2)	1·1053	1·4292
Illuminating gas	0·438±	0·566±

125. Extension of the Principle of Pascal to Gases. — Since a gas is a perfect fluid, it transmits pressure, and when it is at rest the pressures at any point within it are always the same in every direction. (Consider the spherical form of soap bubbles in still air as a delicate proof of this statement.)

Since gases have mass, however, pressures must increase from above downwards, owing to the action of gravity; the pressure at any level being that due to the weight of the superincumbent column of gas. It is possible, by means of a device due to the Italian physicist, Torricelli (1644), who was a disciple of Galileo, to show the existence of this pressure in the case of the earth's atmosphere, and to measure the amount of it.

126. Torricelli's Experiment. — Torricelli filled glass tubes closed at one end with mercury, and inverted them in a cistern of the same liquid.

He found that when the tube had a length of less than about 76 cm., in the case of an experiment performed at the level of the sea (*a*, Fig. 120), or when it was so inclined that the vertical height above the level of the cistern (*b*, Fig. 120) was less than 76 cm., the tube remained full. The mercury which it contained was held in place by the atmospheric pressure upon the surface of the cistern. When, however, the tube rose to a greater height than the above (*c*, Fig. 120), the mercury receded in the tube, and came to rest in such a position that the column, sustained above the general level of the cistern, precisely balanced the atmospheric pressure. This occurs, as indicated above, when the column has a vertical height of about 76 cm. The amount varies slightly in a given locality from time to time through a range of several centimeters.

Torricelli's apparatus affords the most direct and satisfactory means of measuring atmospheric pressure. When used for this purpose it is known as the **mercury barometer**.

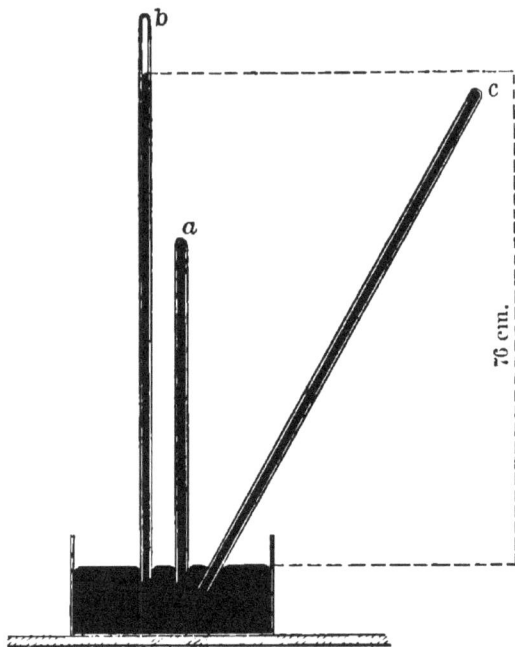

FIG. 120.

127. Construction of the Barometer. — In order that Torricelli's apparatus may serve as an instrument of precision, certain conditions have to be very carefully fulfilled.

(1) The mercury must be pure and clean.

(2) The vacuum above the mercury must be as nearly as possible complete; *i.e.* free from air, aqueous vapor, etc.

(3) The bore of the tube must be so large that the depression of the column, due to the action of the surface film, will be inappreciable, or the diameter must be known, and a correction for the depression must be applied.

The first condition is attained by distilling the mercury

before placing it in the tube; the second, by boiling the mercury in the tube before inverting it in the cistern, thus driving off all gases and moisture.

Barometers are provided with a scale by means of which the height of the top of the mercury column, above the level of the mercury in the cistern, may be read in centimeters and decimal parts of a centimeter (or in countries where the British system of measurement is used, in inches and fractions of an inch).

In order to maintain the mercury in the cistern at a constant level, the latter is commonly provided with a flexible bottom of leather which can be raised and lowered by means of a screw from below (Fig. 121), and with a pointer the apex of which is rigidly adjusted to the zero point of the scale. Before making a reading the mercury is raised until its surface makes contact with the pointer.

Barometers thus arranged are known as *Fortin* barometers.

Sometimes barometers are given the form shown in Fig. 122, in which a bent tube, with one open and one closed arm, is filled with mercury. In this case the difference in the height of the two mercury columns indicates the pressure, and there is no correction for capillary depression.

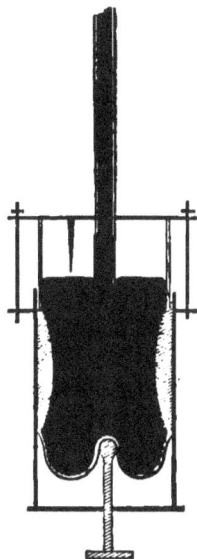

Fig. 121.

128. Nature of the Atmosphere. — From the phenomenon presented by the barometric column certain important conclusions may be drawn concerning the constitution of the atmosphere. A layer of air everywhere of the same density as the atmosphere at the level of the sea, and capable of sustaining a mercury column 76 cm. high against gravity, would need to have a height H, such that,

$$H : h :: d_m : d_a, \text{ or } H = h\frac{d_m}{d_a},$$

where h is the height of the mercury column, d_m is the density of mercury, and d_a is the density of air at 76 cm. pressure. Now 1 cc. of mercury weighs 13.596 grams, and 1 cc. of air at the same temperature (0°) weighs 0·001293 gram.

The value of H is, therefore,

$$H = 76 \text{ cm.} \times \frac{13\cdot596}{0\cdot001293} = 799094 \text{ cm.}$$

$$= 4\cdot95 \text{ miles nearly.}$$

The air being a compressible fluid, however, does not lie in a layer of uniform density enveloping the earth, but diminishes rapidly in density from the surface upwards. There is abundant evidence of the existence of atmosphere at much greater heights than the above. Indeed, direct observations with the barometer have been made by aeronauts at a height of 9600 meters = 31,500 ft.,[1] and there are reasons to think that, for at least fifty miles above the surface of the earth, there is an appreciable atmosphere.

FIG. 122.

---

[1] At that distance above the sea the barometric column was less than 23 cm. long. Life could be sustained, even temporarily, only by supplementing the scanty supply of air by inhalations of oxygen from a cylinder of the compressed gas provided for such an emergency. (See account of the ascent made by Dr. Berson, Dec. 4, 1894; *Nature*, Vol. 51, p. 491.)

129. Manometers. — The barometric column may also be used to measure changes of pressure artificially produced. If, for example, the cistern of a barometer be placed upon the plate of an air pump with the tube projecting through the neck of the receiver, as shown in Fig. 123, we can reduce the pressure upon the surface of the cistern at will. Every stroke of the pump will then lower the mercury column until the minimum pressure is reached.

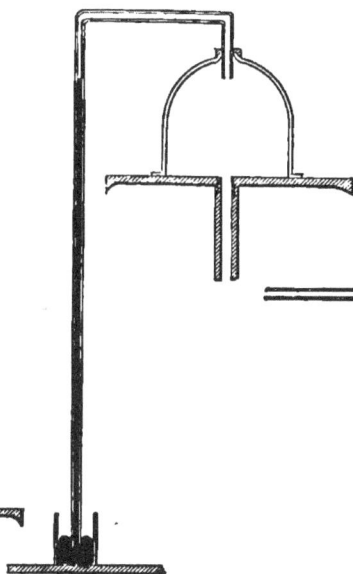

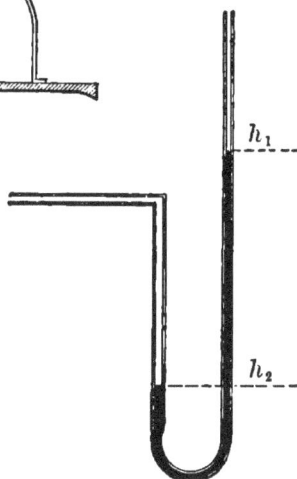

FIG. 123. FIG. 124. FIG. 125.

To any such device for measuring artificial pressures, the name *manometer* is applied.

For the measurement of pressures less than one atmosphere, the form of manometer shown in Fig. 124 is frequently used. It is a barometer in which the vacuum above the mercury column is variable.

Where pressures both greater and less than one atmosphere are to be measured, the *open tube manometer* (Fig.

125) is employed. The difference of level ($h_1 h_2$) of the mercury in the two arms indicates the amount ($+$ or $-$) by which the pressure varies from one atmosphere.

For pressures so great that an open tube manometer would be of inconvenient length, the end of the tube is

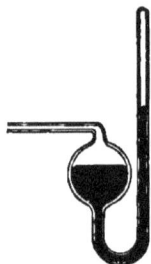

sealed as in Fig. 126, and advantage is taken of the compressibility of air (see Boyle's law) to secure a manometer of compact form. To this type the name *closed tube manometer* is given.

Sometimes, finally, the bending of the flexible top of a metal capsule or box, or the distortion of a curved and flexible tube to which pressure is applied, is magnified me-

FIG. 126.

chanically, giving motion to a pointer which moves along a dial. This principle is applied to barometers (aneroid

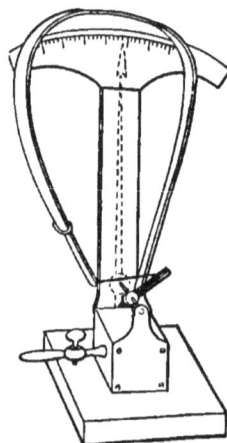

FIG. 127.   FIG. 128.

barometers) and to pressure gauges. The stiffness of the moving part is adapted in each case to the pressures which it is intended to indicate.

Figures 127 and 128 show the essential features of two well-known forms.

130. EXPERIMENT 34. — **Measurement of the Vacuum produced by a Mechanical Air Pump, and Determination of the Rate of Leakage of the Pump and Receiver.**

Apparatus :

(1) An air pump with open neck receiver.

(2) A wooden scale divided to centimeters; a watch or other time-piece.

(3) A barometer.

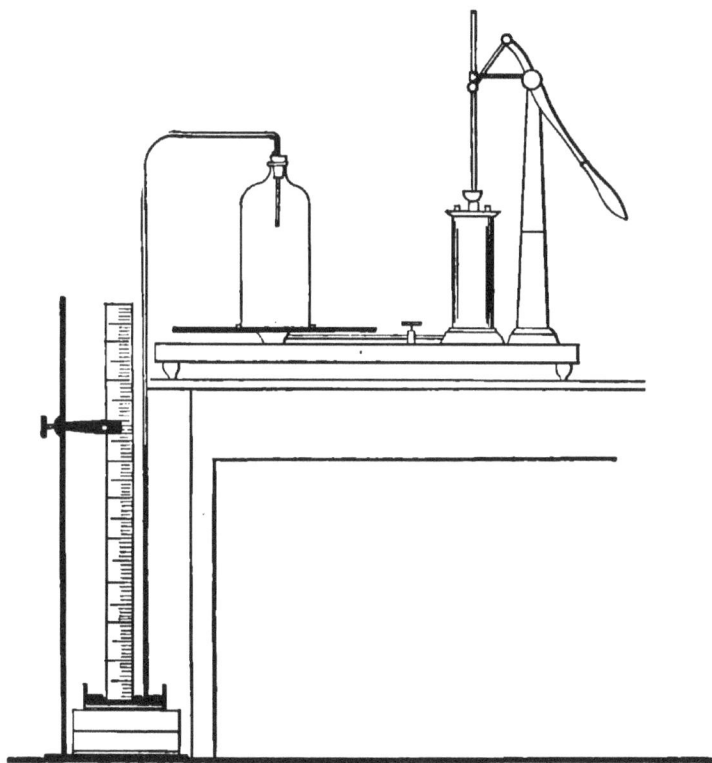

FIG. 129.

Procedure :

(a) Bend a piece of glass tubing, about 120 cm. long and 0·5 cm. inner diameter, so as to form a manometer of the form shown in Fig. 124. Fit the short arm to the neck of the receiver with a good cork or rubber stopper. Mount the receiver, with its manometer

L

tube, upon the plate of the air pump. Bring a flat dish of mercury from below, raising it until the open lower end of the manometer is well submerged and supporting it in position with blocks. Set up the scale vertically with its edge touching the tube, the lower end of the scale submerged in the mercury of the cistern (Fig. 129).

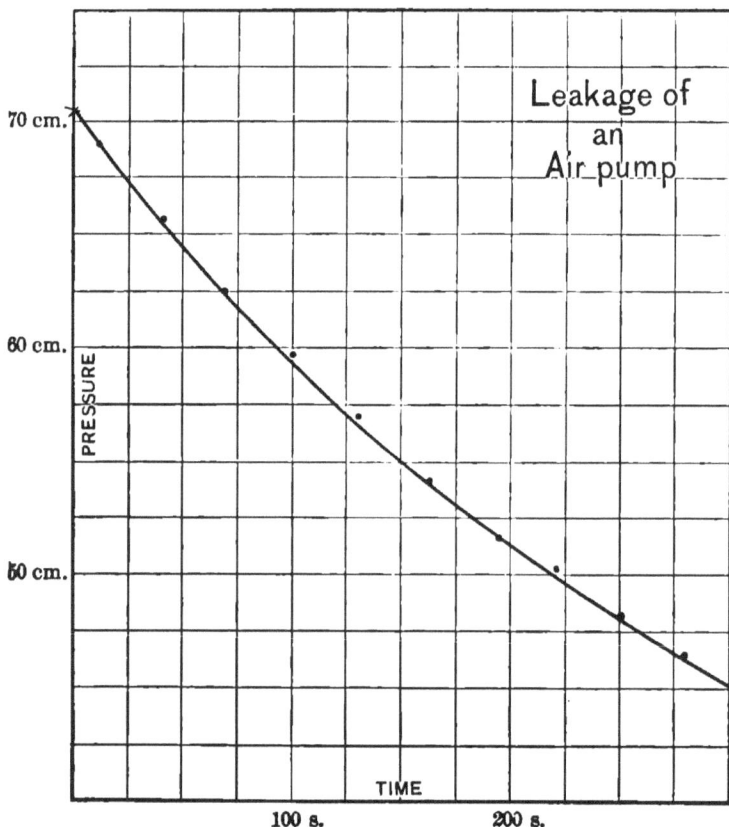

FIG. 130.

(*b*) Take fifty strokes with the pump, timing them to about two-second intervals. Ten seconds after the last stroke read the position of the mercury in manometer column. Immediately thereafter note the height to which the mercury in the cistern stands upon the scale. Every 30 seconds, for at least 5 minutes after the first reading of the height of the manometer column, repeat that reading. (If the leakage is so rapid as to render readings uncertain, apply beeswax.)

The rate of leakage will vary. Occasionally the loss of vacuum will be too slow to give marked change within the time assigned to the readings. In such a case readings may be made at longer intervals.

(c) Read the barometer,[1] and convert the readings into centimeters if necessary.

(d) From these data, which should be tabulated as below, find graphically the pressure at the completion of the last stroke. For this purpose construct a curve with times as abscissas and manometer readings as ordinates. This curve can be drawn backwards so as to cut the line corresponding to 0 seconds (or to the time of the last stroke). The ordinate at that point is the required value. (See Fig. 130.)

TABULATION OF DATA. — EXPERIMENT 34.

Times of observation.			Time after 50th stroke.	Manometer reading.	Height of column.[2]
h.	in.	s.	sec.		
4	30	10	10	70·0	68·7
"	30	40	40	66·8	65·5
"	31	10	70	63·9	62·6
"	31	40	100	61·2	59·9
"	32	10	130	58·5	57·2
"	32	40	160	55·4	54·1
"	33	10	190	53·3	52·0
"	33	40	220	51·7	50·4
"	34	10	250	49·7	48·4
"	34	40	280	47·5	46·2

[1] In case no barometer is available, assume the height to be 76 cm. and use that value in subsequent computation. There will be a systematic error in values for the vacuum, equal to the difference between the assumed and the true barometric height.

[2] Obtained by subtracting the height of the mercury in the cistern above the zero of the scale ; *i.e.* 1·3 cm.

Barometer reading, 74·3 cm.

The curve corresponding to these readings (Fig. 130), when extended to the ordinate for 0 sec., gives for the *initial* height 70·5 cm. The pressure was therefore reduced to 74·3 − 70·5 = 3·8 cm.

PART II — HEAT

CHAPTER XVI

NATURE AND EFFECTS OF HEAT

131. Heat is the name given to a form of energy due to some motion of the particles of a body among themselves, and not of the body as a whole.[1]

Of the various effects of heat, which we shall have occasion to consider, one of the most general is, to increase the temperature of the body. This increase of temperature is usually accompanied by expansion.

For the purpose of indicating the condition of a body with respect to heat energy, also the changes as loss or gain of heat which it undergoes, the above effect, *i.e.* the increase in length or in volume, is the one chiefly employed. This effect is illustrated by means of the following experiments.

132. EXPERIMENT 35. — **Expansion of a Bar.**

Apparatus: (1) A bar of brass or iron about 50 cm. long, and mounted as shown in Fig. 131. The supports s_1 and s_2 are strips of sheet metal cut out in V form to receive the bar, and fastened to wooden blocks.

FIG. 131.

[1] See *Elements of Physics*, Vol. I, p. 151, introductory remarks concerning heat.

148

(2) A small plane mirror. This is fastened to a metal disk (Fig. 132), across the back of which is soldered a thin strip of steel, or of spring brass, about 0·2 cm. wide and 8 cm. long. Just below

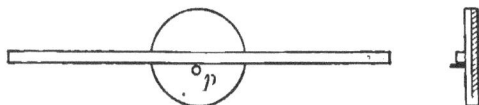

FIG. 132.

the center of this strip, and touching the edge of it, a metal pin p (Fig. 131) is inserted, or is soldered to the disk. The ends of the metal strip are tacked to the arms of a wooden yoke, between which the mirror swings (Fig. 133), and this in turn is clamped to a wooden block at such height that the pin will coincide nearly with the axis of the bar, the expansion of which is to be measured.

Procedure:

(*a*) The bar is mounted with one end in contact with the tack-head

FIG. 133.

(Fig. 131). The block carrying the yoke and mirror, m, is then brought into position at the other end of the bar and adjusted until the pin, which is parallel to the axis of the bar, is in firm contact with the end of the latter.

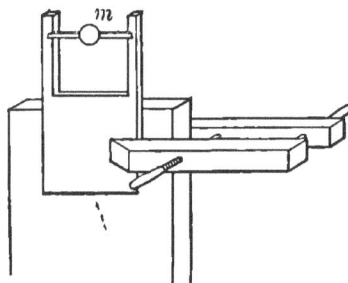

(*b*) A beam of light, either of sunlight directed by means of a mirror, or the rays from a projecting lantern, is thrown obliquely upon the mirror m, and thence to screen. This arrangement of the apparatus once perfected, every increase in the length of the bar will tip the mirror and move the spot of light along the screen.

(*c*) The bar is heated by passing the flame of a Bunsen burner, held in the hand of the experimenter, back and forth between the supports S_1 and S_2. The movement of the spot of light which follows this heating is observed. Note that the direction of the motion is that which corresponds to an elongation of the bar. (With the above described arrangement of the apparatus, the bar in cooling will not return to its original position, a fact which in no way interferes with the success of the experiment.)

133. Expansion of Liquids. — In the study of the expansion of liquids various indirect methods are employed.

To demonstrate merely the fact of expansion, the following device suffices:

A glass bulb, *b* (Fig. 134), is filled with some liquid (glycerine, oil, mercury, or water), until the contents rise part way into the long neck. Around the neck is placed a small ring of rubber,[1] which is to serve as a mark of the height at which the liquid stands.

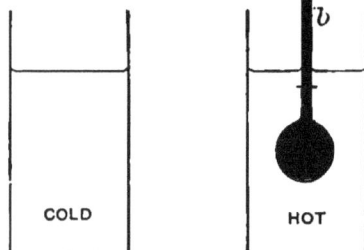

The bulb is plunged in a vessel of cold water, where it remains for several minutes.

The ring is slipped into place to mark the height of the liquid, and the bulb is transferred to a vessel of hot water. The rise of the liquid within the neck indicates not only that the liquid expands with heat, but also that it expands more rapidly than does the containing vessel.

Fig. 134.

The experiment is really a demonstration of what is termed the *relative expansion of a liquid*. When the containing vessel expands more rapidly with rise of temperature than the contained liquid does, the relative expansion is negative. With few exceptions, however, liquids expand more rapidly then solids.

134. Absolute Expansion of Liquids. — To obtain the expansion of a liquid, independently of the expansion of the

[1] Another marker, which is easily constructed, consists of a strip of sheet iron, the length of which is slightly less than the outer circumference of the neck of *b*. If this be rolled into a ring (Fig. 135), it may then be opened and sprung upon the neck, to which it will cling, and along which it may be moved to any desired position.

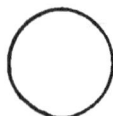

Fig. 135.

containing vessel, the method described in the following experiment is employed.

It depends upon the fact that the height of liquid columns which balance one another are inversely as their respective densities, and dependent only upon the densities.

In this experiment, which is due to the French physicist Regnault, two connecting vertical tubes of glass (Fig. 136) are filled with the liquid to be tested. One of these is packed with ice, while the other is surrounded with steam. The difference in the height of the two columns affords a measure of the relative density of the liquid within the two arms,

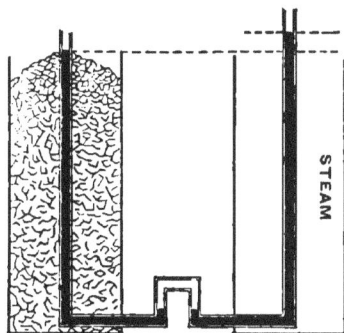

FIG. 136.

and indirectly of the expansion which the former undergoes when heated.

135. Expansion of Gases. — The mere fact of the expansion of gases may be demonstrated by means of apparatus

FIG. 137.

analogous to that used for the relative expansion of liquids; *i.e.* the glass bulb *b*, Fig. 137. This is slightly heated, and the bent end of the long neck is plunged into mercury. Upon cooling, the liquid rises into the neck, which should be slightly inclined from the horizontal position, as shown in the figure.

The gas, when even slightly heated, as by the application of the hand to the bulb, drives the mercury outward along the neck; and when the gas cools, the mercury returns up the neck towards the bulb. Some of the earliest thermometers, devised by Galileo and by Newton, were based upon the same principle as this simple apparatus.

136. Thermometry. — In order to make use of the expansion of any substance for the measurement of temperature, it is necessary to select at least two fixed temperatures of reference and to have some scale by means of which to express the changes which occur. Common experience has led to the adoption of the temperature of melting ice, and that of steam rising from water which boils under a pressure of 76 cm., for the constant temperatures of reference.

Many thermometric scales have been devised, but the one used in scientific work in all parts of the world is the *centigrade scale* devised by Celsius.

In this scale the temperature of melting ice is designated as zero (0° C.), and the interval between the ice point and the steam point is divided into one hundred equal parts, called *degrees*. The designation of the steam temperature at 76 cm. pressure is therefore 100° C.

As the result of common experience, likewise, the relative expansion of mercury in glass has been adopted as the most convenient and trustworthy means of indicating ordinary changes of temperature. The instrument most widely used in thermometry is therefore the *mercury thermometer*. It is constructed as follows:

A tube with capillary bore is sealed at one end, and the sealed end is blown out into an elongated bulb. The bulb is filled with mercury, and is then heated to a temperature above the highest at which the instrument is to be used. When thus heated the expanded mercury must entirely fill the tube. The open end is then sealed by melting, and sometimes a small secondary bulb (v) is blown.

Figure 138 shows the typical form of the mercury thermometer as constructed for scientific measurements.

137. Graduation of a Thermometer. — When the thermometer, constructed as described in the preceding para-

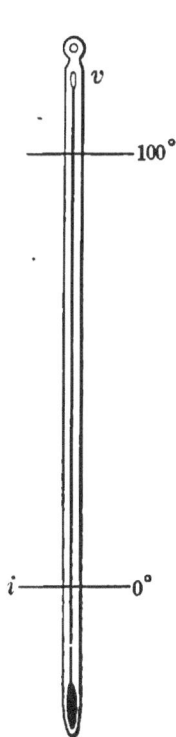

FIG. 138.

FIG. 139.

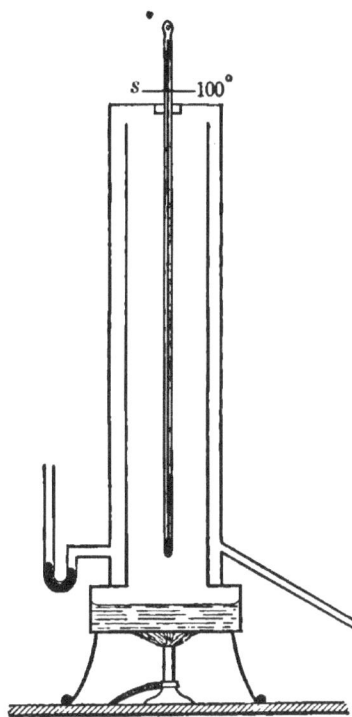

FIG. 140.

graph, is to be provided with a scale, it is placed with its bulb closely surrounded with melting ice (Fig. 139), and the point within the bore to which the top of the mercury column recedes is marked. This is the ice point i, and is taken as the zero of the scale.

The thermometer is then placed in a steam bath (Fig. 140), at a pressure of 76 cm., and the height reached by the mercury is noted. This is the steam point s, and becomes the 100° point of the scale. It now remains only to divide the interval between i and s into one hundred equal parts. Were the bore of the thermometer a true cylinder, these would be divisions of equal length. Owing to the irregularities of the tube, they must, however, be made of unequal lengths, but in such a manner that the successive intervening spaces shall contain equal volumes. The process by means of which the proper lengths are determined is called the *calibration of the tube*. It consists in detaching within the tube a thread of mercury (Fig. 141) and moving it stepwise through the tube, noting its length at each step. The reciprocals of the various

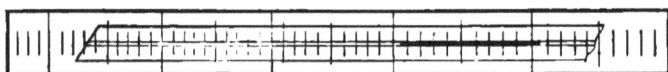

FIG. 141.

lengths thus determined afford a measure of the varying cross-section.

The portions of the tube lying below the ice point and above the steam point are divided into parts equal to those between 0° and 100°. The former are distinguished by the negative sign.

138. EXPERIMENT 36. — **Calibration of the Bore of a Capillary Tube.** The method described above affords the means of an interesting and instructive exercise, which for the sake of convenience may be made with a piece of open tubing of small bore. This method of calibration is of especial importance, because it is applicable to many instruments besides the thermometer.

Apparatus:

(1) A piece of capillary tubing, about 30 cm. long, and open at both ends; a tube of bore so small that a short thread of mercury will maintain its position within, with the tube vertical, but moving freely when the tube is tapped, is to be preferred.

(2) A strip of cross-section paper; some mercury.

Procedure:

(*a*) Introduce into the tube a sufficient amount of mercury to make a thread about 2 cm. long. This is brought nearly to one end by tapping the tube.

(*b*) Lay the tube horizontally upon the cross-section paper, its axis perpendicular to one of the sets of lines, and determine the length of the thread in terms of the number of spaces (or scale divisions) which it covers, also the distance of the nearer end of the thread from the end of the tube, or from a mark thereon which has been placed as a zero of reference. Move the thread by about its own length, towards the middle of the tube and repeat the measurements. Continue thus until nearly the whole length of the tube has been traversed.

(*c*) Tabulate the readings as below, making out the remainder of the table by computation.

The result of these measurements will be to show that the bore is not uniform. Generally speaking, it will be found conical in form, but with more or less marked irregularities. The character of the bore should finally be indicated by plotting a curve in which abscissas are the successive distances of the middle of the thread from zero, *i.e.* from the end of the tube, and ordinates are the reciprocals[1] of the relative lengths of thread. For a truly cylindrical tube this curve would be a horizontal line, the distance of which above the base line is proportional to the cross-section of the bore. For a regularly conical tube, the line would be oblique. In the case of most actual tubes the curve takes a form of which that in Fig. 142 is typical.

139. Mercury in Glass Temperatures. — The desirability of the thermometric scale just described, regarded as a

[1] Reciprocals are plotted instead of lengths, because the latter are *inversely* as the cross-section, or as the volume per unit length of tube, whereas the reciprocals are *directly* proportional to the volumes. The curve therefore gives a sort of picture, very much exaggerated, of the irregularities of the bore.

TABLE.

CALIBRATION OF A CAPILLARY TUBE.

No.	READINGS. End a (distance in scale divisions from zero).	End b (distance in scale divisions from zero).	Length of thread in scale divisions (l).	Position of middle of thread (= distance of a from zero + $\frac{1}{2}l$).	Relative lengths in terms of mean length $\left(=\frac{l}{1\cdot0227}\right)$.	Reciprocals of relative length of thread $\left(=\frac{1\cdot0227}{l}\right)$.
1	0·00	1·04	1·04	0.520	1·017	0·983
2	1·19	2·23	1·04	1·710	1·017	0·983
3	1·98	3·01	1·03	2·495	1·007	0·993
4	3·04	4·06	1·02	3·550	0·997	1·003
5	4·00	5·03	1·03	4·515	1·007	0·993
6	4·88	5·90	1·02	5·390	0·997	1·003
7	6·01	7·03	1·02	6·520	0·997	1·003
8	7·12	8·17	1·05	7·745	1·027	0·972
9	8·21	9·23	1·02	8·720	0·997	1·003
10	9·20	10·22	1·02	9·710	0·997	1·003
11	10·20	11·23	1·03	10·715	1·007	0·993
12	11·15	12·17	1·02	11·660	0·997	1·003
13	12·13	13·14	1·01	12·635	0·988	1·013
14	13·19	14·20	1·01	13·695	0·988	1·013
15	14·17	15·19	1·02	14·680	0·997	1·003
16	15·07	16·09	1·02	15·580	0·997	1·003
17	16·12	17·14	1·02	16·630	0·997	1·003
18	17·22	18·24	1·02	17·730	0·997	1·003
19	18·31	19·32	1·01	18·815	0·988	1·013
20	19·81	20·83	1·02	20·320	0·997	1·003
21	20·95	21·98	1·03	21·465	1·007	0·993
22	21·97	22·99	1·02	22·470	0·997	1·003
23	22·99	24·00	1·01	23·495	0·988	1·013
24	24·07	25·08	1·01	24.575	0·988	1·013

Mean length = 1·0227.

device for measuring temperature, depends upon the accuracy with which its successive intervals correspond to equal accessions of temperature. They would correspond precisely, provided the relative expansion of mercury were uniform. That liquid, however, expands more rapidly at high than at low temperatures, and the same thing is

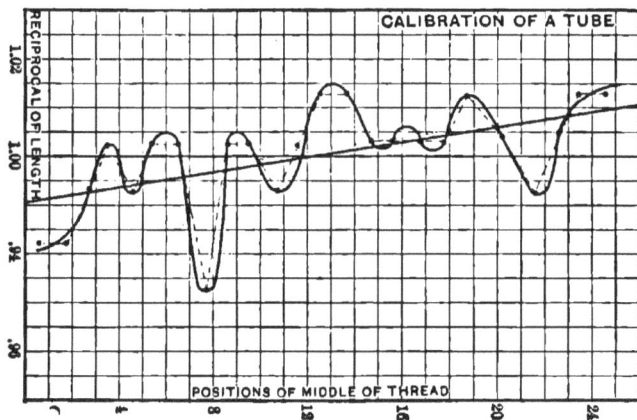

Fɪɢ. 142.

true to a much greater extent of the glass bulb. Between 0° and 100° the errors due to this cause are slight, amounting to only a fraction of one degree. In scientific work of such accuracy that these small errors are of importance, it is customary to indicate the nature of the scale by designating the temperatures as *mercury in glass temperatures*.

140. Charles's Law; the Air Thermometer. — All solids and liquids show variations in the law of their expansion. Perfect gases, however, expand uniformly, obeying the following laws:

(1) *When various gases are equally heated under constant pressure, they all suffer the same expansion.*

(2) *For every degree centigrade which a gas is heated, it expands by an amount equal to* $\frac{1}{273}$ *of its volume at* 0° *C.*

This law is known as Charles's Law and also as Gay Lussac's Law. It was enunciated by the former physicist in 1787, and more completely developed by the latter in 1802.

On account of this uniformity, the expansion of gases has been made the basis of the *standard scale of temperatures*, to which all other scales are referred for comparison or correction. The two standard temperatures of reference, as in the case of the mercury thermometer, are the ice and steam point (see Art. 137), and the interval is divided into one hundred degrees. When the ice point is taken as zero, the scale is called the *centigrade scale of the air thermometer.*

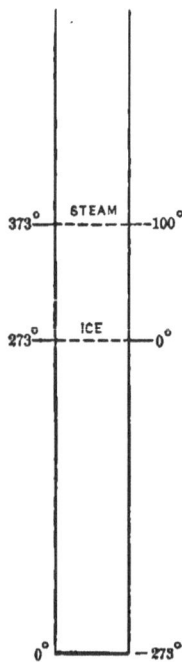

It is sometimes more convenient to use as the zero of the scale a hypothetical point called the *absolute zero*. This may be defined as follows. Imagine an air thermometer cylindrical in form (Fig. 143).

Let the volume of a contained gas, which we must suppose separated from the outer atmosphere by a movable film which it drives before it when heated, occupy the tube up to the line marked *ice*, when brought to the temperature of melting ice, and let it expand to the line marked *steam*, when heated to the temperature of the vapor of water which boils at 76 cm. pressure.

FIG. 143.

If we mark the ice point 0° and the steam point 100°, the scale will be the centigrade scale. If we continue to

divide the tube, using degrees equal to $\frac{1}{100}$, the distance
between the ice point and steam point, the bottom of the
tube will coincide with scale division $-273°$ C. If we
adopt this mark as the zero of our scale and divide the
entire tube into degrees of the same size as before, the
ice point becomes $+273°$ and
the steam point $+373°$. The
zero thus chosen is called the
absolute zero, and the scale is
called the *absolute scale of the
air thermometer.*

141. The Air Thermometer.
— It is not practicable to give
the air thermometer the simple form suggested in the
previous paragraph. The essential features of the actual
instrument are shown in Fig.
144. These are a bulb of
glass or porcelain (*A*), which
contains the dry gas (usually
air or hydrogen), the tendency of which to expand is
to be utilized. The neck of

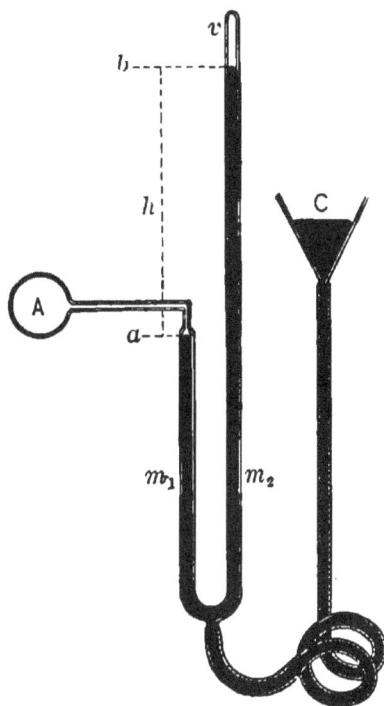

FIG. 144.

this bulb is very narrow, and it ends in one arm of the
manometer m_1, m_2. The other arm (m_2) is long, and is
closed at its upper end. The space (*v*) above the mercury in this arm contains no air, so that the manometer
may be considered as a siphon barometer in which the
vertical distance *h*, between *a* and *b*, measured in centimeters, gives the pressure exerted upon the gas in *A*.

When the gas is heated it expands, driving the mercury

downwards at a and upwards at b. This movement is counterbalanced by raising the cistern C until the level at a is restored. The pressure, as indicated by the difference of level h, necessary to maintain the mercury at the fixed point a, when the temperature of the gas is $t°$, as compared with that necessary to give the gas the same volume at the temperature $t_0°$, measures the change of temperature, $t - t_0$.

This is called the *method of constant volumes.*

The advantage of the air thermometer over other devices for the measurement of temperature, is that it fulfills more nearly the conditions necessary to an absolute standard or system. Mercury thermometers are so much more convenient, however, that they are almost always used, and the air thermometer is reserved for the study of the errors of reading of such instruments.

CHAPTER XVII

CALORIMETRY

142. Heat a Form of Energy. — When the temperature of a body rises, energy of the form which we call *heat* is stored very much as potential energy is stored in bodies which are elevated against the force of gravity to some higher position. Rise of temperature is, therefore, always an indication that work is being done. A body, in other words, cannot be heated without the expenditure of energy. When it cools, the energy stored in it is given up again, either in the way of raising the temperature, and thus increasing the stock of stored energy in neighboring bodies, or in producing motion.

Since heat is a form of energy, it follows that it is capable of being measured. The art of thus measuring *quantities* of heat, as distinguished from the mere measurement of *temperature*, is called *Calorimetry*.

143. Capacity for Heat. — The amount of work which it is necessary to do to raise different substances in temperature varies with the nature of the material. The amount of heat required to raise any given body 1° C. in temperature measures its capacity for heat (thermal capacity). In order to compare the thermal capacities of different substances, water is selected as a standard. Water is a convenient substance for this purpose for many reasons, the most important of which is that its capacity for heat is very nearly uniform through whatever range of tem-

M

perature it may be carried. The uniformity of the thermal capacity of water may be demonstrated by means of the following experiment.

144. EXPERIMENT 37. — The Uniformity of the Thermal Capacity of Water.

Apparatus:

(1) Three beakers, two of which are of the same size (about 400 cc.), the other rather more than twice as large.

(2) Two flasks.

(3) Two thermometers.

(4) Some crushed ice.

(5) A Bunsen burner.

(6) A balance and weights.

Procedure:

(*a*) Note the temperature of the room, as indicated by one of the thermometers, and select two temperatures, one as many degrees above the room temperature as the other is below it (for example, if the temperature of the room be 20° C., 5° C. may be selected as the lower, and 35° C. as the higher temperature).

(*b*) Place the small beakers upon the pans of the balance and add shot or small weights to the lighter pan until equilibrium is produced.

(*c*) Cool a flaskful of water to the lower temperature selected (say 5°) by adding bits of ice and stirring. At the same time warm the other flaskful of water to the higher temperature (say 35°).

(*d*) Nearly fill one of the beakers upon the balance with the warm water, then pour cold water enough into the other beaker to exactly counterbalance it.

(*e*) Remove the beakers simultaneously from the balance, insert the thermometers, one in each beaker, stir, and read the temperatures. Transfer both thermometers to the large beaker, and, without delay, pour the cold and warm water simultaneously into the same. Stir the mixed contents, using both thermometers as stirrers, until the temperature comes into equilibrium. Read the temperature, using both thermometers.

(*f*) Subtract the final temperature as indicated upon the thermometer previously used for determining the temperature of the cold water from the temperature which that water had before mixing, and subtract from the temperature of the hot water the final temperature of the mixture as indicated by the second thermometer.

Since the thermal capacity of water is constant, and since the cold water is heated by the expenditure of energy derived from the cooling of the warm water, the final temperature should be midway between the initial temperatures. The fact, on the other hand, that when this experiment is properly performed the final temperature is thus related to the initial temperatures demonstrates the constancy of the thermal capacity of water.

145. Specific Heat. — The thermal capacity of a body, as compared with that of water, or, in other words, the amount of heat necessary to raise a gram of any substance 1° C., as compared with the amount of heat necessary to raise 1 gram of water from 0° C. to 1° C., is called the *specific heat of the substance.*

146. Heat Units. — The quantity of heat necessary to raise 1 gram of water 1° C. is termed a *heat unit.* The name given to this unit is the *calorie.* A larger heat unit is sometimes desirable, as, for example, in cases where the expenditure of large quantities of heat energy is to be considered. The larger unit generally selected is the amount of heat necessary to raise 1 kilogram of water 1° C. This is equal to 1000 calories.

147. Methods of determining Specific Heat. — *The Method of Mixtures.* The simplest and most direct method of measuring specific heat consists in ascertaining how great a rise of temperature the body under investigation is capable of imparting to a known mass of water, when the former is cooled through a known range of temperature. This method may be best illustrated by means of the following experiment:

148. EXPERIMENT 38. — Specific Heat of a Metal.

Apparatus:

(1) A cylindrical vessel of sheet copper about 10 cm. high and 10 cm. in diameter. (This vessel, which constitutes the calorimeter, that is to say, the apparatus for the measurement of heat quantities, should be polished, in order to reduce so far as possible its radiating power.)

(2) A piece of lead, brass, or copper, preferably spherical or cylindrical in form, the mass of which is about 500 grams.

(3) Two thermometers, one of which should be accurate and sensitive.

(4) A balance and weights, a Bunsen burner, and a small quantity of crushed ice.

Procedure:

(a) Weigh the calorimeter and the metal body, the specific heat of which is to be determined. To the latter a strong bit of twine about 50 cm. long, or a fine wire, should have been attached, and the same should be weighed along with the mass of metal.

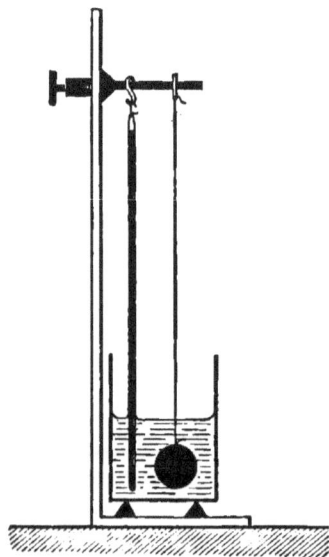

(b) Place the weight in a beaker or can of water over the Bunsen flame; bring the liquid to boiling, and allow it to remain at that temperature for several minutes. In the meantime place the calorimeter upon the scale pan of the balance, half fill it with water, and weigh again.

(c) Remove the calorimeter to the laboratory table, and mount it upon wooden wedge-shaped blocks at the foot of a support, as shown in Fig. 145. Determine carefully the temperature of the water which it contains; also, by means of the other thermometer, determine the temperature of the boiling water which surrounds the metal weight.

FIG. 145.

(d) Remove the weight from the boiling water by means of the twine or wire, the free end of which should have remained outside the vessel for this purpose, and transfer it as promptly as possible to the calorimeter. It may be suspended from an arm of the support, as shown in the figure. Then, by means of the thermometer, which should have remained immersed in the liquid of the calorimeter, stir the contents of that vessel, continuing to do so as long as the temperature rises. Note carefully the highest temperature which the thermometer reaches.

(e) From the rise of temperature which the water has undergone, its mass, the fall of temperature suffered by the hot body after its introduction into the calorimeter, and its mass, the specific heat of the latter may be computed. This quantity, which we may call Q, will be expressed by means of the following equation,

$$Q = \frac{W(t'' - t')}{S(t - t'')},$$

in which W is the weight of the water in grams, S the weight of the metal, t the original temperature of the metal, t' that of the water, and t'' the final temperature which they reach in common, after having been brought together within the calorimeter.

A small part of the heat energy, derived from the cooling of the metal, goes to warming the walls of the calorimeter. This quantity, which has been neglected in the computation indicated above, may be found by multiplying the mass of the calorimeter by the specific heat of copper (0·093) and by the rise of temperature which it underwent during the experiment. In all exact determinations of specific heat this correction is made.

149. Specific Heats of Solids, Liquids, and Gases. — In the table on the following page are given the specific heats of various substances.

150. Method of the Ice-Block Calorimeter. — As will be shown in a subsequent article (see Art. 153), a perfectly definite amount of heat is required to melt a gram of ice and convert the same into the liquid state. Advantage is sometimes taken of this fact for the determination of specific heats. Indeed, this, which was one of the earliest

TABLE.

(1) Specific Heats of Some of the Common Elements.

Substances.	Temps.	Sp. Heat.	Substances.	Temps.	Sp. Heat.
Aluminium . . .	0°–100°	0·2185	Manganese .	14°– 97°	0·1217
Antimony . .	0°–100°	0·4950	Mercury . .	20°– 50°	0·0331
Bismuth . . .	20°– 84°	0·3050	Nickel . .	14°– 97°	0·1091
Cadmium . . .	0°–100°	0·1804	Phosphorus :		
Carbon (diamond)	0°–100°	0·1450	(Red) . .	15°– 98°	0·1698
" (graphite)	0°–100°	0·1860	(Yellow).	13°– 36°	0·2020
Copper . . .	0°–100°	0·0933	Platinum .	0°–100°	0·0323
Gold	0°–100°	0·0316	Silver. . .	0°–100°	0·0568
Iron	0°–100°	0·1130	Sulphur . .	15°– 97°	0·1844
Lead	0°–100°	0·0315	Tin . . .	0°–100°	0·0559
Magnesium . . .	20°–51°	0·245	Zinc . . .	0°–100°	0·0938

(2) Specific Heats of Compounds.

Substances.	Temps.	Sp. Heat.	Substances.	Temps.	Sp. Heat.
Bell metal . . .	15°– 98°	0·0850	Glass :		
Brass	0°	0·0890	(Plate) .	10°– 50°	0·186
Bronze	20°–100°	0·1043	(Crown) .	10°– 50°	0·161
(88·7 Cu + 11·3 Al)			(Flint) .	10°– 50°	0·117
German silver. .	0°–100°	0·0946			
——			Paraffin . .	10°– 15°	0·562
Quartz	20°– 50°	0·1860	Wax . . .	26°– 42°	0·820
Granite	12°–100°	0·1900	Vulcanite .	20°–100°	0·331
Marble	18°– 99°	0·2080	Ice. . . .	–30° –0°	0·505

methods, was described by Black, an English physicist, in 1772. The form of the experiment is as follows:

151. EXPERIMENT 39. — Specific Heat by the Method of the Ice-Block Calorimeter.

Apparatus:

(1) A block of clear ice 20 × 20 × 20 cm.; also a slab of ice 20 × 20 cm., and about 5 cm. in thickness. In the upper surface of the block a cavity is made. The walls of this cavity should be as smooth as possible. The most convenient method for producing such a hole consists in cutting it out in the rough with a knife, and then finishing it by inserting a cylindrical weight of metal previously heated, or a small bottle containing hot water. After a cavity has been made, the size of which is sufficient to receive the mass of metal the specific heat of

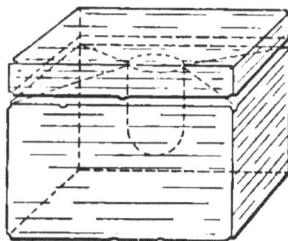

FIG. 146.

which is to be determined, the interior is dried by wiping with filter paper, and the slab of ice is replaced. The ice-block calorimeter thus formed will appear as in Fig. 146.

(2) A balance and weights.

(3) The piece of metal used in Experiment 38; also a beaker of boiling water, two thermometers, and a sponge.[1]

Procedure:

(a) The mass of the metal is determined by weighing, and it is heated as in the previous determination.

(b) Having been subjected to the action of boiling water for a sufficient time, namely, several minutes, the weight is removed by means of the attached twine, and is placed in the cavity of the ice block which just previously must have been newly dried with the filter paper. The slab is instantly replaced, and a sufficient amount of time is allowed to elapse to cool the hot metal to the ice temperature. Not less than ten minutes should be allowed for this portion of the experiment; and if large masses of metal are dealt with, a longer amount of time is required.

(c) The slab is removed, and the weight is carefully withdrawn

[1] Instead of the sponge, several sheets of filter paper may be used.

from the ice cavity, and is dried by contact with the sponge, or with a wad of filter paper which has been previously weighed. The same sponge or filter paper is then employed in removing from the cavity of the ice block all the liquid which has resulted from the melting of the ice therein. The paper, with the water which has been taken up, is placed in the beaker, the weight of which had also been previously determined, and is weighed again. In this way the amount of ice melted is determined. The success of the experiment depends upon fashioning the ice block in such a way that none of the liquid formed by the exterior melting will fall into the cavity. This end may be secured by slightly beveling off the upper face of the block so that the drainage from that face will be outward, and by giving the under face of the slab which is used as a cover a slight degree of concavity.

(*d*) The computation of the result of this experiment is as follows: The amount of heat energy required to melt 1 gram of ice is accurately known; it is 79·25 calories. (See Art. 153.) The mass of the hot metal being known, and the fall of temperature which it suffered, namely, from the boiling point of water to the melting point, we may compute the number of calories which one gram of similar material would have given up in cooling 1° C. This is obviously the number of calories required to melt the ice, divided by the product of the mass of metal into the fall of temperature. The number of calories which a gram of the material would give up in a fall of 1° C. measures, however, the specific heat of the material.

The formula by means of which these statements are to be expressed, and which serves, therefore, for the computation of the specific heat, is

$$Q = \frac{79·25 \ W}{S \, t};$$

where W is the mass of ice melted,
             S is the mass of the metal,
             t is the temperature of the hot bath.

152. Specific Heat of Gases and Liquids. — The specific heats of gases and liquids are determined by methods based upon the principles just described. Many modifications in the procedure are necessary, however, and it will not be possible to describe these in detail in the present work.

153. Heat of Fusion. — The amount of heat necessary to melt a unit mass of any substance, converting it into liquid form without rise of temperature, is called the *heat of fusion*. This is a perfectly definite quantity, always possessing the same value in the case of a given liquid, but differing much with different substances.

Heat of fusion of water is much larger than that of other liquids. It amounts to nearly eighty calories (79·25 calories). To convert ice at 0° C. into water at 0° C., in other words, requires as much heat as to raise water at 0° C. to 80° C. If we take 500 grams of water at 80° C., for example, and place in it the same amount of ice, the heat liberated by the water on cooling will be just sufficient to melt all the ice. The final temperature, instead of being midway between the two initial temperatures (Art. 144), will be 0° C. The experiment is a somewhat difficult one to carry out in this especial form, owing to the loss of heat to surrounding objects. The following is a better method:

154. EXPERIMENT 40. — Heat of Fusion of Water.

Apparatus :

(1) The ice-block calorimeter described in Experiment 39.

(2) A balance and weights, thermometer, Bunsen burner, beakers, sponge, etc.

Procedure :

(a) A clean, dry beaker is placed upon the scale pan and weighed. A quantity of water is then added, sufficient to half fill the cavity of the ice block, and its weight is determined.

(b) The beaker of water is removed from the balance and heated to 80° C., care being taken during this operation that the beaker is covered with a plate of glass to reduce losses by evaporation to as small a quantity as possible. When the water has reached exactly 80° C., as indicated by the thermometer used as a stirrer within the liquid, the slab is removed from the ice-block calorimeter, and the water is poured without delay into the cavity. The latter must have been previously dried by the application of filter paper. The slab is at once replaced, and a sufficient amount of time is given, namely fifteen minutes, for the hot water to have delivered all its heat energy to the surrounding ice. It is obvious that when this condition has been reached, the temperature of the water within the cavity will have fallen to 0° C.; also that since its initial temperature was 80° C., that the amount of ice that it should be capable of melting is equal to its own mass. It is the object of the experiment to verify this latter point.

(c) After the lapse of the requisite time, remove the slab from the ice-block calorimeter. Withdraw the water from the cavity by the use of a sponge to the beaker previously weighted, care being taken to lose none. Weigh the water thus accumulated, and compare its mass with that of the hot water applied to the melting of the ice. Two small corrections are necessary.

(1) That due to the fact that a minute quantity of water clings to the beaker, and cannot be transmitted to the ice cavity. This amount may be quite closely estimated by weighing the beaker filled with water as described under section (a), pouring it out as nearly as possible in the same manner as when the hot water was transmitted to the ice cavity, and weighing the beaker again. The increase of weight will indicate the amount of this correction.

(2) This correction arises from the fact of the loss by evaporation during the process of heating the liquid. To determine this the beaker of water is weighed; it is then heated as nearly as possible under the same conditions as in the actual experiment already described, and is then allowed to cool. It is afterwards weighed again. Its loss of weight will equal (nearly) twice the correction due to evaporation. The student should perform these operations, and thus determine whether these corrections are of sufficient magnitude to affect his result. It is interesting to compare the heat of fusion of ice with that of other substances, and to note that the heat of vaporization of water, given on p. 178, is likewise greatly in excess of that of other substances.

155. Numerical Values of the Heats of Fusion. — The heat of fusion of various liquids is given in the following table:

TABLE.

HEAT OF FUSION OF VARIOUS SUBSTANCES.

Bismuth	12·640 calories	Mercury	2·820
Bromine	16·185	Paraffin	35·100
Cadmium	13·660	Phosphorus	4·744
Glycerine	52·500	Platinum	27·180
Ice	**79·250**	Silver	21·070
Iron	33·50	Sulphur	9·365
Lead	5·858	Tin	13·314

156. Heat of Vaporization. — When liquids are converted into the state of vapor, either by boiling or by quiet evaporation, a certain definite amount of work has to be done upon them. Energy is stored in the process of bringing about this change of state, analogous to that which is stored when a body is caused to rise in temperature. When, on the other hand, a vapor is condensed, the stored energy becomes available again and may be made to do work either in raising the temperature of other bodies or by the application of proper devices, as in the steam engine, in the production of motion. The heat required to vaporize a liquid is, generally speaking, a very large amount. Water, for example, when converted into steam at its ordinary boiling point (100° C.) requires 535 heat units for vaporization. The exact measurement of the heat of vaporization of water is an operation of considerable difficulty, but the phenomenon may be illustrated and the numerical value of the heat of vaporization may be approximately determined by means of the following very simple experiment. With proper care, indeed, excellent results may be obtained.

157. EXPERIMENT 41. — The Heat of Vaporization of Water.

Apparatus:

(1) A flask containing about 1000 cm.³ of water, a cork which fits it well, a piece of glass tube about 60 cm. long and about 1 cm. in external diameter.

(2) A balance and weights, a thermometer, and a beaker.

(3) Two Bunsen burners.

(4) Crushed ice.

Procedure:

(*a*) The glass tube is twice bent through an angle of 90°, giving it the form shown in Fig. 147. Its shorter arm is then fitted to a

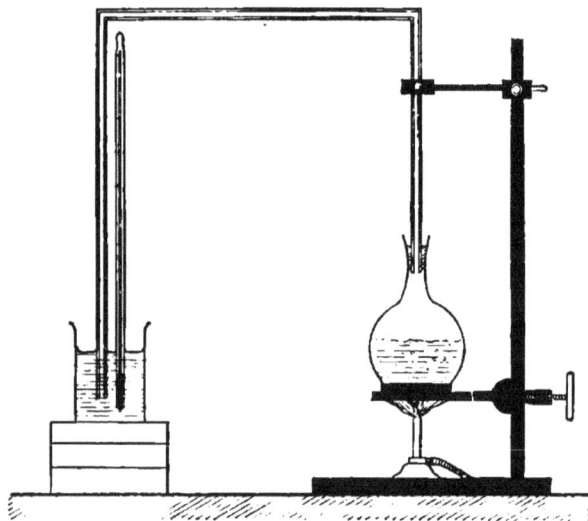

FIG. 147.

cork which is inserted in the mouth of the flask. Water is brought to the boiling point by means of one of the Bunsen burners.

(*b*) The beaker is filled two thirds with cold water chilled as near as convenient to 0° by the use of ice. The mass of this water is determined by weighing.

(*c*) By means of the second Bunsen burner the horizontal portion of the glass tube, through which the steam from the boiling liquid is being delivered, is now to be cautiously heated until the dropping of condensed vapor from its open end ceases. As soon as this condition is brought about, the temperature of the cold water within the beaker is to be read and the beaker is to be inserted under the open mouth of the delivery tube so that the latter will extend to a depth of three or four centimeters below the surface of the liquid. During the remainder of the experiment the beaker must be supported in this position by means of blocks placed beneath it. Under these conditions, the dry steam issuing from the tube will be entirely condensed within the liquid and will give up heat energy in proportion to its mass and to the heat of vaporization, thus raising the temperature of the water which receives it.

It is necessary during the remainder of the determination to watch the delivery tube, and, upon the reappearance of the moisture within the vertical arm, to repeat the heating of the delivery tube as already described. The moment the inner wall becomes dry, however, this heating should be interrupted.

(d) Under the heating effect of the condensing steam, the water within the beaker will rise rapidly in temperature. It should be stirred from time to time and its temperature noted, and when it reaches a temperature as far above the temperature of the room as its initial temperature was below, it should be removed from its position beneath the delivery tube. To determine the amount of steam which has been condensed in the process of heating the liquid through this range of temperature, the beaker is weighed again.

(e) The formula for the computation of the results of this experiment is as follows:

$$V = \frac{W\,(t' - t) - S\,(100 - t')}{S}.$$

In this equation, W is the mass of water in the beaker, S the mass of the condensed steam (both in grams), and $t' - t$ is the rise of temperature caused by the condensation of the latter.

The following table contains the heat of vaporization of various liquids:

TABLE.

THE HEAT OF VAPORIZATION OF VARIOUS LIQUIDS.

Alcohol (ethyl) .	208·92 calories	Chloroform . .	91·11
Ammonia (NH_3) .	294·21 (at 7°·8)	Ether (C_4H_{10}) .	23·95
Benzol	93·45	Iodine . . .	62·00
Bromine	45·60	Mercury . . .	91·70 (at 0°)
Carbon dioxide .	56·25 (at 0°)	Sulphur dioxide	58·49
Carbon disulphide	86·67	Water . . .	535·90

CHAPTER XVIII

PHENOMENA ACCOMPANYING FUSION AND LIQUEFACTION

158. Changes of Volume due to Fusion. — When a solid body is melted it undergoes a change of volume. Substances may be divided into two classes according to the nature of this change. To the first class belong substances the volume of which is decreased by fusion. The chief member of this class is water. The second class, which contains by far the larger number of substances, possesses a larger volume in the melted than in the solid state. The character of these changes may be readily shown by means of the following experiment:

159. EXPERIMENT 42. — **Changes of Volume resulting from the Fusion of Paraffin and of Water.**

Apparatus:

(1) Two pieces of glass tubing 50 cm. long and about 0·3 cm. or 0·4 cm. of inner diameter.

(2) Two beakers each containing about 500 cm.⁸ and a thermometer.

(3) Some crushed ice and salt.

(4) A small piece of paraffin and about 50 cm.⁸ of mercury.

Procedure:

(*a*) One end of each glass tube is heated in the flame of a blast lamp or of a Bunsen burner until it closes. It is then blown into a bulb about 2 cm. in diameter. Care should be taken that the walls of this bulb are not too thin. After the glass has cooled, small chips of paraffin are dropped into one of the bulbs until the latter is about half full. The bulb is then carefully heated in a flame until the paraffin is melted. It is then set aside in a vertical position, with the open end uppermost, to cool. After the paraffin has solidified in this bulb, the tube is bent through 180° in a luminous gas flame, giving it the form indicated in Fig. 148 (3), and after it has again had an opportunity to cool, it is filled with mercury to a point considerably beyond the elbow. If care has been used in the bending of this tube the paraffin will not have been melted. The upper end of the bulb will contain a cap of solid paraffin below which the contents of the bulb and tube will be filled with mercury. Care should be taken to drive out all air bubbles from the space between the mercury and the paraffin. This can be done by continually turning the tube and tapping the outer surface of the glass. We now have a simple form of apparatus in which every change in the volume of the paraffin within the bulb will be indicated by a rise and fall of the mercury column in the bent arm of the tube. The other bulb is bent into the same form as the one just described. It is then filled with water and mercury in such a manner as to afford the same conditions as those which exist in the tube which contains paraffin; *i.e.* water in the upper portion of the inverted bulb with mercury below it and extending a considerable distance around the elbow. Since water is a liquid at ordinary temperatures, this filling can readily be performed after the tube has reached its final shape.

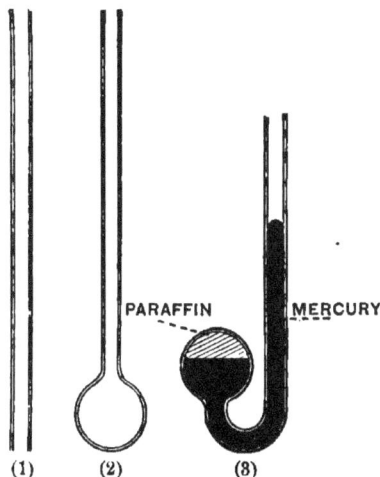

(*b*) The two bulbs having been prepared as described under section (*a*), it remains to bring each of them to the melting point of the substance which it contains, to allow the contents to pass from the solid to the liquid state, or *vice versa*, and to note the modification of the mercury column in the open arm of each tube. For this purpose the tube containing paraffin is placed in a beaker of water previously heated to a temperature of 60°; a temperature which is slightly above the melting point of that substance. It will be noted that when the temperature of the surrounding liquid has been transmitted to the contents of the bulb, and the paraffin begins to melt, the mercury rises in the open arm of the tube, showing a decided expansion as the result of fusion. To bring about the corresponding phenomenon in the case of water, the other bulb is placed in a freezing mixture consisting of crushed ice mingled with fine salt.

This is to be packed around the bulb in the other beaker to such a height as to entirely surround the water. The first effects of subjecting this bulb to the influence of the freezing mixture is a shrinkage, both the mercury and the water diminishing in volume. Even before the water begins to freeze, the downward movement of the mercury in the open arm of the tube ceases, and it begins to rise. When freezing takes place, this upward movement becomes more rapid, and it continues until every drop of the liquid has gone over into the solid state. When the water is completely frozen, a second reversal in the motion of the mercury column takes place, and it begins to fall again, continuing to do so until the contents of the bulb has reached the temperature of the freezing mixture with which it is surrounded.

(c) Upon removing the water bulb from the freezing mixture and the bulb containing paraffin from the hot bath within which it has been melted, the reverse processes may be observed.

160. Influence of Pressure on the Melting Point. — It is a general law that substances of the first class, namely, those which shrink upon fusion, have their melting points lowered by pressure, while substances the volume of which in the liquid state is greater than before fusion may be solidified at temperatures above their normal freezing points by the application of pressure. In the case of water, which is the substance thus far most extensively experimented with, this change of the melting point is exceedingly small. It amounts to 0·007° per atmosphere of pressure. It has been found possible, by the application of very great pressures (about 3000 atmospheres), to lower the melting point of ice to − 20° C. Paraffin, spermaceti, and most wax-like substances belong to the second class. The German chemist Bunsen raised the melting point of paraffin 3·6° by a pressure of 100 atmospheres.

161. Phenomena accompanying Vaporization.—(*a*) *Change of Volume by Vaporization.* When a liquid goes over into the gaseous condition, the change of volume is very much greater than when it is converted from a solid to a liquid. Water, for example, when changed into steam at 100° C. increases in volume in the ratio of 1 to 1578. There is no exception to the statement that vaporization is accompanied by great increase in volume. The precise ratio in each case can be determined only by the indirect method of measuring the density of the saturated vapor produced at the temperature and pressure in question.

(*b*) *Ebullition.* The change from the liquid to the vapor state takes place either by quiet evaporation at any temperature or by the process known as boiling (ebullition). Ebullition consists in the gathering of vapor particles below the surface of the boiling liquid into the form of bubbles. These bubbles are able to maintain themselves against the pressure to which they are subjected only when a certain temperature, which is called the temperature of ebullition, has been reached. At temperatures lower than this all bubbles forming within the mass of the liquid are compressed and condensed, but as the temperature rises, the mass of vapor which constitutes a bubble becomes capable of reacting against greater and greater pressures. Finally the bubbles are able to resist compression, and as soon as this condition of equilibrium is passed, they grow rapidly in size and rise to the surface. Each one of them, as it passes through the surface film, carries its contents of vapor into the outer atmosphere.

The liberation of these masses of vapor is accompanied

N

by the transfer of energy from the liquid mass to the atmosphere above it, and the transfer of energy tends to keep down the temperature of the liquid. If the temperature fall below the condition which has just been described, ebullition ceases. The liquid, on the other hand, cannot rise above that temperature, because the slightest tendency to do so calls forth correspondingly increased violence of boiling; so that the rate at which energy is transferred almost instantly becomes sufficient to lower the temperature again. Thus the temperature of a boiling liquid is in a state of equilibrium, and it remains *constant* as long as the pressure to which the liquid is subjected remains constant.

162. Influence of Pressure upon the Boiling Point. — The pressure to which a liquid is subjected will obviously alter the condition of equilibrium which has just been described; so that the temperature at which the bubbles formed within the liquid mass are able to maintain themselves is higher as the pressure rises and diminishes with the pressure. That under low pressures water will boil at low temperatures, may be readily demonstrated as follows:

EXPERIMENT 43. — The Thermal Paradox.

Apparatus :

(1) A strong flask the bottom of which should be convex (a Florence flask).

(2) A cork well fitted to the flask.

Procedure :

(a) The flask is half filled with water, which is made to boil for several minutes. After a sufficient length of time has elapsed to drive out the air from the atmosphere above the level of the liquid, and also to exclude the air contained by the liquid itself, the cork is inserted into the mouth of the flask and the flame is extinguished.

(*b*) The flask is then inverted. We now have a body of liquid and an atmosphere consisting altogether of its vapor. As the temperature falls this vapor is condensed gradually, so that the pressure under
which the liquid exists diminishes.
By causing a sudden diminution of
pressure, which may be readily done
by pouring cold water over the flask,
as in Fig. 149, the liquid will burst
into violent ebullition. The result of
this boiling is to add new vapor to the
inclosed atmosphere, when the boiling
will diminish in intensity. Fresh applications of cold from without will
cause a repetition of the phenomenon,
and the experiment may be carried on
until the liquid is quite cold.

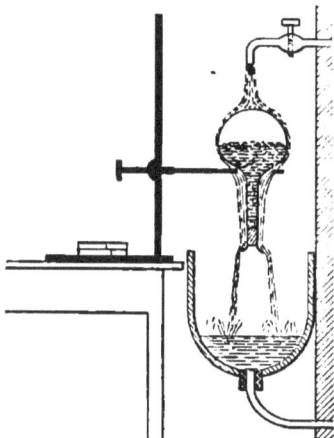

FIG. 149.

163. The Freezing of Water by its Own Evaporation. — By modifying the method of reducing pressure just described, the interesting phenomenon of a liquid boiling at its freezing point may be
shown. The apparatus
necessary for this ex-
periment consists of an
air pump (Fig. 150),
a bell jar, a flat metallic
dish containing a small
amount of water, and
a larger dish of porce-
lain, also very shallow.
The labor of pumping
may be greatly reduced
by having the water
previously cooled nearly to the freezing point. To succeed in the demonstration it is necessary to place within

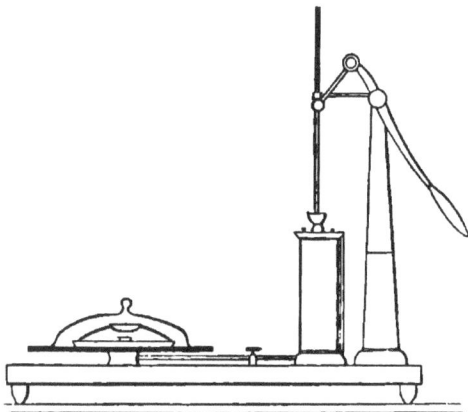

FIG. 150.

the bell jar some substance which will unite with or absorb the aqueous vapor as fast as it is formed by the evaporation of the liquid. The best drying material is phosphorus pentoxide (P_2O_5). Very strong sulphuric acid freed from water by boiling may, however, be used. A very convenient form of apparatus for this experiment is that shown in Fig. 151. A flat dish, which contains the dryer, is surmounted by a flat tripod of wire.

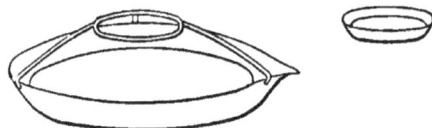

FIG. 151.

Into the central ring of the latter fits the smaller dish, containing the water to be frozen. The latter vessel should be of thin sheet metal and about 5 cm. in diameter. It is only necessary to fill the larger dish with the dryer and to put about 5 cm.³ of cold water into the smaller one, and then to exhaust the air by means of the pump, keeping up the action for a few minutes. The temperature of the liquid will fall to the freezing point, and just before it freezes the liquid in the flat dish will boil. Ebullition will continue vigorously even under the layer of ice first formed, and will not have ceased until the water has all become solid.

164. EXPERIMENT 44. — Relation between Boiling Point and Pressure.

Apparatus:

(1) Two flasks with well-fitted corks.

(2) Three pieces of glass tubing, each about 100 cm. long and 0·3 cm. to 0·5 cm. in diameter; also a T-shaped tube.

(3) A filtering pump (see Appendix V).

(4) A thermometer reading to 100°; a support; a Bunsen burner; rubber tubing.

(5) A glass containing mercury.

(6) A meter scale.

(7) A barometer.

Procedure:

(*a*) Bend one of the glass tubes into the form *abc* (Fig. 152). Cut another in two, and bend one of the pieces into the form *de*. Bore both corks, each with two holes.

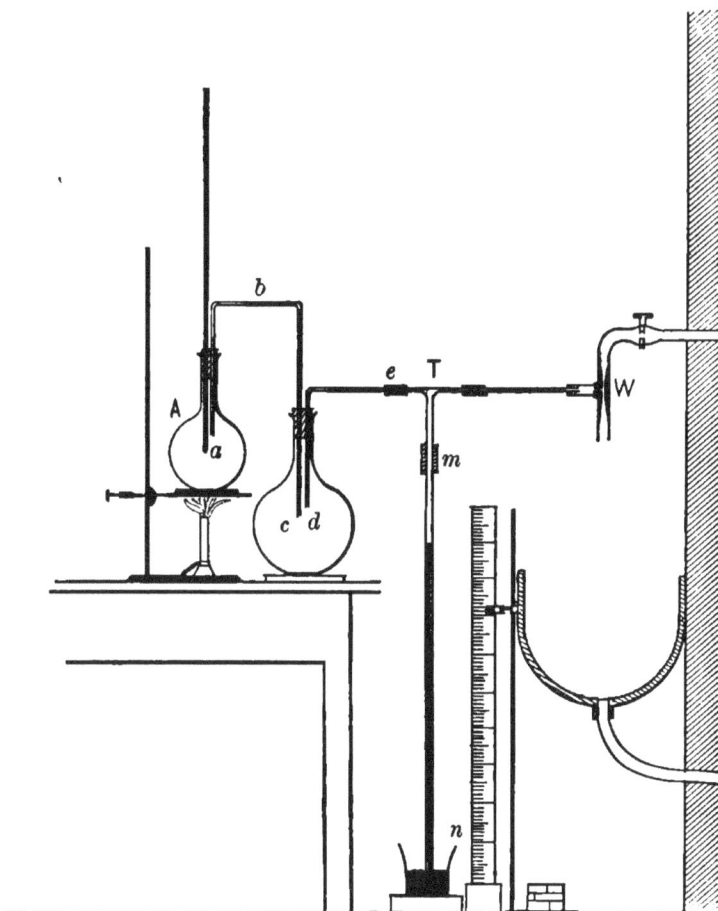

FIG. 152.

Set the apparatus up as shown in the figure, connecting the uncut piece of tubing (*mn*) to the lower end of the T tube by means of a few centimeters of rubber tubing. The right and left hand arms of the T are to be similarly attached to the filter pump *W*, which is fastened to the water faucet, and to the tube *de*. The tube *mn*, the

lower end of which is to be immersed in the glass of mercury, serves
as a manometer. Before the corks are finally inserted in the flasks,
A should be one quarter filled with water. All joints and the corks

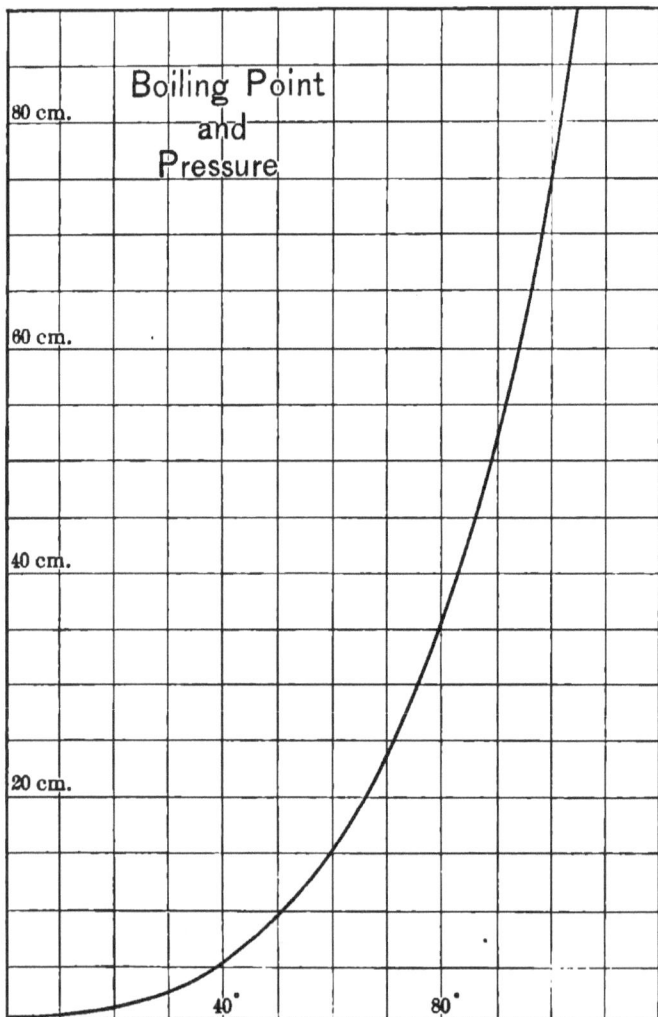

Boiling Point and Pressure

FIG. 153.

should be rendered air tight by the application of rubber cement, or
of beeswax and rosin.

(*b*) Bring the water in *A* to a temperature of about 95°, then extinguish the flame and start the filter pump. Turn on the water gradually until boiling begins in the flask *A*. Then leave the pump in action, and read the height of the mercury in *mn* and also the thermometer.

(*c*) Repeat the readings of manometer and thermometer at intervals of about five minutes, increasing the action of the pump from time to time as may be necessary to produce ebullition. Continue thus until the pump is no longer capable of causing the water to boil.

(*d*) Find the height of the barometer in centimeters.

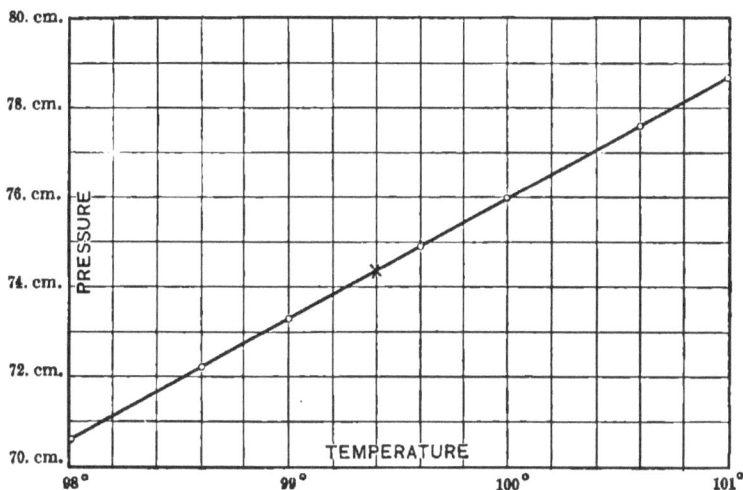

FIG. 154.

(*e*) Plot the curve of your results, with temperatures as ordinates and the height of the mercury in *mn*, subtracted from the barometric reading, as abscissas. Compare your curve with that in Fig. 153, which gives the accepted relation between boiling points and pressure.

(*f*) Instead of reading the barometer as directed in section (*d*), the tube *a* may be withdrawn from the flask *A*, the water may be made to boil by the action of the flame, and the thermometer may be read. From this reading the height of the barometer may be obtained by direct reference to the curve in Fig. 154. This is a small portion of the curve of the preceding figure, reproduced upon a much larger scale. If, for example, the boiling point is 99·4°, the corresponding pressure will be found to be 74·4 cm., which is the barometric reading.

165. Influence of Evaporation upon Temperature. — In addition to the process of ebullition, there is always going on an escape of liquid particles into the atmosphere by the process known as *evaporation*. This process consists in the passage of individual particles of liquid through the surface film and out into the gaseous atmosphere beyond. It occurs more rapidly at high than at low temperatures. Evaporation occurs because the particles of a liquid are ever in motion among themselves. The more rapidly the particles move, the higher is the temperature. The particles of a liquid do not, however, all move with the same velocity, and it is chiefly those of high velocity which escape through the surface film in evaporation. Evaporation, therefore, deprives the liquid of those particles which possess the greatest energy of motion, consequently it produces fall of temperature.

This fact may be shown in many ways. If the hand, for example, be plunged into warm water and then withdrawn, the result will be a distinct sensation of chilliness due to evaporation.

If two thermometers, which give the same reading when placed side by side, are taken, and the bulb of one of them be wrapped in filter paper, or other porous material, and then be moistened with a liquid, the temperature of which is as high as that of the room, or even higher, a difference in the reading of the two thermometers will soon become noticeable. The one with a dry bulb will show a higher temperature than the other. The rate of evaporation of the liquid surrounding the moistened bulb will depend upon the amount of moisture present in the surrounding atmosphere, and since rapid evaporation lowers the temperature to a greater degree, this comparison of wet and dry bulb thermometers affords a means of determining the humidity of the air.

A striking illustration of the lowering of temperature by evaporation is as follows. A test tube of water, surrounded with cotton waste or other porous material which has been well saturated with ether, is placed under the bell jar of an air pump, and the air is removed. The reduction of pressure hastens evaporation, and a sufficient amount of heat is abstracted from the liquid by the surrounding ether to reduce the temperature below the freezing point. In a few minutes the water will freeze.

166. The Spheroidal State. — When a drop of water falls upon a hot surface, the temperature of which is very far above the boiling point of the liquid, it does not come into immediate contact with the surface, as it would do if the temperature of the latter were not so high, but is gathered by the action of the surface film into the form of a spheroid. The globule of liquid thus formed is separated from the surface of the hot solid by a layer of steam. This layer of steam is constantly renewed by rapid evaporation from the lower side of the drop, and it issues with a force sufficient to enable it by its reaction to support the drop out of contact with the hot surface. Steam, like other vapors, is a very poor conductor of heat, consequently the drop remains at a comparatively low temperature (below its boiling point), being kept cool by its own evaporation. Liquid existing in this condition is said to be in the *spheroidal state*. The most convenient form of experiment for illustrating the phenomena connected with the spheroidal state is as follows:

167. EXPERIMENT 45. — **The Spheroidal State of Water.**

Apparatus:

(1) A metal plate composed of some material which is capable of being heated beyond the red heat without being melted. The best

thing to use for this purpose is the inverted cover of a platinum crucible, or a piece of platinum foil carefully flattened and bent into
slightly concave form (a piece of sheet iron will answer, or a porcelain crucible cover).

(2) A *flat* piece of platinum or sheet iron.

(3) A blast lamp.

(4) A beaker of water and a pipette.

Procedure:

(*a*) The metal plate is placed upon the iron ring of a retort stand,
care being taken to have it as nearly level as possible. It is then
heated to a bright red heat from below by means of the blast lamp,
and a globule of the liquid is carefully discharged upon the surface
of the hot plate. It will be seen to assume at once the spheroidal
form. The spheroid will settle over the lowest point of the hot surface, where it will remain without boiling, wasting away rapidly by
evaporation until it finally disappears.

(*b*) Substitute the flat piece of foil for that used in the previous
section. Take a piece of iron wire, and plunge the point of it into
the globule of liquid. The surface film will make contact with the
rod or wire, and the globule can then be moved to any part of the
plate. In this way, holding the eye on a level with the heated surface, and moving the globule to some portion of the latter which is
slightly convex, it will be found possible to look through the layer
of steam between the drop and the
metal, and thus to obtain direct vis-
ual evidence that contact between
them does not exist. Downward
pressure with the rod will flatten
the globule without bringing it appreciably nearer to the hot surface.
Figure 155 shows a drop of water
thus pressed against the layer of
steam which supports it from the hot surface of an inverted crucible. The rod used in this case was simply the pipette by means of
which the liquid had been dropped upon the metal. The figure,
which is from a photograph, shows the layer of steam which serves
as a cushion holding the liquid and solid apart. The flattening of
the drop beneath, caused by the attempt to press it closer to the
crucible, is very evident.

FIG. 155.

CHAPTER XIX

RELATIONS BETWEEN HEAT AND WORK

168. The Mechanical Equivalent of Heat. — We have already spoken of heat as a form of energy consisting of some motion of the individual particles of a substance among themselves. The particles of a body do not come into this condition of motion without being brought into it by the action of outside forces. The transfer of energy which takes place under these conditions manifests itself generally by a rise of temperature. If, for example, a body possessing kinetic energy be brought to rest through impact with some other body, a part of its energy of motion appears to be lost, because the body with which it made impact has not acquired so large an amount of kinetic energy as the body which has been brought to rest has delivered to it.

If we look more closely into the phenomena which accompany impact, we find, however, that both bodies show a rise of temperature. This rise of temperature indicates that the motion, apparently lost, has really been transferred to the *particles* of both bodies, throwing them into motion among themselves. We cannot see these molecular motions, even with the aid of the microscope, but there is no reasonable doubt of their existence. If we measure the rise of temperature and compute the heat energy in calories, we find that it always reaches a certain definite amount per unit of kinetic energy destroyed in producing it (Art. 174). The relationship between the work which disappears and the heat energy, measured in calories, which its disappearance calls forth, is termed the *mechanical equivalent of heat.* This transfer of energy, from the form of kinetic energy to that of heat energy, may be illustrated by numerous familiar examples. The heat developed in the process of boring a hole in wood or metal, or in breaking a wire or rod by bending it back and forth, illustrates this principle. It was through observations of the large amount of heat produced in boring cannon, which first led Count Rumford, who was employed in the supervision of such work by the king of Bavaria, to the conception that heat was a mode of motion and was produced by the expenditure of work. Energy expended in resistance to motion, *i.e.* in friction, is nearly always entirely converted into heat. The following experiment, due to Tyndall, illustrates this transfer of energy, together with the reverse process of converting heat energy into energy of motion :

169. EXPERIMENT 46. — Production of Steam directly from the Heat of Friction (Tyndall's Experiment).

Apparatus :

(1) A whirling table upon which is mounted a brass tube about 2 cm. in diameter and about 10 cm. long. (See Fig. 156.)

(2) A cork which fits the open end of the tube, and a wooden clamp of the form shown in the figure.

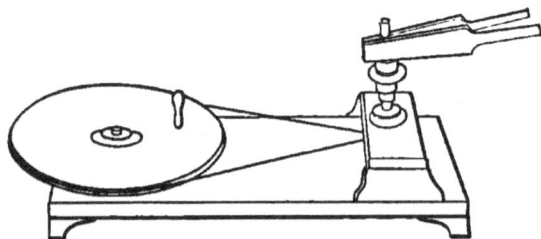

FIG. 156.

Procedure :

(a) The tube is partly filled with water, and the upper end is firmly closed by means of the cork.

(b) The whirling table is put into motion, the tube being held in the jaws of the clamp with sufficient pressure to produce as large an amount of friction as can be obtained without causing the belt to slip. After driving the tube rapidly for a few minutes, a sufficient amount of heat will be developed to cause the water to boil. Steam will be produced, and the pressure will rise until the cork is blown from the mouth of the tube. Here we have the transfer of the energy of motion, through the agency of friction, into heat energy. The result is a rise of temperature, together with a change of state from the liquid to the gaseous form. The energy of motion of the cork is far from being the equivalent of the energy of the work done in driving the whirling table, because a large portion of the heat energy evolved is transferred to surrounding bodies, producing molecular motions in them and rise of temperature.

170. Changes of Temperature as the Result of Compression and Expansion. — One of the many ways in which energy may be stored is by the compression of a gas in a closed vessel. A gas possesses perfect elasticity of volume. It tends to react, therefore, and when it is allowed to escape it is capable of doing work. One of the phenomena which accompanies this process is the rise of temperature as the result of the compression, and a fall of temperature when the gas is released and allowed to expand again. The rise of temperature of a gas in process of compression may be readily observed in the case of the small compression pumps used for inflating the pneumatic tires of bicycles. The tube which connects the pump with the tire grows very warm, and in some cases, where the pumping is rapid, it becomes too hot to be held in the hand.

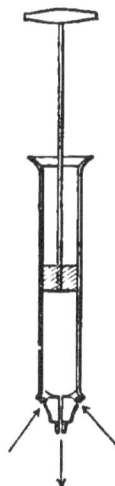

FIG. 157.

The heating due to compression may readily be illustrated by means of the simple form of compression pump shown in Fig. 157. This is an air pump with valves reversed. A convenient method is that described in the following experiment:

171. EXPERIMENT 47.— Heat developed by the Compression of a Gas.

Apparatus:

(1) A simple compression pump.

(2) A thermopile and galvanometer.   (See Appendix VI; also Chapter XXXII.)

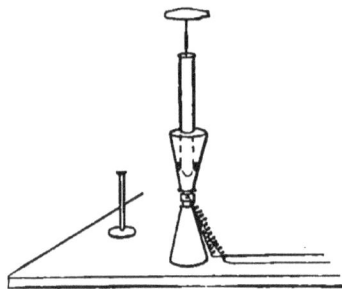

Procedure:

(a) Fit the nozzle of the pump into the hollow cone of the thermopile, as shown in Fig. 158.

(b) Connect the thermopile to the galvanometer, and set the former up in the position shown in the figure, with the pump inserted vertically in the upper cone. By crowding the nozzle of the pump firmly down into the cone, a small volume of air is trapped

FIG. 158.

above the upper face of the pile.   This air will be compressed by the action of the pump.

(c) Work the pump rapidly, but with care, observing the galvanometer. It will be found that the latter responds to every stroke, and that the deflection is in the direction which corresponds to the heating of the upper face of the pile.

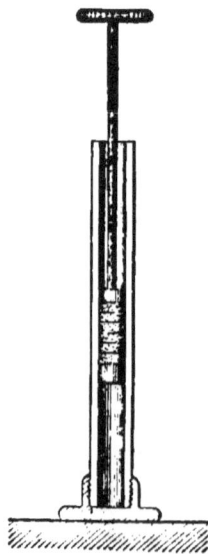

172. The Fire Syringe. — Another form of apparatus for the demonstration of the heat produced in compressing a gas is the fire syringe. This is simply a small compressing pump without valves (Fig. 159). The experiment consists in placing a shred of gun cotton (not more than will go loosely into the bottom of an ordinary thimble) in the bottom of the cylinder of the syringe. The piston is then inserted, and is driven suddenly downwards. If the stroke be sharp and vigorous, the heat generated will ignite

FIG. 159.

the gun cotton explosively, with a puff of smoke. The
result does not follow a slow stroke, because of the rapidity with which the heat is imparted to the walls of the
cylinder.

173. Experiment 48.—**Cooling by the Expansion of a Gas.**
Apparatus:
(1) The compression pump described in Exp. 46 (Art. 169).

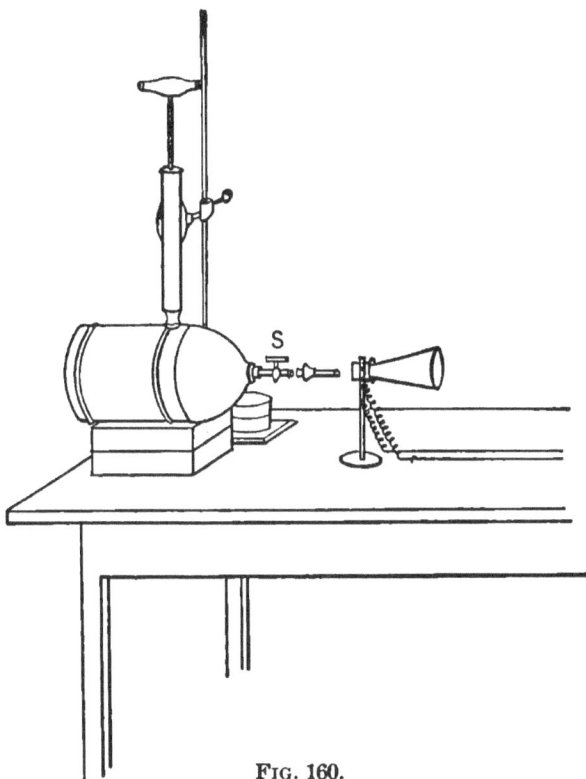

Fig. 160.

(2) The metal reservoir which is usually furnished with such
pumps.
(3) The thermopile and galvanometer.
Procedure:
(*a*) Attach the pump to the reservoir, by the opening in the side
of the latter.

(*b*) Connect the thermopile to the galvanometer, and remove the cone from one face of the former. Set up the pump and reservoir as shown in Fig. 160, and place the thermopile close in front of the nozzle.

(*c*) Fill the reservoir with air by several rapid strokes of the pump.

(*d*) Open the stopcock *S* and note the deflection of the galvanometer, which will be such as to indicate a cooling of the face.

174. The Mechanical Equivalent of Heat. — The first experimenter to measure the exact amount of heat energy in calories, produced by the expenditure of a definite amount of work, was the English physicist Joule. The principle of one of his forms of apparatus is shown in Fig. 161.

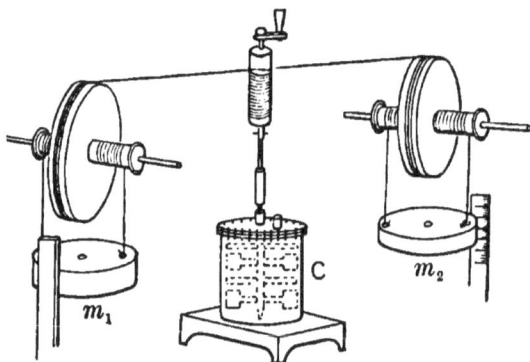

FIG. 161.

The two weights m_1, m_2 fall through measured distances under the action of gravity. They turn a paddle within the calorimeter C, thus stirring and thereby warming the water which the latter contains. The rise in the temperature of the water multiplied by its mass gives the heat developed. Joule measured the work done by the falling weights in foot-pounds, the water in pounds, and temperature in degrees of the Fahrenheit scale.

He found that 772 foot-pounds were expended to warm

one pound of water one degree. Translated into scientific terms, *i.e.* into those of the C.G.S. system of units, this means that to produce one calorie of heat (to raise 1 gram of water 1° centigrade), about 42,000,000 ergs of energy must be exerted.[1]

---

[1] Very large numbers of this kind are more conveniently written in the form $4 \cdot 2 \times 10^7$.

o

CHAPTER XX

TRANSMISSION OF HEAT

175. Conduction, Convection, and Radiation. — There are three methods by which heat is transferred. In the first of these the motions which constitute heat energy are imparted gradually, from layer to layer, of the substance in the path along which transmission takes place. In the second, energy is carried by means of a current, or stream of gas, or liquid. In the third method (radiation), heat energy is transmitted at a very high velocity by means of a wave motion.

The phenomenon of conduction may be illustrated by means of the following simple experiment:

176. EXPERIMENT 49. — **Relative Conducting Powers of Copper, Iron, and Glass.**

Apparatus:

(1) A rod or tube of glass, an iron rod, and a copper rod. These should be nearly of the same diameter.

(2) A metal dish containing paraffin, a Bunsen burner, and a gas burner with a flat flame. [The ordinary "batswing" or "fishtail" burner.]

(*a*) Melt the paraffin, and apply it by means of a brush or cloth to each of the rods which have been previously warmed in the Bunsen flame. Then lay the rods aside to cool to the temperature of the room.

(*b*) Clamp the rods in a horizontal position side by side, as shown in Fig. 162. [The inconvenience, arising from the difference in their diameters, may most readily be overcome by mounting each one in a cork, and then bringing the clamps to bear upon the corks and not directly upon the rods themselves.]

(*c*) Light the batswing burner, and bring it carefully towards the exposed ends of the rods with the plane of the flame parallel to the vertical plane in which these ends lie. Move the flame' up until all three rods are within it.

(*d*) Note the following points:
(1) The transfer of heat through the rods is a slow process.

(2) The rate at which the melting of the paraffin extends from the flame outward is most rapid in the copper, and is much slower in glass than in the metals.

(*e*) After a considerable time the distribution of temperatures in the three rods will have reached a permanent condition; that is to say, the position of the limit of the melted paraffin upon each rod will have come to rest. Note that the position which is thus reached is furthest from the flame in case of the copper, less in the iron, and least in the glass.

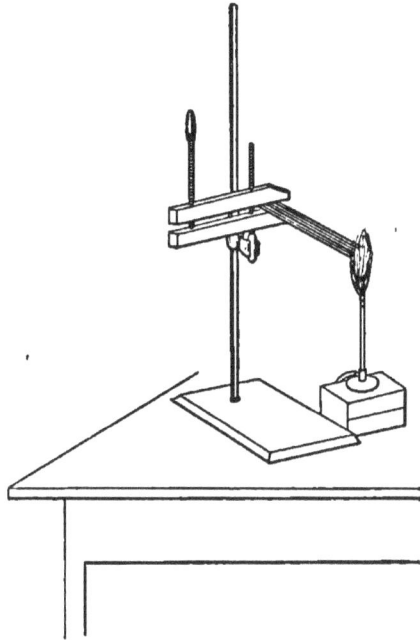

FIG. 162.

It is evident that in order to maintain those portions of the rods which are above the temperature of the room, at that temperature permanently, in spite of the fact that heat energy is being continually given off from the surface of the rods to the surrounding air, a transmission of heat energy from the flame outward must continue to take place. The amount of heat transferred will necessarily be greatest in those materials which are delivering the largest amount of heat energy to the surrounding atmosphere, and least in those which are losing the least heat. Evidently the copper rod is the one which loses the most, because a greater portion of it is heated above the temperature of the room than in case of the others.

This property of transferring heat is called *conductivity*. Of the three substances experimented with, copper is the best conductor, glass by far the poorest.[1]

177. EXPERIMENT 50. — Conductivity of Liquids.

The conductivity of liquids, with the exception of mercury, is very much smaller than that of most solids. How very slowly liquids convey heat may be shown as follows:

Apparatus:

(1) A funnel, or a bottle without a bottom, through the neck of which is inserted a glass tube ending in a bulb, as shown in Fig. 163.

(2) A small quantity of benzine.

Procedure:

(a) The funnel or bottle is mounted vertically as shown in the figure, the bulb of the glass tube which it contains being a few millimeters below the level of the mouth.

(b) Water is poured into the funnel until it rises high enough to nearly cover the bulb. The flame of the Bunsen burner is applied to the surface of the latter so as to slightly warm it, and the lower end of the glass tube is then inserted in a beaker of water. In consequence of the heating of the bulb a small amount of air will have been expelled, and when this cools again water will rise in the tube. If the bulb was not too strongly heated, this liquid column will come to rest at a point below the neck of the funnel or bottle, and the bulb will then serve as a simple form of air thermometer.

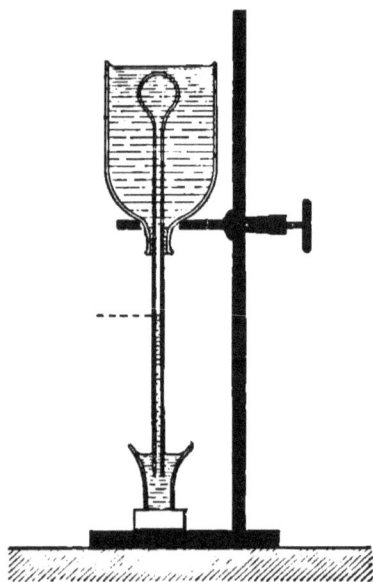

FIG. 163.

(c) Add water to the funnel until

[1] The quantitative study of thermal conductivity is beyond the range of the experiments given in this book. For a description of the methods by which such measurements are made, see Preston's *Theory of Heat*, p. 505, etc.

the same is brimming full, in which condition there.should be a depth of liquid over the bulb of three or four millimeters. From a very small flask or bottle pour about a cubic centimeter of benzine upon the surface of the water in the funnel and ignite. The benzine will burn freely and will generate much heat.

(*d*) Watch the liquid column in the air thermometer. It will be found that the latter remains almost completely stationary in spite of the fact that the layer of liquid which separates it from the flame is very thin. It follows, therefore, that water is a very poor conductor of heat. The demonstration of the conductivity of gas is a more difficult matter. It has been found, however, that gases conduct heat even more slowly than liquids do.

178. Thermal Conductivity a Slow Process. — The chief characteristic of thermal conductivity, as compared with other methods by which heat energy is transferred, is the slowness with which it takes place. A striking example of this occurs in the distribution of temperatures in the earth's crust. Geologists have been much interested in measuring the temperatures at different depths below the surface of the earth by means of instruments placed in the shafts of mines, and also in the narrower bore holes which are often carried to great depths in searching for oil, gas, etc. By means of the thermo-element and of other electrical devices, it is possible to make records of the temperatures in such localities, within the crust of the earth, day by day for very long periods of time, and it has been found that the heat of the summer weather upon the surface of the earth penetrates surely, but very slowly indeed, into the interior. At a distance of several hundred feet, for example, the hottest days may be found to occur not in midsummer but several months later, the fluctuations of heat and cold being reproduced, although with greatly diminished range of temperature within the rock masses below. The greater the distance from the surface, the longer will be the time which must elapse before the effect of a hot spell of weather, for instance, will make itself felt, and the smaller will be the variations of temperature.

179. Convection. — If the condition of the air surrounding any heated body be observed, it will be found to be in motion. Heated air rises from the surface, and other currents of cooler air come in to take its place. These are called *convection currents.* Since air possesses thermal capacity, the result of this circulation is that each molecule of the moving gas carries away with it a certain definite amount of energy which it has obtained by contact with the hot surface.

Convection currents occur in liquids also, and their presence may be shown in a variety of ways. The following experiment is one of the best adapted to this purpose, and is easily performed, provided a storage battery or other source of electrical current is available:

180. Experiment 51. — Convection Currents in a Liquid.

Apparatus:

(1) A cell consisting of two plates of glass clamped between a piece of rubber tubing as described in Appendix VII.

(2) A short piece of fine platinum wire or iron wire, also some heavier copper wire.

(3) A storage battery of two or three cells, or a primary battery capable of giving a considerable current.

(4) A small amount of colored water (a solution of an aniline dye will do, or even a few drops of ink).

Procedure:

(a) From the platinum wire make a spiral coil of the form shown in Fig. 164, and of such size as to fit the cell. In the figure, *ab* and *dc* are pieces of the copper wire bent as shown, and the ends set through two small corks, which should be slightly greater in diameter than the inner thickness of the cell. At b and c join the ends of the coil e, either by soldering or by wrapping the ends of the platinum or iron wire snugly around the copper wire.

Place this arrangement in the cell with the coil as near the bottom and as well centered as possible. It will be held in place by the two corks, which are to be crowded in between the glass plates at the top of the cell when the adjustment is made.

(*b*) Nearly fill the cell with water, and by means of a pipette introduce at the bottom a layer of coloring .matter (water tinted with red ink). This layer should be about 1 cm. in depth. If this operation be delicately performed, there will be a sharply marked line between the clear and the stained liquid.

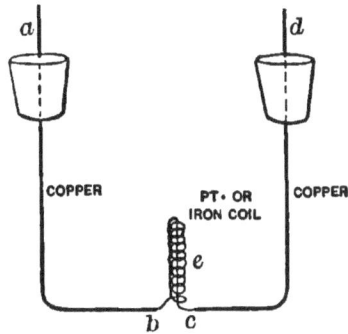

FIG. 164.

(*c*) Connect the terminals of the wire coil with the poles of the battery. The result will be to heat the wire to a temperature considerably above that of the surrounding liquid. Particles of liquid which are in contact with the metal will receive heat from the same, will expand, and owing to their diminished density will rise toward the surface of the liquid within the cell. These will be replaced by cold particles from the layers of liquid at the bottom of the cell. These in turn having been heated by contact with the coil, others again will take their place. Since the region from which the ingoing convection current comes is filled with colored liquid, and since the particles which form this ingoing current a moment later constitute the upward current, the nature of the movement will be plainly seen. It will be seen to consist of a movement of the colored matter or liquid through the mass of the surrounding water. In the course of a few minutes a large portion of the ink solution will have been transferred from the bottom of the cell to the surface of the same, where, having cooled, it will begin to return, falling as a slow current on either side along the walls of the vessel. It will be seen by means of this experiment that in convection we have a circulatory movement of the liquid or gas, the liquid particles moving in closed curves and carrying heat energy with them each time from the surface of the coil to the cold regions lying at a distance from the same. That convection currents serve to transfer heat may be shown in the following manner:

181. EXPERIMENT 52. — The Cooling of a Heated Body in Air and in Vacuo.

Apparatus:

(1) An air pump with an open-necked receiver.

(2) A small test tube which will pass easily into the neck of the receiver.

(3) A thermometer reading to 100°.

(4) A clock or watch indicating seconds of time.

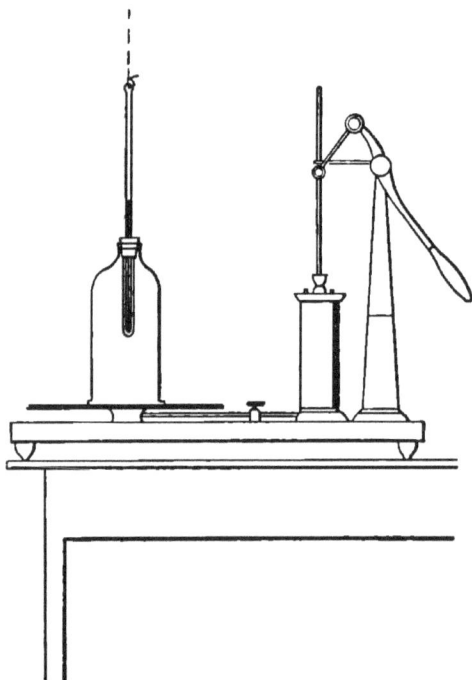

FIG. 165.

Procedure:

(a) Fit to the neck of the bell jar a cork containing a hole of sufficient size to admit the test tube, as shown in Fig. 165, and make the joint air tight by the application of rubber cement or of rosin-beeswax cement. Insert the thermometer in the test tube, holding it in place by means of a loosely fitting cork, and suspending it likewise from above, as in the figure.

(*b*) Smoke the outer surface of the test tube, and fill the tube with hot water to a height sufficient to cover the bulb of the thermometer. Place the apparatus thus adjusted upon the plate of the air pump. At intervals of sixty seconds, beginning as soon after the apparatus has been arranged as convenient, read the thermometer, noting the times corresponding to each reading. Continue these readings of the thermometer for ten minutes.

(*c*) Refill the test tube with hot water, and exhaust the bell jar by means of the air pump. Great care should be taken in this experiment to have the air pump placed upon a firm foundation, otherwise the thermometer will be endangered. The pump should be clamped to a laboratory table. As soon as the vacuum is complete, make a new series of temperature readings as described under section *b*, beginning these at the same temperature as before. The two sets of readings should be tabulated side by side as follows:

TABLE.
Curves of Cooling.

In Air.			In Vacuo.[1]		
Time.	Time from first reading.	Temp.	Time.	Time from first reading.	Temp.
h. m. s.	m.		h. m. s.	m.	
12 42 0	0	64·20°	12 59 0	0	64·30°
— 43 —	1	61·30	1 0 —	1	62·20
— 44 —	2	58·90	— 1 —	2	60·40
— 45 —	3	56·60	— 2 —	3	58·80
— 46 —	4	54·50	— 3 —	4	57·20
— 47 —	5	52·60	— 4 —	5	56·00
— 48 —	6	50·90	— 5 —	6	55·40 [2]
— 49 —	7	49·20	— 6 —	7	53·20
— 50 —	8	47·70	— 7 —	8	52·10
— 51 —	9	46·30	— 8 —	9	51·30
— 52 —	10	45·00	— 9 —	10	50·50

[1] Twenty strokes of the pump between each reading to maintain the vacuum.

[2] Reference to Fig. 166 shows that this reading was probably just 1° too high. The value 54·20° falls upon the curve; the position is marked by means of a cross.

(*d*) From the data obtained plot two curves upon the same sheet of cross-section paper, with temperatures as ordinates and times as abscissas. A comparison of these will show that cooling occurs much more slowly *in vacuo* than when the heated flask is surrounded by air.

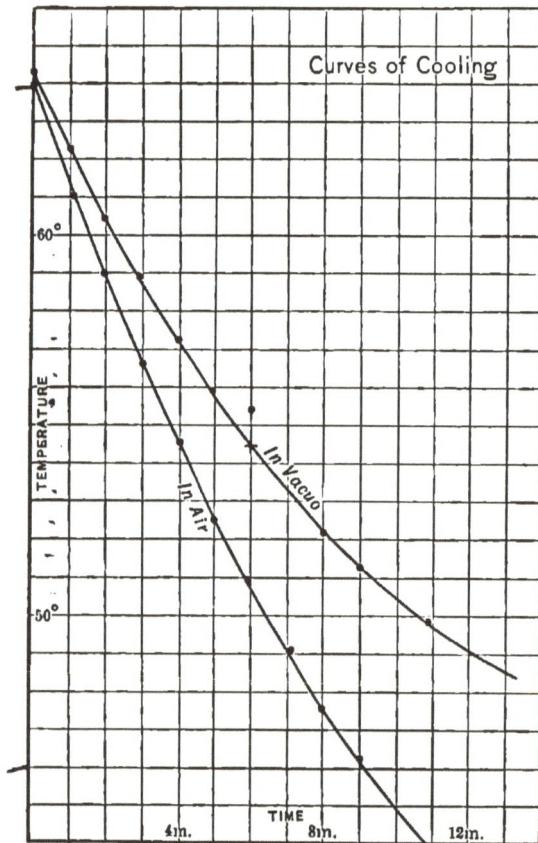

Fig. 166.

Figure 166 is drawn from the results of such a set of measurements. It indicates that in air the tubeful of water becomes 5·4° cooler, at the end of ten minutes, than is the case *in vacuo*.

182. Radiation. — The third method by which heat energy is conveyed through space is called *radiation*. This term is applied solely to the transfer of energy by

the face of the thermopile and about 20 cm. long. One end of this rod must be cut squarely and smoothly off at right angles to its length.

(3) A luminous gas flame (without chimney).

(4) A block of ice.

Procedure:

(*a*) Set up the apparatus as shown in Fig. 168. Connect the thermopile with the galvanometer, and place the copper bar with its smooth end in contact with one face of the former.

[This face of the thermopile, from which the cone has been removed to admit the bar, may

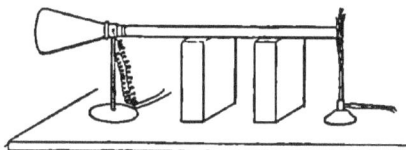

FIG. 168.

be protected from draughts of air by a wrapping of adhesive tape.]

(*b*) When the galvanometer needle has come to rest, bring the gas flame under the free end of the bar. Note roughly, by counting, how many seconds elapse before an appreciable deflection of the galvanometer indicates the arrival of heat at the end of the bar which touches the face of the thermopile. Remove the flame and notice that the effect on the galvanometer continues to increase for a considerable time after the free end of the bar has begun to cool.

(*c*) After equilibrium has again established itself (*i.e.* after the bar has had time to cool throughout) repeat section (*b*), removing the flame, however, after half the interval necessary to produce the first effect upon the galvanometer. Note the result and compare with *b*.

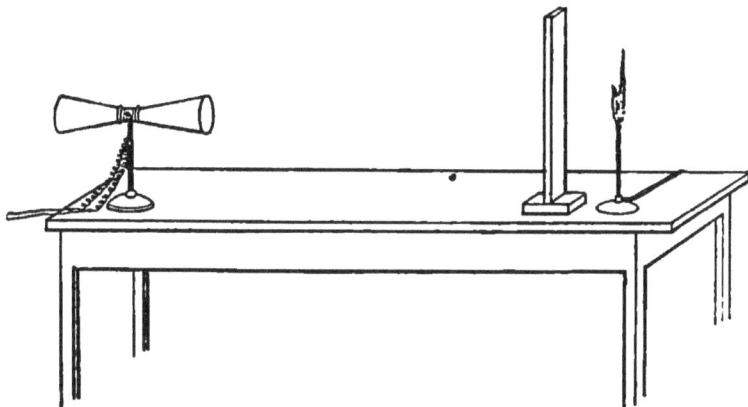

FIG. 169.

(*d*) Repeat (*b*), using the block of ice at the free end of the bar instead of the flame.

(*e*) Remove the bar and place the cone on the pile. Set up a wooden screen (Fig. 169) 100 cm. from the face of the pile, and behind it the gas flame. Remove the screen suddenly and note that the galvanometer begins to show the effect of the exposure instantly.

185. Influence of the Surface upon Radiation. — The intensity of radiation depends upon the character of the surface (*i.e.* upon the material of which it is made and upon the quality of the surface as regards polish, etc.).

It also depends upon the temperature of the surface. The former statement may be verified as follows:

186. EXPERIMENT 55. — Leslie's Cube.

Apparatus:

(1) A hollow cubical vessel of sheet metal (Leslie's cube), the edges of which are about 10 cm. long. This contains an opening in the center of one face through which it may be filled with liquid, and in which a thermometer may be inserted. The four vertical faces of the cube should be treated as follows:

Face 1 should be of bright polished metal, namely, either nickel-plated, or, if the cube be made of brass or copper, rubbed to a polish.

Face 2 should be smoked with lampblack.

Face 3 should be painted with red lead.

Face 4 should be smoked in the flame of a burning magnesium ribbon. The result will be to coat the metal with a thin layer of oxide of magnesium. (It may be painted with zinc white instead of being smoked.)

(2) A galvanometer and thermopile.

Procedure:

(*a*) The cube is filled with hot water, and is placed upon a revolving stand in front of the thermopile as shown in Fig. 170. The stand must be at such a distance that the radiation from the vertical faces of the cube may produce a considerable deflection of the galvanometer needle.

(*b*) Present faces Nos. 1, 2, 3, and 4 in succession to the face of the thermopile, and note the behavior of the galvanometer. It will be found that when the radiation is that received from the lamp-blacked face, the deflection is greatest, and when from the polished metal face of the cube the deflection is least. The other two faces give intermediate values. Since these faces are all of them subject to the temperature of the hot water within the cube they will all radiate

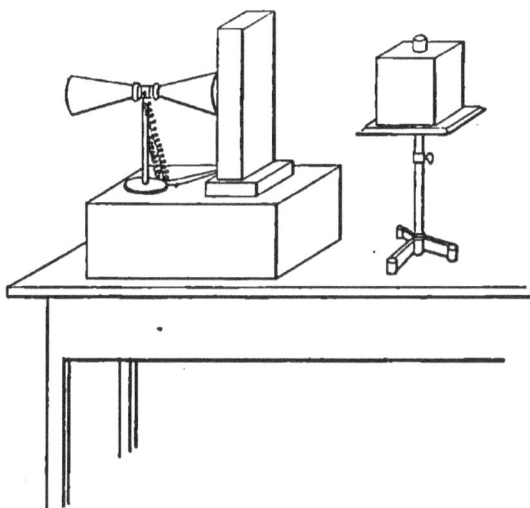

FIG. 170.

energy to the face of the pile at the same temperature. The changes of temperature which the cube undergoes during the progress of the experiment will be extremely small, — so small that they need not be taken into account. It will be seen from this experiment that at the same temperature various surfaces radiate very different amounts of heat.

187. Kirchhoff's Law. — The differences which various surfaces exhibit, as regards radiating power, depend upon what is known as *Kirchhoff's law.*

The statement of this law is as follows :

Every body radiates those waves which it is capable of absorbing, and in the same proportion.

The significance of Kirchhoff's law will appear from the following illustrations :

(1) Imagine a body polished so as to be an almost perfect mirror. Of the rays which fall upon its surface nearly all, being reflected, have no heating effect. Were the body a good radiator, it would continually give off more energy than it received, and would grow colder and colder instead of remaining at the temperature of its surroundings.

(2) A black body, on the other hand, which absorbed nearly all the rays which reached its surface and converted them into heat, would grow continually hotter than the environment, unless it were a correspondingly powerful radiator.

The fact that all known bodies tend to approach the temperature of their surroundings, instead of becoming hotter or colder, affords the basis for the law.

That good reflectors are bad radiators may be shown as follows:

188. EXPERIMENT 56. — **Kirchhoff's Law.**

Apparatus:

(1) A porcelain crucible containing a blue or black trade-mark burned in the glaze. The porcelain of Meissen or of the royal Berlin potteries have such a mark. (Any thin-patterned porcelain will do.)

(2) A blast lamp.

Procedure:

(a) Note the appearance of the porcelain by reflected light, the mark showing dark upon a bright background.

(b) Heat the porcelain in the flame of the blast lamp, beginning very cautiously and raising the temperature gradually until the whole piece is incandescent. Note that the trade-mark glows more brilliantly than the rest, so that the appearance now is of a bright mark upon a darker ground.

(c) Extinguish the flame and watch the return to the original appearance.

189. Radiation and Temperature. — It has already been stated (Art. 185) that the intensity of radiation depends upon the temperature. As the temperature of the radiating surface rises, the radiation increases, but much more rapidly. Its growth between 0° and 80° may be demonstrated by means of the following experiment:

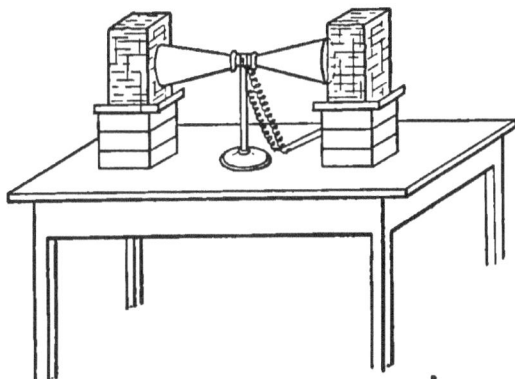

FIG. 171.

190. EXPERIMENT 57. — Increase of Radiation with Temperature.

Apparatus:

(1) The thermopile and galvanometer.
(2) A Leslie cube.
(3) Two blocks of ice; hot water.
(4) A thermometer.

Procedure:

(*a*) Mount the thermopile and galvanometer as in Exp. 54, etc. Set a block of ice in front of each cone (Fig. 171). After equilibrium has been reached, read the galvanometer.

(*b*) Replace one of the ice blocks by the Leslie cube. Fill the cube with water at about 20°, and insert the thermometer (Fig. 172). After equilibrium read both thermometer and galvanometer.

(*c*) Fill the cube with water successively at 40°, 60°, and 80°, and read temperatures and deflections.

P

(*d*) Tabulate your data, and plot a curve with temperatures as abscissas and deflections as ordinates.

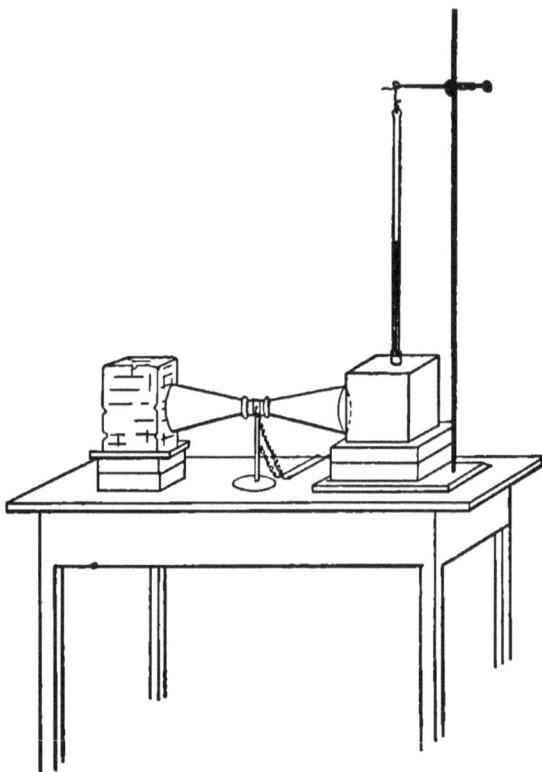

Fig. 172.

It will be found that the trend of the curve is upwards; *i.e.* that the deflection increases faster than the temperature. The deflection, however, is proportional (or very nearly so) to the radiation. This experiment, therefore, serves as a demonstration of the statement made in Art. 189.

PART III — ELECTRICITY AND . MAGNETISM

CHAPTER XXI

INTRODUCTION TO ELECTROSTATICS

191. The Relation of Electricity to other Portions of Physics. — The phenomena to be studied under the head of electricity and magnetism are very closely related to those which have been described in the foregoing chapters. It is still with the action of forces upon matter, and with the expenditure, storage, dissipation, and transformation of energy that we shall have to do; but the phenomena generally spoken of as electrical or magnetic, respectively, have certain features in common which make it desirable to consider them in a separate group or class by themselves.

192. EXPERIMENT 58. — Electrostatic Attraction.

Apparatus :

(1) The wooden support described in Art. 38.

(2) A strip of wood about 60 cm. long, 2 or 3 cm. wide, and 1 cm. thick.

(3) A piece of hard rubber (vulcanite) about 30 cm. long (an ordinary black rubber ruler or straightedge will do); a glass rod or tube about 30 cm. long and 2 cm. in diameter; a catskin or other piece of fur; a silk handkerchief.

211

Procedure: (*a*) Balance the wooden strip upon a pointed wire which has previously been mounted in a wooden block (Fig. 173).

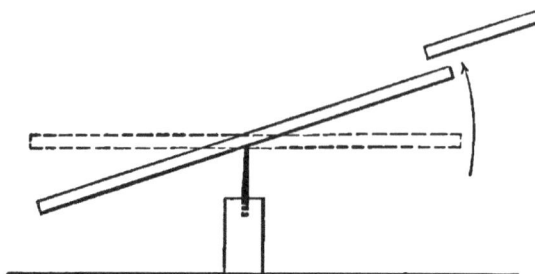

FIG. 173.

FIG. 174.

(*b*) Rub the vulcanite ruler vigorously with the catskin and bring it near one end of the balanced strip, first above, then below the latter, then on either side. Note the attractive force between the wood and the vulcanite.

(*c*) Rub the glass rod with the silk handkerchief and repeat the operations described in (*b*). Note that the wood is attracted towards the glass, although perhaps not so strongly as towards the vulcanite. (It is necessary to have both glass and silk quite dry.)

(*d*) By means of a silk thread, and a stirrup made by bending a bit of wire as shown in Fig. 174, hang the vulcanite ruler, previously well rubbed with fur, from the support. (See Fig. 175.) Bring near it the wooden strip, any metal tool, the hand itself, in fact any object not previously rubbed, and note that strong attractive forces exist between the body thus brought near and the suspended ruler.

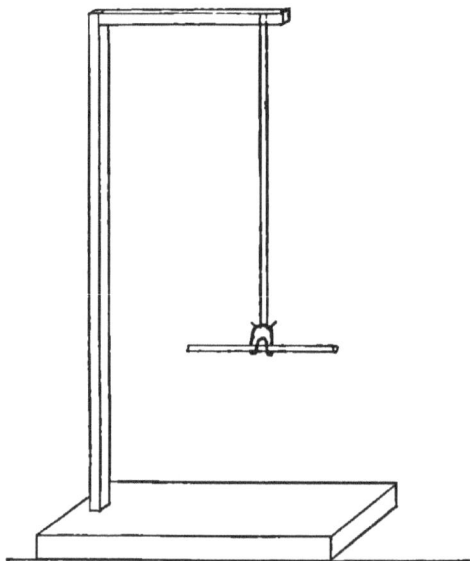

FIG. 175.

(*e*) Rub the glass rod with silk and repeat operation (*d*).

193. The preceding experiment (58) affords the basis for the following statement:

Various bodies, such as vulcanite and glass, can be brought by friction (i.e. *by doing work upon them*) *into a condition such that they attract and are attracted by whatever bodies may be near them.* This condition is called **electrification.**

An extension of the inquiry would lead to the conclusion that all substances are capable of being electrified by friction, provided that they are rubbed with some material differing in nature from themselves; also that the process of electrification is always one which involves the expenditure of energy.

194. EXPERIMENT 59. — Electrostatic Repulsion.

Apparatus:
(1) The support used in Experiment 58.
(2) Two rods of vulcanite and two rods or tubes of glass.
(3) Catskin and silk.

Procedure:
(*a*) Electrify one of the vulcanite rods by friction with the catskin, and·suspend it as in the previous experiment. Rub one of the glass rods with silk and bring it near. Note the attraction between the two.

(*b*) Electrify the other vulcanite rod and bring it near the suspended rod. *Note that they repel one another strongly.*

(*c*) Repeat with one of the glass rods electrified and suspended. Note that, whatever combination may be made, there is repulsion between electrified glass and electrified glass, and attraction between electrified glass and electrified vulcanite.

By means of experiments of this kind upon a great variety of substances, it has been shown that all bodies, however electrified, fall into one of two classes. They either repel glass rubbed with silk, and attract vulcanite rubbed with fur, or *vice versa.*

Bodies of the first class, *i.e.* those which repel glass, are said to be positively charged. Bodies which attract glass, as above, are said to be negatively charged.

In general, electrified bodies either attract or repel each other. *Those which repel each other* are said to be similarly charged; *those which attract each other* are said to be oppositely charged.

195. The results of the two experiments (58 and 59) just described may be stated as follows:

(1) *Bodies, however electrified, attract neutral bodies.*

(2) *Bodies similarly charged repel each other.*

(3) *Bodies oppositely charged attract each other.*

(4) Any charged body, however electrified, may be classified either as *positively charged* (or vitreously charged), or as *negatively charged.* In the former case, it repels glass rubbed with silk; in the latter, it attracts that body.

196. The Electroscope. — The study of the phenomena of electrification is greatly aided by the use of the elec-

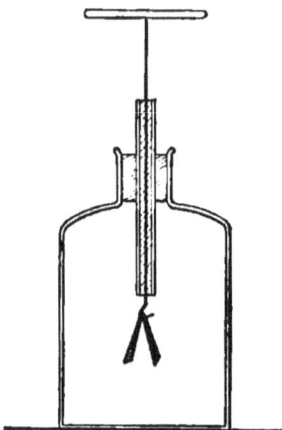

FIG. 176.

troscope. This is an instrument designed for the purpose of indicating the presence of an electric charge, and of ascertaining whether the same be positive or negative. The form generally employed is that shown in Fig. 176. It is called the gold-leaf electroscope, and it consists of a metal rod, one end of which is inserted in a glass flask or receiver, the object of which is to protect it from draughts of air, etc. To the lower end of this rod are attached two strips of gold foil of equal size. These hang, side by side, from the lower extremity of the rod. The outer end of the rod, which projects into the open air, is surmounted with a metal plate or ball. The action of the electroscope is best understood by means of the following experiment: [1]

197. EXPERIMENT 60. — The Gold-Leaf Electroscope.

Apparatus:
(1) The electroscope described in Appendix VIII.
(2) Rods of vulcanite and of glass, a stick of sealing wax, a piece of roll brimstone, a lump of rosin.
(3) A catskin, a silk handkerchief, a piece of flannel or other woolen cloth.

Procedure:
(*a*) Electrify the vulcanite rod very slightly by rubbing it with the catskin. Bring it into contact with the plate of the electroscope for an instant, and remove it. The gold leaves which previously had hung side by side now repel each other. They stand at such an angle that the force of gravitation and the repellent force between them balance one another.

(*b*) Touch the plate of the electroscope with the finger, and note that the gold leaves immediately come together into their original position.

These observations (*a*) and (*b*) indicate, among other things, that a body may be charged by contact with an electrified body; also that the charge may be dissipated by contact with a neutral object.

(*c*) Repeat observations (*a*) and (*b*), using glass electrified by rubbing with silk.

(*d*) Bring the vulcanite rod, or the glass rod more strongly electrified than before, gradually into the neighborhood of the plate of the electroscope, and then withdraw it. Watch the leaves. It will be seen that the latter begin to repel each other while the electrified body is still at a considerable distance, and that the effect increases as it is brought nearer and nearer. As the rod is withdrawn from the neighborhood of the plate, the leaves come gradually together again, and, finally, they return entirely into their original position. The electroscope, under these circumstances, is said to receive a *temporary charge*, and the process is called *charging by induction*.

[1] For details of construction of a simple but serviceable electroscope, see Appendix VII.

In section (a) of this experiment, the electroscope received a *permanent charge*, *i.e.* it continued to exhibit repulsion of the gold leaves after the removal of the electrified rod. The temporary charge, however, exists only so long as the electrified body remains in the neighborhood, and the effect is evidently dependent upon the distance between the rod and the plate of the electroscope.

(e) Bring the electrified vulcanite rod into the neighborhood of the plate of the electroscope as in (d). Touch the plate with the finger for an instant; then withdraw the hand, and, finally, remove the electrified rod from the neighborhood. Note that the effect upon the leaves is now a permanent one. They have received a permanent charge by induction.

(f) The electroscope being still charged as a result of the previous operation, bring the vulcanite rod near and watch the movement of the gold leaves. Note that they move gradually towards one another as if losing charge upon the approach of the vulcanite rod, and that, when the latter is withdrawn, they return to their former position. Repeat this operation, using the glass rod electrified by rubbing with silk, and note that, in this case, the gold leaves are more and more strongly repelled as the glass rod approaches, and return to their former condition when the same is removed.

We have in these last operations a more complex phenomenon. The electroscope has a permanent charge given to it by means of operation (d), and in addition to this it receives temporary effects from the subsequent proximity of the electrified rods. These effects are dependent upon the character of the charge of the body which is brought near them. A body negatively electrified, as is the case with the vulcanite rod, produces a temporary reduction of the repulsion of the gold leaves. A body positively electrified like the glass rod increases the angle between the leaves while it remains in the neighborhood of the electroscope.

198. Further Discussion of the Results obtained from Experiment 59 ; Hypothesis of Two Fluids. — To explain the phenomena observed in the course of Experiment 59, we may make use of the following very convenient but entirely artificial assumption concerning the nature of an electric charge. Let us suppose that the electric charge consists in the existence upon the surface of the electrified body of an invisible fluid, the presence of which gives the charged body the power of attraction and repulsion, as described in Experiment 57. Since there are two kinds of charge, positive and negative, we may suppose that there are two distinct fluids capable of producing these charges, and that these fluids are present on the surface of all bodies. When they are present in equal quantities, they neutralize each other, and the body in question shows none of the properties of an electrified body. It is said to be neutral or unelectrified. When either fluid is in excess the body shows a static charge of the character which that fluid is capable of producing. The process of electrification, according to this theory, consists in the separation of the two fluids. Whenever two objects which differ from one another are rubbed together, such a separation tends to take place and one of the objects will become positively, the other negatively charged. (See Experiment 61.) This is the explanation of electrification by friction, according to the assumption of which we are making use.

In order to account for the phenomena of electrification, it is also necessary to assume that each of these two fluids exerts a repellent action upon itself, and that each attracts the other. By means of this assumption it is easy to account in an artificial way for the phenomena of temporary and permanent charging by induction, as well as for the process of permanent charging and discharging of the body by contact.

199. Explanation of the Foregoing Phenomena by Means of the Hypothesis of Two Fluids:

(*a*) *Charging by contact.* When the electrified body is brought into contact with the plate of the electroscope, the fluid which constitutes its charge is immediately distributed, because of its repellent action upon itself, to all portions of the surface of the metal portion of the instrument. (The reason why this distribution is confined to the metal portions of the electroscope will be considered in Art. 200.) The repellent action of the fluid upon the leaves of the electroscope results in driving these apart. The same repellent forces exist between all neighboring portions of the charged body, but the forces of repulsion are so small, compared with the molecular force between the particles of the metal, that no appreciable change of form or volume takes place. The two gold leaves, however, are not held together by molecular forces, and the electric forces due to the charge manifest themselves in driving the leaves asunder. The sphere of action of the molecular forces, as was shown in Chapter X, is exceedingly small. The electrical forces of attraction and repulsion, on the other hand, like that of gravitation, act at all distances in inverse ratio to the square of the distance.

(*b*) *Discharge by contact.* Although the fluid, which is assumed to constitute the electric charge, repels itself and is thus driven into all portions, however remote, of the surface of the body upon which it exists, it cannot leave that body excepting by traveling along the surface of other bodies, or by being carried away by the latter. The result is that a body charged, and then isolated from other matter, would remain charged indefinitely. If, however, we bring a body that is capable of carrying away the fluid into contact with any portion of the charged metal, the fluid, on account of its repellent action towards itself, will immediately flow away over the surface of that body. The electrified body thus loses its charge by contact.

(*c*) *Temporary charging by induction.* When the vulcanite rod previously electrified is brought near to the plate of the electroscope, but not into contact with the latter, the two fluids already existing upon the surface of the electroscope in equal amounts are acted upon. The fluid of positive charge is attracted by the neighboring fluid of negative charge upon the vulcanite rod. The negative fluid, being similar in character to that which constitutes the charge of the vulcanite rod, is repelled and flows downward. It takes up its position chiefly in the leaves of the instrument. The leaves are therefore repelled, and the electroscope shows a temporary charge by induction.

Neither of the fluids in this operation has been allowed to escape from the surface of the electroscope, nor has any charging fluid been added from without. The only thing that has taken place has been the re-arrangement of the fluids upon the surface of the instrument. The positive charge gathers upon the plate of the instrument as indicated in Fig. 177, while the negative charge is driven downward to the gold leaves. When the inducing body, namely, the vulcanite

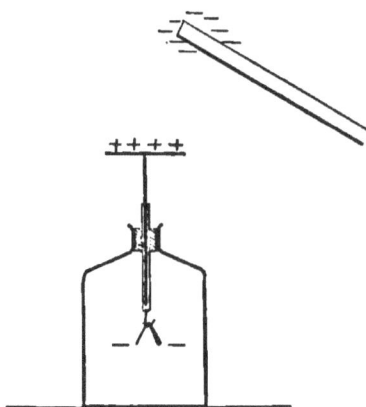

FIG. 177.

rod, is removed, these fluids, which attract each other, return to their former distribution. They there neutralize each other on every portion of the metallic surface, and the charge disappears.

(*d*) *Permanent charging by induction.* The first portion of this operation is that described in the previous section. As has already been stated, however, it is necessary to bring into contact with some portion of the metal of the electroscope, a body which is capable of conveying the electric fluid away. The fluid which thus escapes is the one which is repelled by the charge of the electrified rod. This fluid passes off to the body of the operator, and so to the earth, leaving behind an excess of the positive charge. When the finger is removed from the plate of the electroscope, and that instrument is thus isolated, and when the electrified rod is also removed to a distance, it is the same as though a permanent charge by contact with a positively electrified body has taken place ; that is to say, there is an excess of the positive fluid which immediately spreads itself over the entire surface and increases the repulsion of the leaves.

The reason why, under these circumstances, the leaves of the electroscope are drawn together when the electrified vulcanite rod is brought near is obvious. Since the fluid which constitutes the charge is positive, it is attracted by the negative charge of the rod, and is drawn out of the leaves of the plate, thus reducing their repellent action upon each other. When, on the other hand, a positively charged body, such as a glass rod rubbed with silk, is brought near, the positive charge repels a like charge on the metal of the electroscope, driving more of it down into the leaves, and increasing their repellent action.

CHAPTER XXII

CONDUCTORS AND NON-CONDUCTORS: ELECTRICAL MACHINES

200. Conductors and Non-conductors. — In discussing the performance of the electroscope, the statement was made that the charge imparted to the body by contact with the electrified body immediately distributed itself over the surface of the metallic portion of the instrument. There is a distinction in this respect which it is important to consider.

Some bodies distribute with great rapidity whatever electric charge they may acquire to all portions of their surface. Such bodies are called *conductors* of electricity. Other substances are not capable of thus rapidly conveying the charge. The latter are called *non-conductors.*

The distinction is not an absolute one, and it would perhaps be more accurate to say that certain bodies are good, while others are bad, conductors. To illustrate the very great difference in this respect between such substances as glass and the metals, the following experiment may be tried:

EXPERIMENT 61. — Conductors and Non-conductors.

Apparatus:

(1) The gold-leaf electroscope.

(2) A naked copper wire about 2 m. long and a glass rod or tube of the same length.

(3) Rods of vulcanite and glass, a catskin, and a silk handkerchief.

Procedure :

(*a*) Suspend the copper wire horizontally as shown in Fig. 178, hanging it by means of two silk threads to which wire stirrups have been attached. One end of the rod should be in contact with the plate of the electroscope. Electrify the vulcanite rod, and bring it near the free end of the wire. Note the effect upon the electroscope. It will be found that the instrument can be charged both by contact

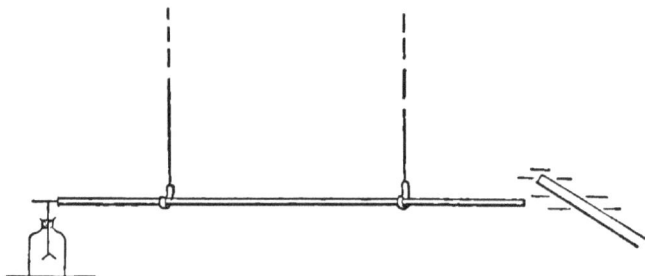

FIG. 178.

and induction, using the wire as an intervening medium. Repeat, using the glass rod as a source of electrification, instead of the rod of vulcanite.

(*b*) Substitute for the copper wire the long glass tube, and attempt to repeat operation (*a*). It will be found that the electroscope does not respond. Whatever movement of the gold leaves is brought about will take place just as well without the intervening glass rod as with it. The body through which electrification takes place, as in the case of the copper wire, is called a *conductor*, one through which it does not take place a *non-conductor*.

201. Equal and Opposite Charges. — In electrification by friction, two equal and opposite charges are always produced; if the body rubbed becomes positively charged, the body by means of which it is electrified invariably takes a negative charge. This fact may be demonstrated by means of the following experiment:

EXPERIMENT 62. — **Production of Opposite Charges by Friction.**

Apparatus :

(1) The electroscope.

(2) The silk and catskin; also the glass and vulcanite rods previously mentioned.

Procedure:

(*a*) Charge the electroscope by induction from the vulcanite rod, and test its condition by afterwards bringing the rod near, but not into contact with, the plate several times and noting that the leaves are drawn together.

(*b*) Electrify the glass rod by friction with the silk handkerchief, and bring the handkerchief, held with the tips of the fingers, so that as small a portion of its surface as possible may come into contact with the hand and thus be discharged near the plate of the electroscope. Note that the leaves are drawn together, which is a sign of the negative charge upon the handkerchief. Bring the glass rod itself into the neighborhood of the electroscope and note the divergence of the leaves. It will be found, however often this operation be repeated, that the glass and the silk handkerchief are always oppositely charged; the former positively, the latter negatively.

(*c*) Repeat this operation using the vulcanite rod and the catskin. It will be found somewhat more difficult to show the presence of the positive charge on the catskin than to show the presence of the negative charge on the silk, since the catskin is by no means so good an insulator. If, however, a small piece of fur or a bit of woolen cloth be tied to the end of the glass rod and then be pushed lightly but briskly over the surface of the vulcanite, it will gather a positive charge which is not able to escape on account of the poor conductivity of the glass. The nature of its charge can thus be readily demonstrated.

The experiment may be extended to various other bodies which are not conductors of electricity.

202. The Electrostatic Series. — The phenomenon of electrification by friction is not confined to a few such substances as glass and vulcanite. It is indeed a perfectly general property which may be stated thus:

Whenever two bodies not identical in their chemical constitution and molecular arrangement are rubbed together, they become charged, one of the bodies assuming a positive and the other an equal negative electrification. In the case of metals and other good conductors of electricity, special precaution must be taken to prevent the immediate dissipation of the charge; in the case of poor conductors, the dissipation is so slow that there is no difficulty in detecting the same.

By means of the electroscope, the fact that a great variety of different bodies can be electrified by friction may be shown. It is only necessary for this purpose to collect bits of sulphur, rosin, sealing wax, glass, quartz, etc., together with pieces of flannel, silk, paper, and various other textile materials and to rub these together pairwise, testing each pair by bringing first one and then the other of the two substances which have been in contact into the neighborhood of the electroscope. It will be found that whatever substances be selected, electrification, more or less marked, will result from rubbing them together, and that the body rubbed and the one with which it has been in contact always take opposite charges. By rubbing every object in this collection successively with every other, it will be found possible to arrange them in order, placing first in the list that body which is positive when electrified by contact with each and every other member of the collection, next after it the body which is positively electrified by friction with all the remaining members, and so on, until at the end of the list is put that body which is found to be negative by contact with every other member of the collection. The properties of substances in this regard are, to a great extent, due to external differences of structure or surface, so that the series does not always take the same order at the hands of different observers. The following is the arrangement given by Faraday:

+

Fur.	Linen.
Flannel.	Silk.
Ivory.	The hand.
Feathers.	Wood.
Quartz.	Shellac.
Glass.	Metals.
Cotton.	Sulphur.

—

203. Electrical Machines. — Any device for facilitating the production of an electrostatic charge is called an *electrical* machine. There are two classes of such machines : those which depend upon friction for their action, and those in which electrostatic induction is made use of. The former are called *frictional* machines, the latter *influence* machines.

The earliest form of frictional machines consisted simply of a ball or cylinder of glass mounted upon a horizontal axis, so that it could be given a rapid motion of rotation by means of a crank. The

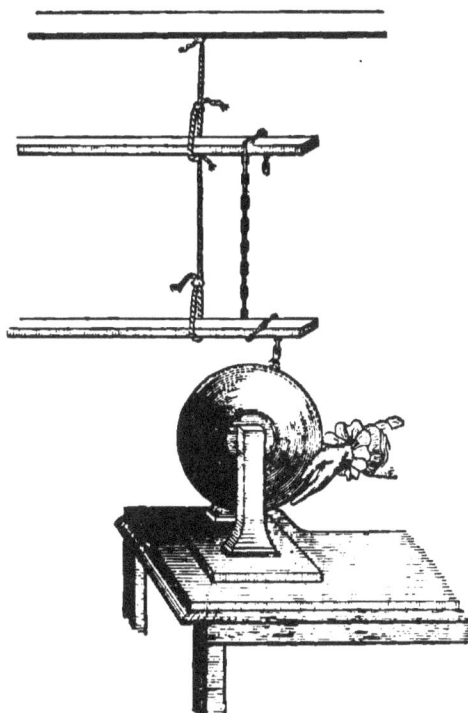

FIG. 179.

hands of the operator served as the rubber. Figure 179 shows such a machine as depicted by the Abbé Nollet in the eighteenth century.[1]

[1] Nollet, *Lettres sur l'Électricité.*

Q

In later machines glass disks were substituted for the spheres and cylinders, and a rubber consisting of silk or of undressed leather was used instead of the hand. It was soon found that an application of an amalgam of sodium to the leather greatly enhanced the activity of the machine. Figure 180 shows the usual form given to these frictional machines, which, to distinguish them from other types, are called *plate* machines. The essential parts of such instruments are:

Fɪɢ. 180.

(1) A glass plate mounted upon a cylindrical axis and turned by means of a crank.

(2) A pair of leather-faced rubbers, *r*, which clamp the machine near the periphery of the glass.

(3) A prime conductor, as it is called, which is a metal body, usually cylindrical in shape with rounded ends, upon which the charge from the glass is gathered by means of a pair of metal combs, the teeth of which come as nearly as possible into contact with the electrified plate at a point 180° distant from the rubbers. The prime conductor, *C*, and also the clamp are supported upon glass posts. The post which carries the rubbers is generally capped by a hollow brass ball, as shown in the figure.

That the action of such a machine is similar to that of a glass rod rubbed by hand with a silk handkerchief, may be shown by testing the character of the charge upon the prime conductor, and upon the ball, which is in contact with the clamp. It will be found that the prime conductor always possesses a positive charge, and the ball a negative one. This fact may be easily ascertained by bringing a proof plane[1] into contact with the conductor and then into the neighborhood of the electroscope, previously charged by induction from the vulcanite rod. The result will be to cause the leaves to diverge more strongly, which, as has already been pointed out, is indicative of a positive charge upon the proof plane. When the electroscope is charged as above, the same proof plane, if discharged and brought into contact with the ball of the machine, will show a negative charge when brought near the plate of the electroscope.

204. The Electrophorus. — The simplest form of machine for electrifying by induction is the electrophorus (Fig. 181). This consists of a disk of solid metal, about 20 cm. in diameter, which rests upon a plate of vulcanite, or other similar material which is capable of being negatively electrified by friction. The best material for this purpose is hard rubber. To the middle of the metal disk is attached a glass handle. Owing to the irregularity in the surface of the disk and of the plate upon which it rests, contact is

FIG. 181.

---

[1] Proof plane: the name given to a small metal disk, 2 cm. or 3 cm. in diameter, with a handle of glass or vulcanite. It is used, as above, for testing the charge of bodies which cannot be easily transported to the neighborhood of the electroscope.

made only at three points, the remainder of the disk being separated from the bed of vulcanite by a thin film of air.

If the vulcanite plate be charged by friction with the catskin, and the metal disk be laid upon it, the latter is charged by induction, positive electricity being drawn to the under side of the plate, as indicated in Fig. 182, and the negative charge repelled. If the finger of the experimenter is then brought for a moment into contact with the plate, thus affording the repelled charge opportunity to flow off, there will remain an excess of the positive charge which has been attracted to the inner surface of the disk. The latter may now be raised from its position upon the vulcanite plate, and it will carry with it a strong positive charge. If the finger be brought near the plate, a spark will pass between the disk and the hand. If the disk be brought near the plate of the electroscope, the presence and also the sign of the charge may be determined. If the vulcanite plate itself be brought near the electroscope, it will be found charged negatively; *i.e.* in a manner opposite to that of the disk.

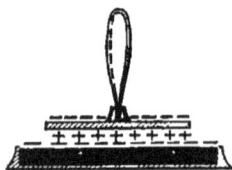

FIG. 182.

Since the disk of the electrophorus is charged by induction and not by contact, no portion of the original charge existing on the surface of the vulcanite plate is carried away in the operation just described. It is possible, therefore, to return the disk to its place and to withdraw it again newly charged, without having electrified the vulcanite plate in the meantime by friction. This process may indeed be repeated as often as desired without in any way exhausting the original charge of the vulcanite.

It is true, nevertheless, that the positive charge upon the disk after each charging represents a certain amount of energy. The spark which is formed when the disk is discharged, for example, is a manifestation of the development of heat, which in itself indicates the expenditure of energy. The source of the energy of these successive charges is *the work done in withdrawing the charged disk from the neighborhood of the oppositely charged vulcanite plate.* As has already been pointed out, there are attractive forces between bodies oppositely charged. These forces can be overcome only by the expenditure of energy. The equivalent of the energy thus expended is stored in the charged body, and when the charge is distributed it is transformed either into heat energy or into energy of motion.

205. Influence Machines. — Machines of this type, of which the electrophorus is the simplest form, are devices for going through the cycle of operations described in the previous article, automatically.

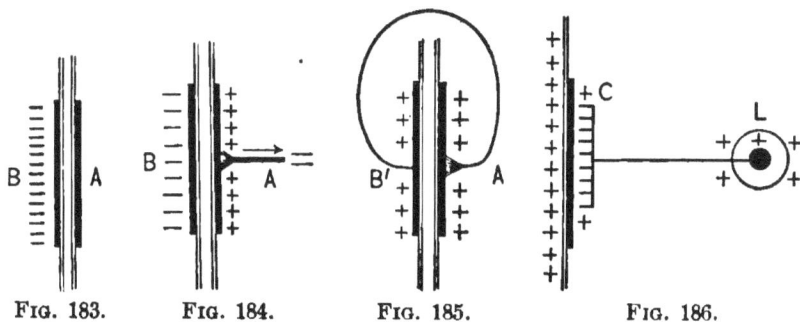

FIG. 183.   FIG. 184.   FIG. 185.   FIG. 186.

(1) A carrier of electric charge, *A* (Fig. 183), corresponding to the disk of the electrophorus, is brought near to a charged body, *B*, corresponding to the vulcanite plate.

(2) The carrier having become charged by induction, the repelled charge is carried away by momentary contact with a conducting body (Fig. 184).

(3) The carrier is made to deliver its remaining positive charge in part to a second charged body, B' (Fig. 185), and in part to a storage reservoir L (Fig. 186).

(4) The second body, B', subsequently induces a negative charge upon A, which is transferred by a similar set of operations in part to B, increasing its original charge, and in part to another storage reservoir. This process is continued indefinitely, until the insulators which separate the stored charges become inadequate, and discharge takes place.

206. The Toepler-Holtz Machine. — The manner in which this cycle of operations is performed may be conveniently studied by means of the Toepler-Holtz machine (Fig. 187).

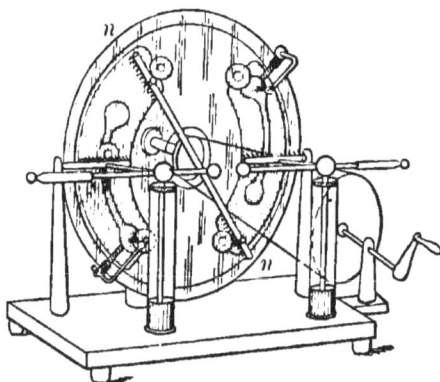

FIG. 187.

This apparatus consists of two vertical glass disks, one of which is stationary while the other revolves in front of it. Upon the back of the stationary plate are two pieces of tin foil, B, B' (Fig. 188). These correspond to the charged bodies B and B', mentioned in the preceding article. Upon the front of the revolving plate are a set of equidistant metallic carriers, each consisting of a small disk of foil surmounted by a brass button (Fig. 189). Each of these disks corresponds to the carrier, A, described in Art. 205. Imagine the revolving plate mounted in front

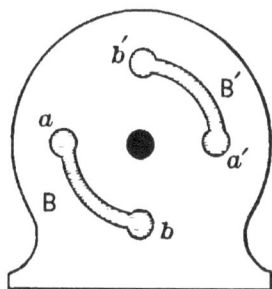

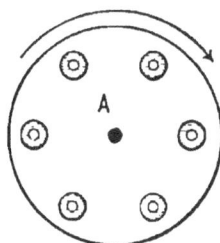

FIG. 188.   FIG. 189.

of the stationary plate, and turning in the direction indicated by the arrow. Let the body B have a small negative charge. At the beginning of the cycle of operations one of the carriers, A, is opposite the point a of B, and a separation of positive and negative charges takes place upon its surface. It then comes into contact with a tinsel brush upon the end of the neutralizing rod nn, which is shown in Fig. 187, and the repelled (negative) charge passes away. The carrier thus charged positively by induction is carried by the rotation of the plate to a point opposite b' upon the other charged body B'. Here it passes under another tinsel brush, called the *charging brush*, by means of which it shares its charge with B'. This brush is mounted at the end of a bent metallic arm, which extends around the edge of both plates

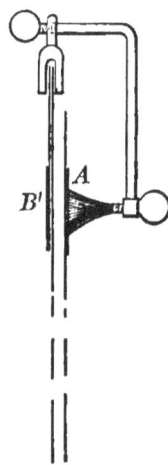

FIG. 190.

from B', as shown in Fig. 190. Finally A passes under the collecting comb and delivers another portion of its charge to the storage reservoir.

In the meantime the carrier diametrically opposite to
A is undergoing a similar cycle of operations, but with
opposite charge, whereby the body *B* receives increase in
its negative electrification, and negative charge is stored
in the reservoir upon that side of the machine. As the
machine revolves, and new pairs of carriers come opposite
B and *B'*, respectively, these cycles of operations are continually repeated.

The storage reservoirs of an influence machine are two
Leyden jars (Fig. 191), the inner coatings of which are
metallically connected with the two collecting combs, and
likewise with two sliding rods tipped with balls. These
rods are mounted horizontally
with their common axis in
front of, and parallel to, the
plates of the machine. The
distance between the balls
may be adjusted at will.

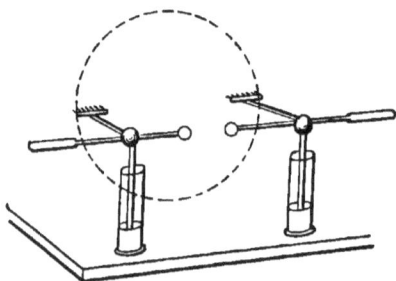

When the revolving plate
is turned, there is a continued
accumulation of positive and
negative charge, respectively, in the two Leyden jars. If
the balls are set a few centimeters apart, the attraction
between the charges becomes so great as to overcome the
insulating power of the intervening air, whereupon a spark
leaps between them. This results in the partial discharge
of the machine; but the accumulation is repeated until
another spark passes, and so on. To test the above statement of the performance of the Toepler-Holtz machine, the
following experiment may be tried:

207. EXPERIMENT 63. — Testing a Toepler-Holtz Machine.

Apparatus:

(1) A Toepler-Holtz machine.

FIG. 191.

(2) The electroscope.

(3) A proof plane.

(4) A vulcanite rod and catskin.

Procedure :

(*a*) Charge the electroscope by induction from the vulcanite rod, and test its charge by noting that the approach of the rod causes the leaves to collapse.

(*b*) Put the machine in motion in such a direction that the buttons pass in a direction from the charging brushes towards the metallic combs of the conductor. Continue turning until sparks pass from between the terminals when the latter are 2 or 3 cm. apart.

(*c*) Stop the machine ; then bring the proof plane into contact for an instant with one of the balls of the machine, and carry it into the neighborhood of the electroscope. Note the movement of the gold leaves. Discharge the proof plane and bring it into contact with the arm of the charging brush on the same side of the machine. Its charge thus obtained by means of the electroscope ought to correspond in sign to that obtained from the ball upon the same side of the machine.

(*d*) Repeat this test for corresponding parts upon the other side of the machine.

(*e*) Test the electrification of one of the carriers before it reaches the charging brush, between the brush and the comb, after passing the comb, and after passing the brush of the neutralizing rod. See whether your results are in accordance with the statements of the action of the machine given in the previous article.

208. Holtz Machines and Wimshurst Machines. — In the influence machine, invented by Holtz (Berlin, 1865), of which the Toepler-Holtz machine is a modification, the charged bodies are simply sectors of paper fastened to the back of the stationary glass plate. The carrier is the revolving plate itself. Instead of charging brushes, pointed strips of paper are glued to the paper sectors. These project through windows in the stationary plate and make contact with surface of the revolving plate. One form of the Holtz machine is shown in Fig. 192.

The Wimshurst machine is an influence machine in which both plates revolve but in opposite direction. Each

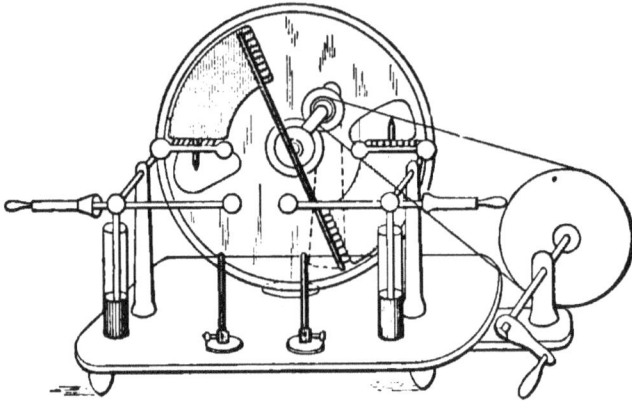

FIG. 192.

plate bears a series of metallic carriers. (See Fig. 193.) For a full discussion of the action of this machine, see *Elements of Physics*, Vol. II, p. 108.

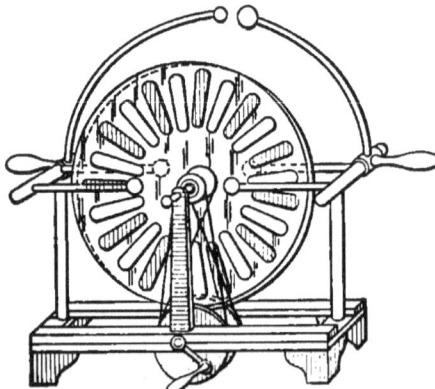

FIG. 193.

CHAPTER XXIII

DISTRIBUTION OF THE ELECTRIC CHARGE UPON CONDUCTORS

209. Electrostatic Charges reside only upon the Outer Surface of Bodies. — This fact may be illustrated by means of the following experiment:

EXPERIMENT 64. — **Faraday's Ice-Pail Experiment.**

Apparatus:

(1) A cylindrical vessel with an open mouth of the general form shown in Fig. 194. Any wide-mouthed metal can will answer for this purpose.

(2) An electrical machine.

(3) The electroscope.

(4) An insulating stand, or a plate of glass or vulcanite, upon which the can may be placed.

(5) A small metallic ball, or button, attached to the end of a silk thread, the other extremity of which is fastened to a short glass rod.

(6) A vulcanite rod and a catskin.

Procedure:

(a) Turn the plate of the electrical machine in the proper direction until it becomes well charged. Connect one terminal to the earth. If a frictional machine is used, the ball connected with the rubbers should be so connected by means of a wire. Connect the other terminal of the machine with the interior of the metal can which has previously been placed upon the insulating stand, or upon a sheet of vulcanite or glass.

Fig. 194.

(b) Charge the electroscope by induction from the vulcanite rod, and test the character of its electrification. Charge the metal can from the machine by turning the plate of the latter for a few minutes; then, by means of the vulcanite rod, withdraw the connecting wire from the interior of the former, leaving it charged and insulated.

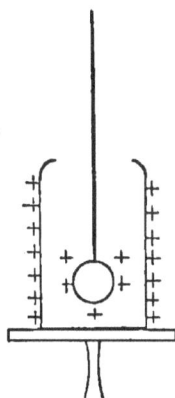

Lower the brass ball which takes the place of the proof plane, being more convenient than the latter for this experiment, into the bottom of the can, and withdraw it without making contact with the opening. Transfer it by means of a glass handle to the plate of the electroscope, and note that it brings no charge to the latter; from this we conclude that the interior of the can has no electrification.

(c) Bring the ball into contact with the outer surface of the can, and remove it into the neighborhood of the electroscope. The movement of the leaves will indicate that the ball is strongly charged.

210. Faraday's Bag. — Faraday, to whom the preceding experiment is due, tested this point in many ways. He constructed a set of buckets, one within another, and showed that when the interior of the innermost was charged, the charge was immediately transferred by repulsion to the outer surface of the exterior vessel. He also made use of a conical bag woven of linen, which material is a conductor of electricity. This is mounted upon an insulated hoop, as shown in Fig. 195. By means of a silk thread attached at the apex of the cone, this bag can be turned inside out without bringing it into contact with any conducting body. When charged, the outer surface shows electrification. The proof plane applied to the interior of the fabric, however, shows no charge. When the bag is turned inside out, by drawing upon the thread which is attached from the interior, the charge is found to have transferred itself altogether to what then becomes the outer surface, and the surface previously without, but which has now become the inner surface of the bag, is discharged. This is an ingenious modification

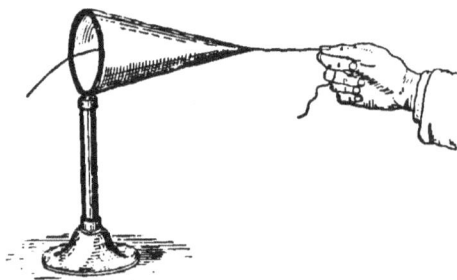

FIG. 195.

of the ice-pail experiment. A considerable degree of dexterity is required to perform the experiment satisfactorily, chiefly because the proof plane is likely to become charged by friction against the linen fabric of which the bag is made.

That an electrostatic charge gathers upon the outer surface of conductors may be also very simply demonstrated as follows: A strip of sheet metal, about 40 cm. long and 10 cm. wide, is bent into the form *abcd*, Fig. 196. It is mounted upon an insulating support, and pith balls attached to short silk fibers are fastened, two to the outside and two to the inner surface, as shown in the figure. When this little apparatus is electrified, the two outer pith balls become charged by contact with the surface and are strongly repelled; those which are attached to the inner surface of the bent sheet of metal remain neutral.[1]

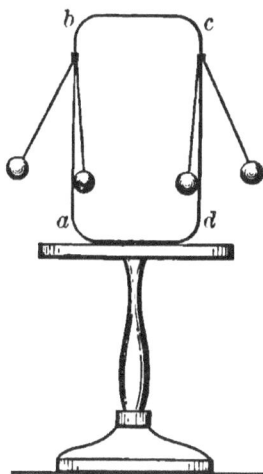

FIG. 196.

211. Quantity of Electricity. — An electrostatic charge being the result of the expenditure of energy, we may use the words *quantity of electricity* in the same way in which we used them in the subject of Heat. (See Arts. 142 and 146.) The unit by means of which, or in terms of which, electrical quantity is measured, is defined in the following manner:

[1] A tinned can of rectangular section, such as is commonly used for packing mustard, spices, etc., may be utilized for this experiment. If the effect is to be shown to a large class, the bottom should be removed from the can as well as the cover, so that the apparatus may be placed in the field of the lantern for projection.

The unit is that quantity which at a unit's distance from an equal quantity repels it with unit force. The unit of distance in this definition is one centimeter, and the unit of force is one dyne. The space between the charged bodies in this definition is supposed to be filled with air at ordinary pressure.

212. Intensity of Charge. — When a charge representing a definite amount of electrical energy is imparted to a body, it becomes distributed over the entire surface of the latter. The intensity of the effects produced will depend upon the area of that surface. If the surface be small, there will be indications of a high degree of electrification upon it; if the surface be very large the same quantity of electricity will not produce manifestations of so high a degree of electrification at any given point. The case is analogous to that of heat. If a given quantity of heat energy be imparted to a small amount of water, the result will be a great rise of temperature; if the same quantity be imparted to a very large body of water, the rise of temperature will be slight. The amount of energy in the two cases is the same. It is represented by the quantity of water multiplied by the rise of temperature which it undergoes. The corresponding relation is expressed in electrostatics by means of the words *intensity of charge*, or sometimes by the words *density of charge*. That the intensity of electrification, or in other words the density of charge, depends upon the surface over which it is distributed, may be shown by the following experiment:

213. EXPERIMENT 65. — **Diminution of Density of Charge with Increase of Charged Surface.**
   Apparatus:
   A sheet of tin foil, mounted as shown in Fig. 197. The figure

shows an ordinary spring curtain roller, over which a piece of glass tubing, about 30 cm. long, has been slipped. The tube is cemented to the roller with sealing wax or with beeswax-rosin cement. One end of a sheet of tin foil is fastened to the glass tube by means of shellac. When dry the remainder of the sheet is rolled around the tube, and the roller is mounted in the usual manner, so that the sheet of foil may be rolled and unrolled like a curtain.

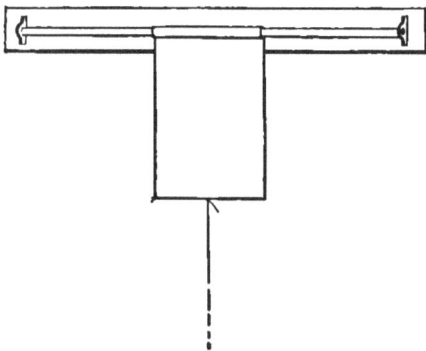

FIG. 197.

A silk thread attached to the middle of a light rod or strip, which is fastened along the free end of the sheet of foil, serves as a handle. By drawing upon the silk thread carefully, the tin foil may be unrolled like a window shade from its roller. When released it will be rolled up again by the action of the springs.

Procedure:

(a) Mount the proof plane above the plate of the electroscope, at the distance of about five centimeters. Attach a fine wire to the proof plane, and carry the same to the foil, to the free end of which it may be fastened by means of a bit of gummed paper (Fig. 198).

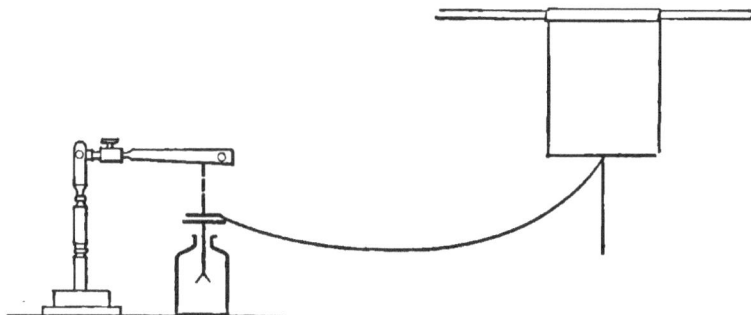

FIG. 198.

(b) The foil being rolled up, electrify it gradually until the leaves of the electroscope show marked divergence. By means of the silk thread unroll the foil carefully and watch the gold leaves. As the outer surface of the foil increases in area in the process of being unrolled, the charge is compelled to distribute itself over greater and greater areas, and it diminishes in density. This change is indicated by the gradual coming together of the leaves of the electroscope. Roll up the foil again and note that this change was not due to loss of charge. The leaves will be seen to dilate nearly to their former position. Gold leaves, attached directly to the curtain, will indicate the change in the density of charge in the same manner as the electroscope.

214. Distribution of Charge upon the Surface of Conductors. — An insulated conductor, when charged, does not assume the same intensity of electrification at all points upon its surface unless it be spherical in form. The density of charge at each point depends, in fact, upon the curvature of the surface at that point.[1]

The relationship between the electrification upon the surface of the conductor and the curvature of that surface, may be stated by saying that the density of charge increases as the radius of curvature diminishes. This statement may be verified by testing the electrification of charged bodies of different forms.

CASE 1. *An ovoid conductor.* If an insulated conductor, ovoid in form, be charged, and the density of charge at different portions of its surface be tested by means of the brass ball described in Experiment 64 and the electroscope, it is found that the greatest charge can be obtained by bringing the ball into contact with the more

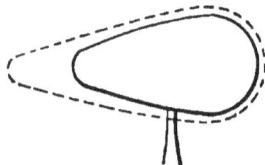

FIG. 199.

---

[1] The curvature of a surface is expressed by means of what is called the radius of curvature. Whatever form the surface may have, we may consider the portion of it which immediately surrounds any given point as part of a sphere, provided we select a sufficiently small portion. The length of the radius which this sphere would have is the radius of curvature of the surface at the point in question.

pointed end of the conductor. We may indicate the distribution of charge upon the surface of the conductor by drawing a dotted line around the conductor, as in Fig. 199, the distance of this line from the surface indicating the intensity of the electrification, or the density of charge.

CASE 2. *A disk.* In this case the radius of curvature becomes very small at the edge of the disk, and it is here that the charge is chiefly gathered. The point may be verified by bringing the ball successively into contact with the edge of the disk and with a point upon its plane surface. Figure 200 gives a diagram of the distribution.

FIG. 200.

CASE 3. *A cup-shaped conductor.* Here the maximum effect will be obtained at the lip of the cup; in the interior scarcely any indication of the electrification can be found. Upon the convex outer surface the presence of charge may be discovered, but the density is less than at the lip.

CASE 4. *A conductor carrying a point.* In this case the charge will be gathered altogether at the point. If the point were a perfect one, and so that the radius of curvature were reduced to zero, the density of charge would be relatively infinitely great as compared with that on other portions of the surface. Even with such a point as can be actually produced, the radius of curvature is so very small that the charge is almost entirely centered at the point.

215. Action of Points. — A high degree of electrification in the neighborhood of a point upon any charged body leads to certain interesting phenomena. The particles of

R

air in immediate proximity to such a point are strongly
attracted. When they come into contact with the surface of the charged body they are electrified and repelled.
Others take their places and are repelled in their turn.
The result is the establishment of a convection current,
similar to that which has been described in Chapter XX
(Art. 180). The direction taken by these currents is
shown in Fig. 201. The existence of this convection
current may be shown by
mounting a point upon
any convenient conduct-
ing body. The latter is
to be placed upon an in-
sulating plate and con-
nected by means of a wire

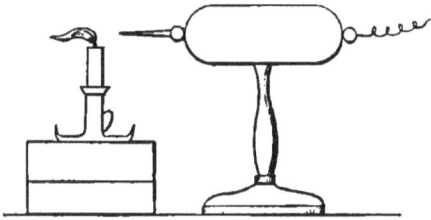

FIG. 201.

with the terminals of the Toepler-Holtz machine. The
flame of a lighted candle held near the point will be
strongly blown by the draft of air. (See Fig. 201.) The
reaction of this convec-
tion current takes the
form of a backward thrust
upon the point, and if
the latter be mounted in
such a manner that it is
free to move, it will be
driven in the opposite
direction from that in
which the current flows.
This phenomenon is illus-
trated by means of the

FIG. 202.

apparatus known as the *electrical tourniquet* (Fig. 202).
This consists of two wires at right angles to one another,
the points of which are bent as shown in the figure. They

are mounted upon a jeweled bearing at their common center. When the tourniquet is charged from an electrical machine, it revolves in the direction indicated by the arrow.

216. The Distribution of an Induced Charge. — The distribution of charge upon bodies, discussed in the previous article, is that which exists when the charged body is not in the neighborhood of another electrified body. In the latter case, the forces between the positive and negative charges modify the character of the distribution. Imagine, for example, a small metal sphere which has not been electrified. The density of its charge is everywhere zero. If this be brought into the neighborhood of a large conductor, positively charged, such as the prime conductor of an electrical machine, it causes a charge by induction, as already described (Art. 199). The portions of its surface nearest the positively charged conductor will become negatively electrified, owing to the attractive force between unlike charges, while the more remote parts of the surface will become positively charged through the action of the forces of repulsion between similar charges.

The distribution will then be of the kind indicated in Fig. 203.

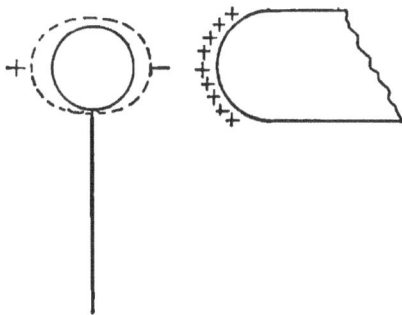

FIG. 203.

To verify this statement, bring a small proof plane or insulated brass ball into contact successively with the positive and negative portions of the sphere, and test the charge imparted to it by means of the electroscope. In case the sphere were already charged before being brought into the neighborhood of the conductor of the machine, a modification of the distribution of the charge upon its surface would likewise take place.

Imagine the initial charge to be positive. So long as the charged sphere remains by itself uninfluenced by neighboring bodies, the distribution of charge upon its surface will be uniform. When it is brought into the field of influence of the charged conductor, however, the same attractive and repellent forces which charge it by induction in the previous example will be found active. The positive charge upon the sphere will be driven to the most remote portions of the surface by the action of repulsion between it and the positively charged conductor of the machine. The distribution of the charge will therefore be a variable one with a maximum of density at that point of the surface which is furthest away from the conductor of the machine, and a minimum at the nearest point upon the surface of the sphere. If the positive charge be small as compared with that upon the conductor of the machine, it may easily happen that the surface of the small sphere nearest the conductor will become negatively charged.

A similar case is frequently observed in the use of the electroscope. If the instrument be given a slight positive charge and a strongly charged vulcanite rod be brought near, the leaves are drawn together, and as the charged rod approaches are repelled again. This second repulsion is due to a residual negative charge which is no longer neutralized by positive electrification.

CHAPTER XXIV

CONDENSERS

217. The Condenser. — When two conducting surfaces are brought near together, and one is charged, the other becomes charged by induction. The process of charging may be carried on as long as the intervening material is capable of resisting the attractive forces which are in action, and which are trying to bring the two charges together. Such an arrangement is called a *condenser*. It may be regarded as a device for storing electrical energy, since, when the two charges are allowed to unite, an amount of energy which is equivalent to the sum of all the energy expended in charging the condenser is obtained.

218. The Leyden Jar. — One of the best-known forms of condenser is the Leyden jar (Fig. 204). It consists of a wide-mouthed bottle of thin glass, the inner and outer surface of which are coated with tin foil from the bottom upwards for about two thirds of its height. The inner coating is connected, by means of a chain or wire, with a rod which is inserted through the cover of the jar. This rod ends above in a ball or plate. The two layers of tin foil, together with the intervening layer of glass, form a condenser. They are commonly spoken of as the inner coating and the outer coating. The object of the rod and ball is simply to give convenient connection between

FIG. 204.

the inner coating and the outer air. The apparatus derives its name from the city of Leyden in Holland, where it is said to have been discovered by a student of electricity named Cuneus, who was experimenting on the charging of a glass of water. He held the glass in the palm of his hand, and connected the liquid with the conductor of an electrical machine. Upon removing the liquid from contact with the machine, and touching it with the other hand to see whether it was charged, he received a severe shock. The water formed the inner coat of the condenser, and the hand in which the glass was held, the outer coat. This stored a considerable amount of electrical energy during the time that the water was in connection with the machine, all of which was instantly released when Cuneus made contact between his other hand and the liquid.

219. Specific Inductive Capacity. — The capacity of a condenser increases as its surface increases, and the distance between its plates is diminished. *The capacity also depends upon the character of the intervening medium.*

The word *capacity* is used in the above statements with reference to the quantity of electricity which the condenser is capable of storing under given conditions. The quality of the intervening medium with reference to the extent to which induction takes place through it, is termed its *specific inductive capacity*. The specific inductive capacity of air is considered to be equal to unity, and the specific inductive capacities of other substances are expressed in terms of this. Thus, paraffin has a specific inductive capacity of 1·99, sulphur of 2·58, shellac of 2·74, and glass of 3·25.

220. Experiment 66. — The Leyden Jar.

Apparatus:

(1) The Leyden jar.

(2) An electrical machine (either a Toepler-Holtz or a frictional machine will do).

(3) The electrophorus.

(4) A discharger. (The discharger consists of a jointed rod tipped with brass balls [Fig. 205]. It is mounted upon a handle of glass. By means of this instrument, the balls of which are adjustable as to their distance apart, the outer coating and the ball of the Leyden jar can be readily connected without danger of sending the discharge through the hand or body of the operator.)

Fig. 205.

Procedure:

(*a*) Charge the electrical machine, using the electrophorus to set it into action if necessary.

(*b*) Place the vulcanite plate of the electrophorus near the machine and upon it a Leyden jar, connecting the knob of the jar with the conductor of the machine, and the other terminal of the machine with the ground (*i.e.* with the nearest gas or water pipe); these connections should be made by means of copper wire, or with chains. Connect also the outer coating of the Leyden jar with the ground.

(*c*) Turn the plate of the machine for about one minute, and then bring one ball of the discharger to the outer coating, and the other to the knob of the jar. If the machine has been acting properly, there will be a heavy spark which indicates that a considerable amount of energy has been stored in the charging of the jar.

(*d*) Disconnect the outer coating from the earth, leaving it carefully insulated by the vulcanite plate upon which the jar stands. Put the machine into motion as before, for about the same length of time, then discharge the jar as in operation (*c*), and compare the character of the discharge with that previously noted. It will be seen that the spark is now very insignificant indeed. The reason for this condition is that in the latter experiment no opportunity was given for the repelled charge upon the outer coating to escape. However strongly, therefore, the outer coating may have been charged by induction, this charge would be temporary. Under these circumstances no considerable storage of electrical energy takes place.

(*e*) Arrange the apparatus as shown in Fig. 206. The wire connecting the machine with the inner coating is to be supported upon a glass rod, and its end is not to make contact with the knob of the jar. The intervening air space should be about 1 cm. The wire between the outer coating and the earth is to be removed from the outer coating by about the same distance. This arrangement perfected, repeat operation (*c*), and note that whenever the spark passes between the machine and the knob of the jar, a similar spark leaps across the air space between the outer coating and the wire which leads to the earth.

FIG. 206.

Under these conditions the jar becomes strongly charged, as may be shown by the use of the discharger.

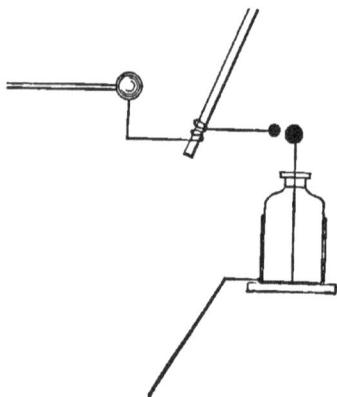

(*f*) The condenser cannot be discharged by contact with one coating at a time. To test this point charge the jar as described in operation (*c*); then remove the wire which connects the outer coating with the earth, thus leaving the jar insulated. Place one hand behind the back, and with the other touch the knob of the jar. Note that while a feeble spark passes between the hand and the jar, the discharge is not in any way comparable with that which occurs when the two coatings are connected together. Let go the knob and then touch the insulated outer coating with the hand. Another slight spark passes between the finger and the coating. The jar, however, is still strongly charged, as may be shown by the use of the discharger. The reason for putting one hand behind the back in performing this experiment is that while the discharge of a single Leyden jar through the body is not in any way dangerous, although rather unpleasant, this is the method which insures safety in experimenting with charged condensers and other electrical apparatus. In some experiments it would be a more serious matter to receive the discharge through the body. The habit of always working with one hand behind the back is easily acquired, and it secures immunity from many unpleasant experiences.

221. EXPERIMENT 67. — **The Capacity of a Condenser increases with the Area of its Coatings.**

Apparatus:

(1) Four Leyden jars.

(2) A Toepler-Holtz machine.

Procedure:

(a) Remove the small Leyden jars which are attached to the machine.

(b) Get the machine into action and drive it without the jars for half a minute at, as nearly as possible, a uniform speed, noting the rapidity with which sparks pass between the terminals of the machine, and noting likewise the appearance of the sparks. The terminals should be adjusted to a distance of 1 cm. by means of a piece of wood of that thickness, used as a gauge.

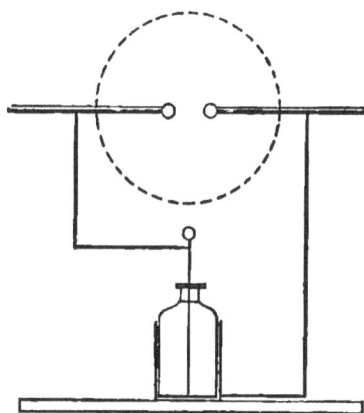

FIG. 207.

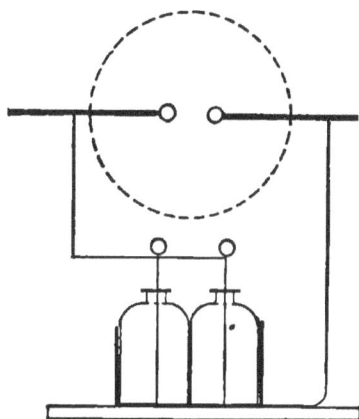

FIG. 208.

(c) Lay a pane of glass in front of the machine and place one of the four Leyden jars upon the glass. Connect the knob of the jar to one terminal of the machine, and the other terminal to the outer coating of the jar, as shown in Fig. 207. Drive the machine as before and note :

(1) That the sparks pass much less frequently and that each is much heavier than before, *i.e.* that it is louder and gives more light. Estimate *approximately* the number of turns of the handle of the machine to each spark.

(*d*) Place an additional jar upon the glass, the outer coatings in contact. Connect the knobs by means of a wire, as in Fig. 208, and repeat observation (*c*). Note that sparks occur at even greater intervals, and that each one represents a greater amount of energy than before. Estimate again the relation between sparking intervals and revolutions of the handle of the machine.

(*e*) Repeat, with three and finally with four jars.

The Leyden jars, whatever their number, when connected as above, form a single condenser, the surface of which is proportional to the number of jars. The increasing length of time necessary to produce an intensity of charge sufficient to give a spark affords a direct *indication* of the increased capacity as each jar is added.

It does not afford an exact *measure* of the capacity, because of the irregular manner in which Toepler-Holtz machines work even when driven at constant speed.

222. EXPERIMENT 68. — **Bound and Free Charges.**

Apparatus :

(1) The gold-leaf electroscope.

(2) The electrophorus.

(3) A sheet of plate glass about 20 cm. square.

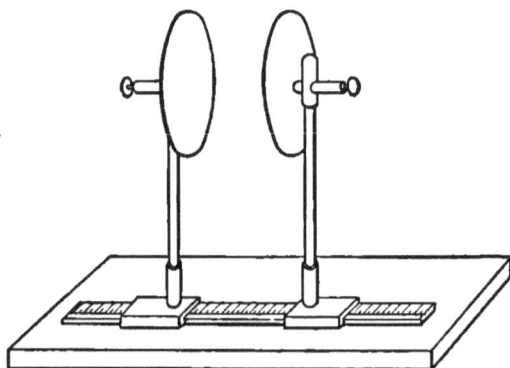

FIG. 209.

(4) A condenser of the form shown in Fig. 209. This instrument is known as an *air condenser*. It consists of two metallic disks, mounted vertically and coaxially upon insulating supports. One or both of them should have freedom of motion in the direction of their common **axis**.

Procedure:

(*a*) Remove one plate of the air condenser to a considerable distance from its neighbor, and attach the other plate to the disk of the

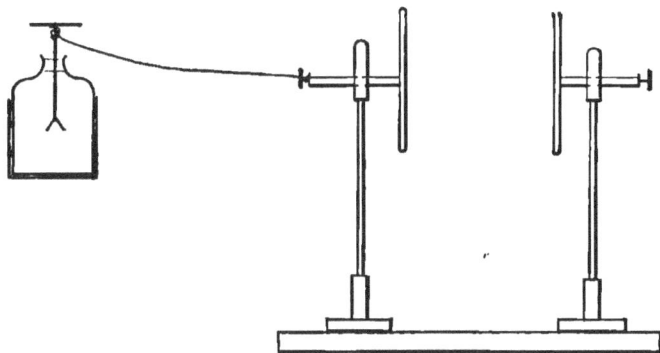

FIG. 210.

electroscope, as shown in Fig. 210, by means of a fine wire. Charge the attached plate until the leaves of the electroscope show a dilation of nearly 90°.

(*b*) Move up the free plate of the condenser until the distance between it and the attached plate is equal to the thickness of the glass. (One corner of the sheet of glass may be used as a gauge, provided care has been taken not to electrify by cleaning it with paper, silk, or wool.) Note that the leaves of the electroscope come together upon the approach of the plates. The charged system has not been discharged, but the charge upon it is said to be "bound" by the condenser action between the plates.

(*c*) Withdraw the free plate and note the gradual freeing of the charge, as indicated by the dilation of the leaves of the electroscope.

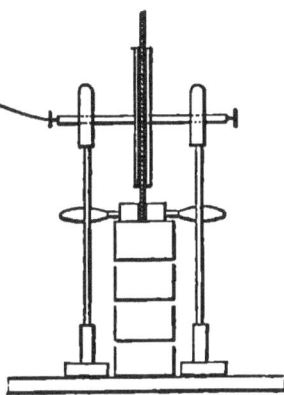

FIG. 211.

(*d*) By means of a joiner's clamp and blocks mount the glass plate vertically in front of, and in contact with, the attached plate, as in Fig. 211. (Care must be taken to handle the glass so as not to electrify it.) Charge the attached plate again to about the same intensity as before. Move up the free plate until it touches the glass, and note the effect upon the electroscope. The condenser action is much more marked on account of the high specific inductive capacity of the glass. By removing the glass without discharging the attached plate, it will be found possible to compare the effects of glass and air when used as a dielectric.[1] It will be found necessary to bring the plates of the condenser together until their distance is about one third the thickness of the glass plate, to obtain the same action.

To have the same capacity as an air condenser, therefore, a condenser with glass as the dielectric will have only one third the surface, provided the coatings are the same distance apart in the two cases. Glass, on account of this property, is one of the best materials with which to insulate condenser plates; where large capacity is desired, however, this may be obtained more compactly by bringing plates of large surface very close together with mica, or paraffined paper, as a dielectric.

This experiment may be performed in a simpler manner by using the metal disk of the electroscope, and that of the electrophorus, as the two plates of the air condenser.

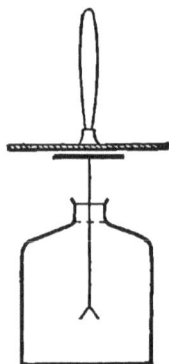

Bring the latter down over the disk of the electroscope as in Fig. 212. Touch the upper disk to connect it to earth, and note the effect upon the gold leaves. Then place the glass plate upon the disk of the electroscope, bring the dilation of the leaves to the same degree by further charging with vulcanite, and bring the electrophorus disk down upon the glass. Connect with earth, and note the greatly increased condenser action due to the presence of the glass. The experiment in this form frequently succeeds when, owing to the state of the atmosphere, difficulty is encountered in carrying it out in the form previously described.

Fig. 212.

223. The Quadrant Electrometer. — For the experiments hitherto described, the gold-leaf electroscope is sufficiently

[1] Dielectric: the name applied to any material through which electrostatic induction takes place.

sensitive. It is possible, however, to construct instruments of far greater delicacy. One of the most useful of these is the quadrant electrometer. It consists of a thin sheet of aluminium, called the *needle* which is shown in Fig. 213. This is suspended by means of two parallel fibers of unspun silk attached to a

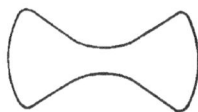

FIG. 213.

platinum wire which passes through the center of the needle. To the other end of this wire, which extends below the needle, is attached a tiny vane of platinum. (See Fig. 214.) The needle hangs in the middle of a flat, cylindrical box of brass which is cut diametrically through in two directions, and is thus divided into quadrants (Fig. 215). These quadrants are slightly separated

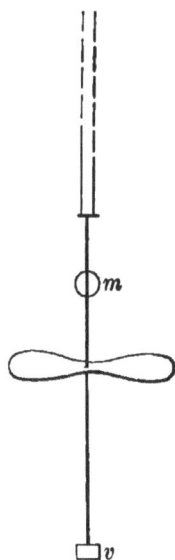

· FIG. 214.

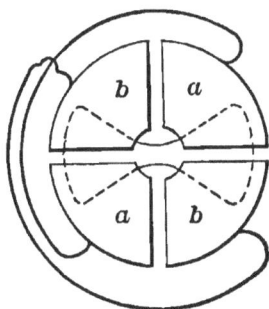

FIG. 215.

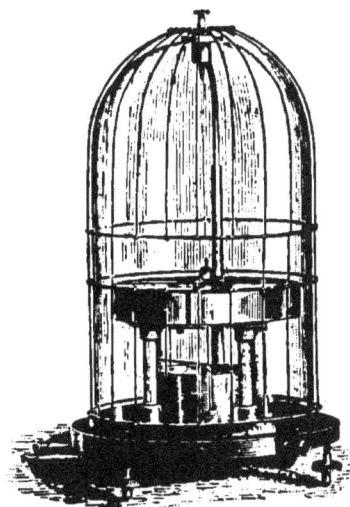

FIG. 216.

from one another, and are mounted upon glass posts which serve as insulators. A glass jar below has an outer coating of foil like a Leyden jar. Within, it is partly filled with strong sulphuric acid. This is placed at such a height that the vane *v* attached to the needle dips into the acid. Figure 216 shows the arrangement of a simple form of a quadrant electrometer.

A small mirror, *m* (Fig. 214), is attached to the platinum wire above the needle. By means of the direction in which it reflects a beam of light thrown upon it, every movement of the needle is indicated.

To use the electrometer, the acid within the jar is charged by means of the electrophorus. The charge is shared by the needle, since the latter is in contact with the acid through the platinum wire and vane. The position of the needle within the box is shown by the dotted line in Fig. 215. If two opposite quadrants as *aa*, or *bb*, be given a small charge, the needle will turn either towards

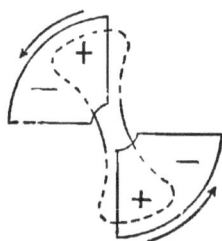
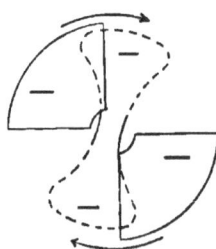

FIG. 217.   FIG. 218.

them, in case the charge is unlike that of the needle (Fig. 217), or away from them, if the charges be similar (Fig. 218). With this arrangement, exceedingly small degrees of electrification can be detected and measured.

CHAPTER XXV

THE ELECTRIC SPARK

224. The Disruptive Discharge. — When two neighboring conductors, such as the terminals of an electrical machine, are oppositely electrified, there is attraction between them which increases as the intensity of electrification increases. The intervening medium resists this pull between the charged bodies, and, in so doing, is strained as a spring is strained when subjected to forces.

In the case of the dielectric, as in that of a spring, energy is stored in the production of this strained condition, to be released when the strained medium returns to its normal state.

If the electrification be carried far enough, the dielectric is no longer able to withstand the strain. An electric spark then passes between the conductors, and they are discharged. Such a discharge is called the *disruptive discharge* because it follows a *breaking down* of the medium under excessive strain.

The discharge manifests itself in two ways:

(1) By means of a flash. The path along which the discharge takes place is heated intensely, and the medium, whether a gas, a liquid, or a solid, is momentarily rendered brilliantly incandescent. Nearly all the stored energy is thus transformed into heat and is dissipated.

(2) By means of a report. A certain portion of the stored energy is dissipated in the form of sound waves.

225. Path of the Spark. — The spark always occupies the path of least resistance. In a perfectly homogeneous medium, this path is a straight line, but in air, the spark usually follows a crooked path, as shown in Fig. 219,

FIG. 219.

which is from a photograph of a spark between the terminals of a Holtz machine. The irregularity of such paths is due to the presence of dust in the air.

226. Energy of the Spark. — The energy set free when disruptive discharge occurs increases with the length of the spark and is directly proportional to the capacity of the discharged conductors.

To illustrate this relation, drive a Toepler-Holtz machine at as nearly uniform rate as possible. Adjust the distance between the terminals until the sparks follow one another at intervals of about one second. Note the character of the spark as regards length of path and apparent thickness. Attach two additional Leyden jars to the machine, as shown in Fig. 220. The outer coating of each is to be in metallic contact with the fixed jar upon the same side of the machine, the inner coatings connected with the respective terminals. Now drive the machine at the same speed and

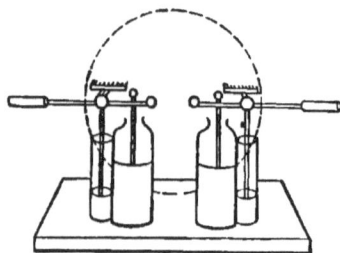

FIG. 220.

readjust the knobs until the sparks occur with the same frequency. It will be found:

(1) that the length of spark is less than before;

(2) that the spark is thicker.

Since the machine is driven at the same speed in the two trials, we may assume, for the purpose of this experiment, that it produces electrical energy at the same rate; and since the sparks occur with equal rapidity in the two cases, they represent the same amounts of energy converted into heat. The two sparks are equivalent, but not identical. We may distribute the energy over a long path, producing a long, thread-like spark, or concentrate it by increasing the capacity of our storage reservoirs and by bringing the terminals of the machines nearer together.

227. Influence of Pressure upon the Spark. — The form of spark described in the foregoing articles occurs in air at ordinary pressures. If we greatly reduce the pressure, the discharge undergoes remarkable changes. These may be conveniently observed by means of the apparatus shown in Fig. 221. This is a glass receiver, fitted with brass caps. Through this are introduced brass terminals, one of which is adjustable; also a tube with a stopcock, which connects the receiver with the air pump. If the upper terminal be pushed down until the air gap is reduced to a few centimeters, and the terminals be connected with those of an electrical machine by means of wires, sparks of the character usual in air will be obtained.

FIG. 221.

Upon exhausting the receiver by means of the air pump,

s

the sparks gradually change their appearance. They pass more frequently and are less intense, showing that the dielectric strength of the air diminishes with the falling pressure. The path of the spark becomes ill defined, and the layers of air immediately surrounding it begin to glow with a bluish or purple light. As the pressure decreases, this illuminated region extends, it becomes brighter, while the original linear path gradually vanishes from view. When the pressure has been reduced to a few centimeters of mercury, the entire atmosphere within the receiver partakes of the purple glow. It will now be found that the upper terminal may be withdrawn to the very top of the receiver without interrupting the discharge. A machine which is capable of producing a spark a few centimeters long in air at ordinary pressure will discharge through much greater distances in rarefied air. This point may be illustrated by means of the following experiment :

228. EXPERIMENT 69. — Equivalent Spark Length of the Discharge in Vacuo.

Apparatus:

(1) A glass tube between 0·5 cm. and 1·0 cm. in diameter and at least 200 cm. long.[1]

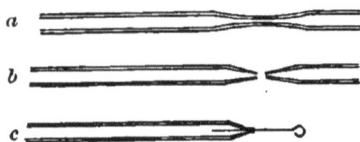

(2) Mercury (enough to fill the closed tube, leaving an excess of at least 100 cm.³), also a glass tumbler.

(3) The Toepler-Holtz machine.

FIG. 222.

(4) A barometer.

[1] This tube is to be drawn out at one end in the flame of a blast lamp, as shown in Fig. 222, *a*, until the bore in the contraction is reduced to about 1 mm. The tube is then to be cut off where the diameter is smallest, *b*, and a piece of platinum wire 3 or 4 cm. long is to be inserted. The end of the tube is then sealed around the wire, *c*, and completely closed by means of the flame.

Procedure:

(*a*) Fill the closed tube with mercury, close the end with the fore-finger, then invert it carefully in the tumbler. The latter should already contain the excess of mercury to insure prompt immersion of the open end. The mercury will flow out until the proper barometric height is reached, leaving a Torricellian vacuum above the mercury column more than a meter in length.

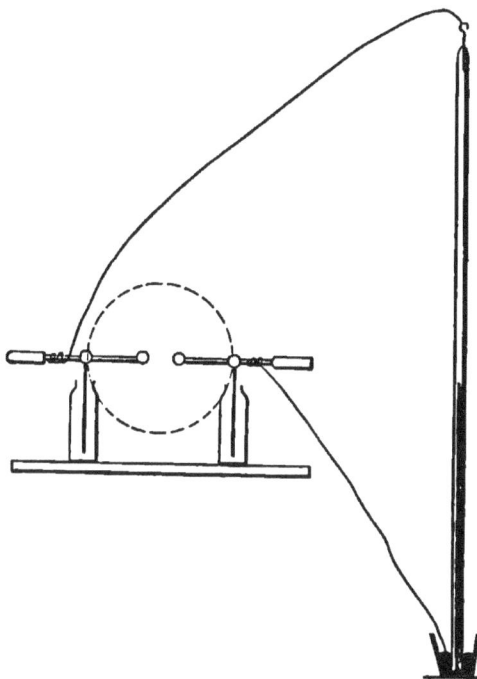

FIG. 223.

(*b*) Support the inverted tube in a vertical position and connect the outer end of the platinum terminal at the top of the tube with one pole of the Toepler-Holtz machine by means of a flexible wire. Attach a wire to the other terminal of the machine and dip the free end into the mercury within the tumbler. (See Fig. 223.)

(*c*) Push the terminals of the machine together, start the machine, then very gradually draw the terminals apart again. Note that at first sparks pass between the knobs. Before the latter are any considerable distance apart, however, the discharge abandons the direct

path between them and passes between the platinum wire at the top
of the tube and the surface of the mercury column. The tube
becomes filled with a whitish or bluish glow. Push the terminals of
the machine together until the spark returns to the space between the
knobs. Measure the distance between them in centimeters. Draw
the knobs apart until the spark again transfers itself to the tube.
Measure the distance again. Repeat these alternate measurements
five times. The average of the entire set may be taken as the *sparking distance* in air, which is equivalent to the discharge within the
tube.

(*d*) Measure the distance between the top of the mercury column
and the platinum terminal within the top of the tube. Express this
length in terms of the sparking distance in air. Finally, measure the
height of the mercury column in centimeters and compare it with
the true barometric height for the time and place. The difference
between these heights will give the pressure, in centimeters of mercury, for which the comparison of sparking distances has just been
made.

This experiment may be repeated, at different pressures, by letting
a bubble of air into the vacuum, repeating the measurements under
(*c*) and (*d*), admitting more air from time to time, until the discharge
can no longer be driven through the tube.

229. Intermittent Character of the Discharge in Vacuo. —
The discharge through the Torricellian vacuum, described
in Art. 227, often appears to be nearly or quite continuous.
It is, however, as truly intermittent as the spark discharge
in air. This may be shown by the use of the revolving
mirror depicted in Fig. 224. It consists of four square,
plane mirrors mounted so as to form the vertical sides of
a cube which revolves upon an axis passing through the
centers of the remaining faces. When we view the image
of the tube through which the discharge occurs in this
mirror, the illuminated region, instead of being spread out
into a continuous sheet, is broken up into a series of parallel, bright images when the mirror revolves. Each of
these appears just as it would in a stationary mirror.

Each is displaced from its neighbor by the angular distance through which the mirror has turned between the successive discharges. The fact that the images in the moving mirror are not appreciably broadened indicates that the discharge is so brief in duration that the mirror does not turn through a noticeable angle during the life of the spark. The same observation may be made upon the ordinary spark in air. This breaking up of the image of the discharge may be observed, although less conveniently, by viewing it in a common hand mirror,

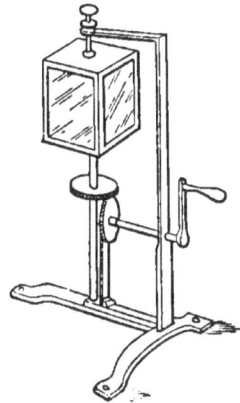

FIG. 224.

and giving the mirror a sudden turn to the right or left. It may even be observed by turning the head suddenly while looking at the discharge, provided the eyes be allowed to follow the motion of the head instead of remaining fixed upon the tube.

By means of mirrors revolving at much higher speeds, it is found that the electric spark is not instantaneous, but has a duration amounting, in general, to a few millionths of a second.

230. Discharge in Higher Vacua. — By means of mercurial pumps, it is possible to carry the exhaustion of a tube or receiver much further than in the experiments just described.

When a tube with two metallic terminals sealed into the glass (Fig. 225) is thus exhausted, we get, as the pressure falls, a suc-

FIG. 225.

cession of further modifications in the discharge. In such a tube, so long as the pressure exceeds about 0·5 cm., the discharge presents the appearance already described. A tube in this condition is termed a *Geissler* [1] *tube.* When the pressure is further reduced, the discharge becomes laterally laminated and undergoes a succession of beautiful and striking changes. Finally, the discharge within the tube becomes nearly invisible. There now emanate from the terminal, which is connected with the negative pole of the electrical machine, rays which travel in straight lines through the tube. Wherever these impinge upon the glass, the walls of the tube glow with a bright green phosphorescence. In this condition the surface of the tube emits, externally, invisible rays which are known as the *Roentgen rays*, or as the *X rays.* These pass through many opaque substances, such as aluminium, hard rubber, wood, flesh, most textile fabrics, etc., while they are absorbed by some transparent materials, such as glass. They affect the photographic film, so that shadow photographs may be taken by means of them; and they render certain substances, such as calcium tungstate, zinc oxide, and many of the double platinum cyanides, luminous. If a screen coated with one of these substances be placed in the path of the *X* rays, the shadow of any body interposed will be visible upon the screen. This will be the shadow of those portions of the body which are opaque to the *X* rays, and not of those opaque to ordinary light. Thus, if the arm be interposed, the two bones will be clearly visible, but the fleshy part of the arm will scarcely cast any shadow.

Photographs taken with the *X* rays possess the same peculiarity. Figure 226 is a reproduction of such a photo-

---

[1] From Dr. Geissler, of Bonn.

graph.[1] It shows the bones of the human wrist; also a bit of broken needle which had become lodged in the membrane surrounding one of the bones. The outline of the wrist is dimly visible.

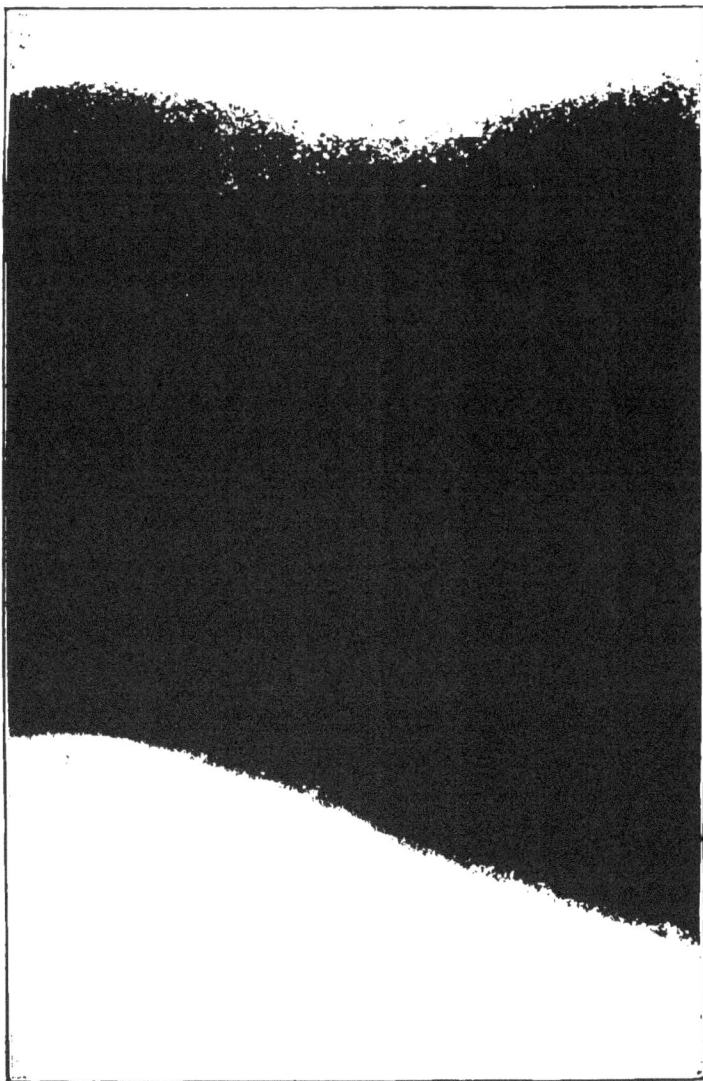

FIG. 226.

[1] From a photograph taken by R. W. Quick.

CHAPTER XXVI

THE ELECTRIC CURRENT

231. Electrification by Chemical Action. — Thus far we have considered the electrification of bodies by friction. It is also possible to electrify two metals, oppositely, by chemical action. It is usual to speak of two conductors thus electrified as possessing a *difference of potential*.[1] If a piece of zinc and a piece of copper, for example, be placed in a glass (Fig. 227) containing an acid, such as dilute sulphuric acid, the copper will become positively, and the zinc negatively, electrified. Such an arrangement is called a *voltaic cell*.[2] The source of electrical energy, in this case, is the chemical reaction between the zinc and the acid.

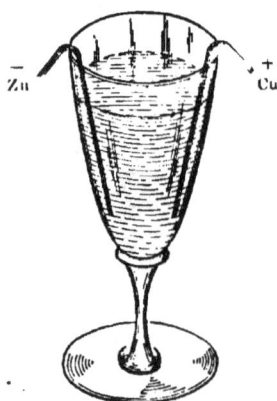

FIG. 227.

This difference of potential is so small that it can scarcely be shown by means of an ordinary electroscope. With the quadrant electrometer described in a previous

---

[1] The term *difference of potential* is frequently used in speaking of the difference of electrical condition of bodies, however electrified. If a positively and negatively charged conductor be metallically connected, there will be electric flow, and their charges will unite. If a charged body and a neutral one are connected, there will be electric flow, and the former will share its charge with the latter. In all such cases, we say that there was *difference of potential*.

[2] From the Italian physicist, Volta, 1745–1827.

article, a considerable deflection will be produced when one pair of quadrants is connected to the zinc and one to the copper. By getting together a large number of such cells, the copper of each connected to the zinc of the next, we are able to add together their individual differences of potential, and to secure a battery, the difference of potential between the terminals of which can be shown even with the gold-leaf electroscope.

232. The Difference of Potential between the Metals of a Voltaic Cell is Independent of the Size of the Cell. — To verify this statement, plunge sheets of copper and zinc into a jar containing a liter or more of dilute sulphuric acid (1 : 20, or thereabout). Connect these with wires to the quadrant pairs of an electrometer of the simple form described in Chapter XXIV. Note the deflection. Then, with shears, cut a very narrow strip from each of the sheets of metal. Dip the ends of these into a drop of acid, remove from the jar, and place in a watchglass or other convenient receptacle. Connect with the electrometer as before, and note that the deflection is as large as in the previous case. This point may be demonstrated, also, and more conveniently, by means of the galvanometer.

233. The Electric Current. — If the terminals of a voltaic battery be connected together by means of a wire, or other conductor, the unlike charges which have been gathered at the terminals as a result of chemical action will tend to neutralize each other precisely as opposite charges upon any two conductors would do in case they were metallically connected. This equalization of charge may, as a matter of convenience, be considered as a flow of electricity through the connecting wire, and we may say that the direction of the flow is from the positively charged terminal to the negatively charged terminal. The result is what we call the *electric current*. One very important difference between the cases of bodies charged by friction and bodies charged by chemical action is as follows:

When two bodies, previously given opposite charges by friction, are brought into contact with one another, there is immediately an instantaneous exchange, and the charges neutralize each other. If the two conductors, however, form terminals of a voltaic battery, they may be connected metallically, and the difference of potential between them will be continually maintained by chemical action going on within the cells.

The flow of current between the positive and negative terminal, in the latter case, is not a matter of short duration, as in the case of two conductors which have been previously charged, and which discharge themselves upon contact, but will continue as long as the chemical action continues. The existence of an electric current shows itself chiefly in the following ways:

(1) By its magnetic action.

(2) By its thermal action.

(3) By its chemical action.

234. EXPERIMENT 70. — **Phenomena of the Simple Cell.**

Apparatus:

(1) A glass beaker or an ordinary tumbler.

(2) Strips of zinc and copper, sulphuric acid.

Procedure:

(*a*) Nearly fill the glass with dilute acid, *i.e.* ten parts of water to one of acid.

(*b*) Bend the strips at one end so as to form a hook, and hang them by these hooks to the lip of the glass. Note the generation of gas at the surface of the zinc. If collected in an inverted test tube and ignited, this gas will be found to be hydrogen.

(c) Lay a bit of naked copper, scraped until bright, across between the strips as in Fig. 228.[1]

Note that a change in the generation of gas immediately follows, and that gas bubbles begin to appear upon the copper strip as well as upon the zinc. If tested, these gases would likewise be found to consist of hydrogen.

As soon as the terminals are connected by means of the copper wire, current begins to flow. The direction of this current is from copper to zinc along the wire, and from zinc to copper through the liquid.

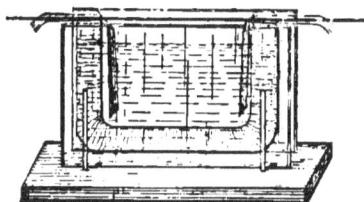

FIG. 228.

In order to have a permanent flow there must be a completed, or, as we say in electrical terms, a *closed circuit*. This current is maintained by means of the energy generated by the chemical reaction going on within the cell. It has just been noted that this reaction differs when the circuit is open and when it is closed.

Reaction on Open Circuit. — The zinc is attacked by the sulphuric acid, giving as a product zinc sulphate, *i.e.*

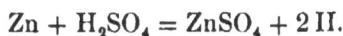

$$Zn + H_2SO_4 = ZnSO_4 + 2 H.$$

The gas is given off at the surface of the zinc.

The presence of the copper is unnecessary to this reaction.

The energy of the reaction is converted into heat.

If the zinc be amalgamated by rubbing with mercury, the reaction will cease altogether.

Reaction with Closed Circuit. — The product of the reaction is the same as before; but hydrogen is evolved both at the zinc and at the copper terminals.

The gas evolved at the zinc pole is due to the continuance of the open circuit reaction.

If the zinc be amalgamated the evolution at the copper continues, while that at the zinc ceases.

[1] This figure shows the apparatus mounted for use in the field of the lantern. It differs from the arrangement indicated above only in that the liquid is contained in a flat cell with glass sides instead of a cylindrical glass. (See Appendix VII.)

The energy of the reaction is converted into electrical energy and is a source of electric current.

(*d*) Test the foregoing statements with reference to the effect of amalgamation of the zinc upon the action of the cell.

235. Polarization of the Voltaic Cell. — The gathering of gas at the terminals of a cell is detrimental because of the greatly increased resistance to the passage of the current. The hydrogen upon the zinc terminal is obviated by amalgamation as already shown. To prevent the formation of the gas upon the copper terminal, a substance containing loosely combined oxygen is added to the liquid. Such materials are called *depolarizers*. The most frequently employed are bichromate of potassium, chromic acid, permanganate of potassium, and manganese dioxide.

A cell with hydrogen upon the copper possesses much less difference of potential between its terminals than would otherwise be the case. It is to obviate this effect, which is called *polarization* and which is more important than the rise in the resistance of the cell, that the depolarizer is employed.

236. Forms of the Voltaic Cell. — The voltaic cells in common use are of two types: *open circuit cells* and *closed circuit cells*.

Properties of Open Circuit Cells. — These cells are used for the production of momentary currents only, with considerable periods of rest (for operating signals, ringing bells, etc.).

They suffer polarization when continued action is demanded of them.

The open circuit cell, although incapable of furnishing strong currents continuously, must be capable of responding when the circuit is closed, without previous preparation, for months at a time.

The way in which the necessary conditions are met in the Le Clanché cell, which is a typical form, is as follows:

The negative terminal is a rod of zinc. This is immersed in a liquid which acts upon it only when the circuit is closed, *i.e.* in an aqueous solution of ammonium chloride.

The positive terminal is of carbon. It is surrounded with dioxide of manganese, which acts as a slow depolarizer. This is sometimes packed into a porous cup together with the carbon ; sometimes it is compressed into blocks, as in the figure, which are attached to the carbon by means of rubber bands.

The jar is closed to hinder evaporation.

Figure 229 is a sketch of the Le Clanché cell.

Properties of Closed Circuit Cells. — These cells must be capable of furnishing strong currents continuously for

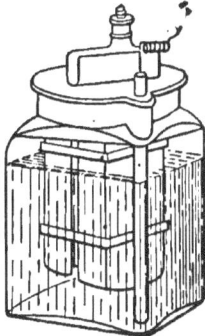

FIG. 229.

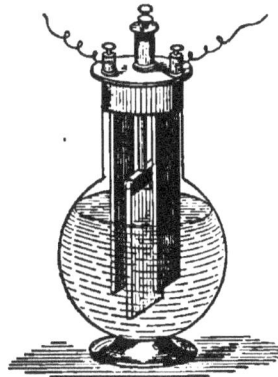

FIG. 230.

considerable intervals of time. The conditions to be met, therefore, are *low resistance* and *prompt depolarization*.

In the bichromate cell (Fig. 230), which is one of the best-known forms of closed circuit cells, the positive terminal consists of two parallel slabs of carbon which are metallically connected at the top. Between these is the negative terminal, a plate of zinc. The large surfaces of these, and their proximity, secure very low resistance.

The liquid is dilute sulphuric acid in which potassium bichromate has been dissolved. The chromic acid thus formed in the solution gives up a portion of its oxygen to the hydrogen, as fast as the latter appears. It thus acts as a very prompt depolarizer.

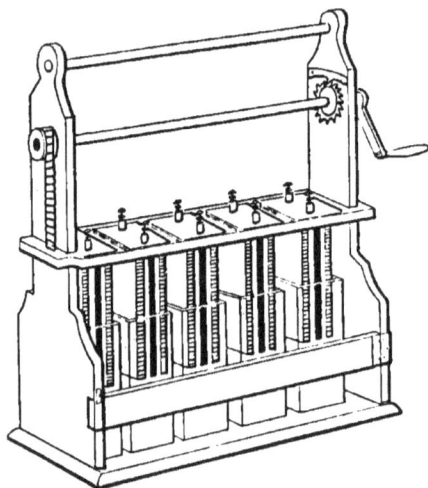

FIG. 231.

This liquid acts upon zinc, even when the circuit is open. It is necessary, therefore, to withdraw the zinc when the battery is not in use; this is accomplished by means of a sliding rod, as shown in the figure. Sometimes a battery consisting of several bichromate cells are arranged so that all the zincs may be raised or lowered simultaneously. (See Fig. 231.) Such an arrangement is called a *plunge battery*.[1]

---

[1] For a full discussion of the voltaic cell and for a description of numerous forms to which no reference is made here, see Carhart, *Primary Batteries;* see also *Elements of Physics*, Vol. II, Chap. V.

CHAPTER XXVII

THE MAGNETIC EFFECTS OF THE CURRENT

237. Preliminary Observations concerning the Magnetic Effect of the Current. — If several cells of the bichromate battery described in Art. 237 be placed in series, and the terminals of the battery thus formed be connected by means of a copper wire, a strong current of electricity will flow through this wire from the positive to the negative terminal of the battery. If this wire passes through a hole in the center of a flat block of wood, as shown in Fig. 232, and if the upper surface of the block which should previously have been covered with glazed paper or cardboard be strewn with fine iron filings, these will be seen to arrange themselves in concentric circles, with the axis of the wire as a

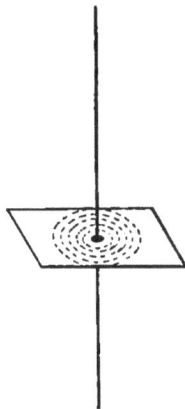

center. Figure 233, which is from a photograph, shows the arrangement of iron filings in such a case.

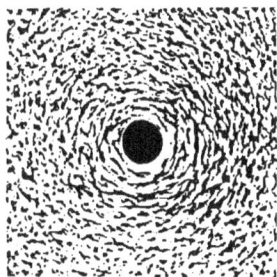

FIG. 233.

FIG. 232.

Each particle of iron in the neighborhood of the electric current becomes a magnet. The entire region around a wire which carries an electric current becomes what is called a *magnetic field*. In this field forces are at work which tend to move the poles of any magnet which may be in the field along certain lines which are called *lines of force*. The minute magnets formed from the iron filings are acted upon by these forces in such a way as to cause them to arrange themselves in the lines of force. The lines of force around a straight wire through which current passes are circles.

238. EXPERIMENT 71.— **Magnetic Field of a Coil of Wire through which Current flows.**

Apparatus:

(1) Six cells of bichromate battery arranged as described in the previous article.

(2) A hollow cylinder about 15 cm. long and 3 cm. in diameter. A cylindrical glass lamp chimney such as is used with the Argand burner may be employed.

(3) Several meters of copper wire. The insulated wire known as annunciator wire or office wire, size No. 18 or 20, is well adapted for this experiment.

Procedure:

(*a*) Wrap the cylinder carefully from end to end with wire. One layer will suffice. Secure the layer in its place by means of threads passed through the cylinder and tied. The coil thus formed is called a *helix* or *solenoid*.

(*b*) Cut two pieces of stiff cardboard 20 cm. square into the form shown in Fig. 234, and trim the tongue of each piece so that it will fit the inside of the cylinder. Touch the tongue of each with muci-
lage, and insert in the cylinder in such a way that the flat portions will afford platforms opposite the mouths of the cylinder as shown in Fig. 235.

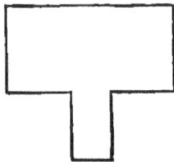

FIG. 234.

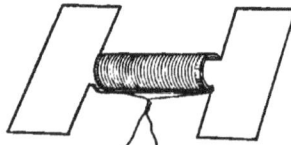

FIG. 235.

(*c*) After the apparatus is dried, mount the coil with its axis horizontal, and the two platforms of cardboard also in a horizontal plane. Send the current through the coil, and strew filings upon the cardboard. Note the tendency of the filings to arrange themselves along lines of force. This tendency may be increased by tapping the cardboard. The arrangement will be that shown in Fig. 236. Note that all the lines of force appear to enter the mouth of the cylinder: in point of fact, if we were to trace the lines of force throughout, we should find that they entered the cylinder at one end, passed through its entire length, and issued from the other; also that they returned through the outer atmosphere, forming a closed curve. It may be laid down as a general law that all magnetic lines of force are closed curves. The arrangement of the filings observed in this experiment indicates the character of a plane section of the field produced by the coil of wire. In the neighborhood of such a coil through which current flows, all iron tends to become magnetized. This fact shows itself most strongly in the case of pieces of that metal which are inserted within the cylinder itself, because in that region the number of lines of force are greatest.

FIG. 236.

239. EXPERIMENT 72. — **Magnetization of Iron by Means of a Solenoid.**

Apparatus:

(1) The battery or solenoid described in the previous article.

(2) A bar of soft Norway iron about 1 cm. in diameter and 5 cm. longer than the solenoid.

Procedure:

(a) By means of two pieces of gummed paper fasten the soft iron bar as nearly as possible in the axis of the solenoid, immediately below the cardboard platforms. Let its ends project equally from the coil.

(b) Send the current through the coil and map the field of force by means of iron filings as in the previous experiment.

(c) Note that the lines of force now nearly all appear to enter the iron. Compare the character of the field — which will be similar to that shown in Fig. 237 with that obtained in the previous experiment. The iron bar has become for the time being a magnet, and the regions in the ends where the lines of force enter the iron from the outer air are called *magnetic poles.*

T

If we consider the lines of force to be closed curves, entering the iron and the core at one end and issuing from the other, we have the means of distinguishing between the two poles of the magnet. One of these is called the *north* pole of the magnet and the other the *south* pole. There is a general agreement among physicists to consider the lines of force as issuing from the north pole and entering at the south pole., The names of these poles are derived from the behavior of magnets when suspended in the magnetic field of the earth.

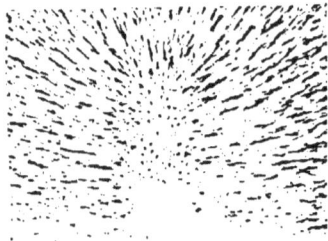

FIG. 237.

(*d*) Turn the solenoid over so that the bar of iron is above the cardboard. Send the current through it again, and strew filings. Note how the filings cling to the iron, especially in the regions which have already been described as poles; note further that the filings tend to attach themselves to the magnetized poles endwise, proceeding from the poles outward. Note that each one tends to set itself in the direction of a line of force.

(*e*) Remove the bar from the solenoid, and notice that the tendency of the iron filings to cling to its poles decreases, and when the bar has been entirely withdrawn ceases altogether. (If the iron used in this experiment is not soft, this last statement will not be found to be quite true. The pole will remain slightly magnetized, for reasons which are described in the following articles.) The magnetization of the soft iron bar, which it has been the object of this experiment to consider, is called *temporary magnetization:* it depends for its existence upon bringing the iron into the magnetic field, and it lasts only so long as the iron remains in that field.

The term *permanent magnetization,* on the other hand, is used to designate the *lasting* effect produced upon the properties of steel, and of various hardened varieties of iron, by the action of the magnetic field. The intensity of permanent magnetization is usually small as compared with that of the temporary magnetization produced by the same field. It varies greatly with the temper of the metal. Permanent magnetization varies with the temperature of the magnet. It diminishes as the temperature rises; and finally, when a certain point called the *critical temperature* is reached, the iron (or steel) becomes non-magnetic. The critical temperature for steel is about 735° C.

240. EXPERIMENT 73. — Permanent Magnetization of Steel by means of a Solenoid.

Apparatus:

(1) The battery and solenoid described in the previous experiments.

(2) Six steel knitting needles, which should be somewhat longer than the solenoid, also several sewing needles.

Procedure:

(*a*) Connect the solenoid with the terminals of the battery, noting the direction in which the current flows through the wire; that is to say, start at the copper or carbon pole of the battery, which is called the *positive* pole, and consider that the current flows from that through the coil to the zinc pole. (See Fig. 238.)

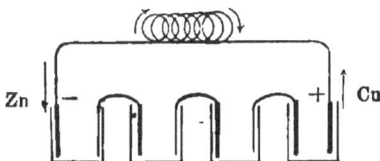

FIG. 238.

(*b*) Insert one of the sewing needles into a small cork, pushing it through until the cork reaches the middle of the needle.

(*c*) Face the solenoid, looking toward that end of it at which the current is traversing the turns clockwise (Fig. 239), and insert the needle, point first, into the axis of the coil. In this position the needle will afford a path for the passage of lines of force through the coil. The region where the lines of force enter will become a south-seeking pole; the region where the lines issue from the steel into the air again will become a north-seeking pole. In short, the needle will become a magnet. These two regions lie, of course, very near the ends of the needle. Steel differs from soft iron in that it is capable of retaining its magnetized condition after being withdrawn from the field of the coil.

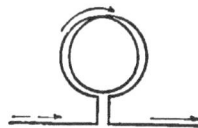

FIG. 239.

(*d*) To test this statement, tap the needle sharply three or four times with the end of a pencil or other piece of wood, and remove it from the coil. Withdraw the cork, and place the needle beneath a piece of stiff paper or cardboard, strew iron filings upon the surface of the paper, and note the arrangement of them. It will be seen that they tend to arrange themselves as shown in Fig. 240.

The pattern indicates the trend of the lines of force. It is that which one would expect from a magnet with poles near the ends of the needle.

(*e*) Drop the magnetized needle lightly upon the surface of a dish of water. If this operation is performed in such a way that the needle strikes the water with its axis parallel to the surface, it will be sustained by the surface film, and will float. Note that it immediately sets itself nearly into a north and south position, with the point northward.

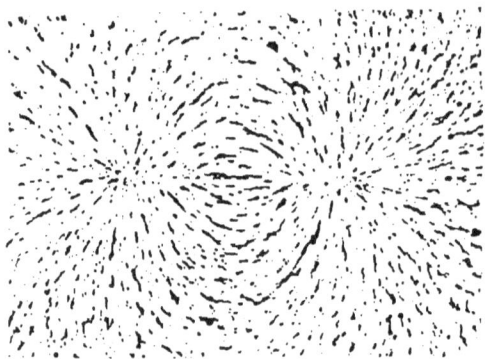

(*f*) Insert into the coil, through which the current is flowing as before, one of the knitting needles. Tap the latter sharply with a block of wood, and withdraw it from the coil: it also has become a permanent magnet. Bring it into the neighborhood of the floating sewing needle. Note that the end of the magnetized knitting needle which corresponded in position, when within the coil, with the point of the sewing needle, and which has likewise been made a north-seeking pole, attracts the eye end of the magnetized needle and repels its point (Fig. 241). Note that the other end of the knitting needle, which has been made into a south-seeking pole, attracts the point and repels the eye end of the floating needle. From this observation we conclude that like magnetic poles repel each other, and unlike attract each other.

To test this matter further, remove the magnetized needle from the dish of water, and replace it with a needle which has not been magnetized. Note that the latter possesses but little if any tendency to set itself in a north and south direction; furthermore, that either end of it will be attracted by the magnetized knitting needle, or by either end of the magnetized sewing needle. (To succeed in this portion of the experiment, it is necessary to take some precautions to prevent the sewing needles from becoming slightly magnetized; in other words, they must be kept away from all magnetizing bodies and from the magnetizing coil.)

(*g*) Magnetize two more sewing needles in the manner described in (*b*) and (*c*) of this experiment. Float these three needles on different parts of the dish of water and watch their behavior. It will be seen that they drift slowly together, turning so as to arrange themselves side by side with unlike poles in contact.

241. Action of a Wire, bearing Current, upon a Magnet. — It has already been shown that any wire which carries an electric current possesses a magnetic field. If we bring such a wire near to a magnet, forces are brought into action between the two. These may be considered to be due to the mutual action between the lines of force of the two fields. It is as though lines of force running in the same direction tended to repel each other. If the magnet have freedom of motion, it will tend to set itself in such a position that its lines of force are parallel to those which surround the current. This action, which may readily be shown by means of the floating needle, forms the basis for the instrument known as the *galvanometer*. It is well illustrated by means of the following experiment:

242. EXPERIMENT 74. — **Magnetic Influence of a Wire, carrying Current, upon a Floating Needle.**

Apparatus:

(1) The floating needle previously described, a cell of battery, three or four meters of copper wire.

Procedure:

(*a*) Connect the wire to the terminals of the battery, and hold about a half meter of the intervening portion of the same in the hands. Bring this part of the wire near to the needle, holding it above the latter in a horizontal position, and as nearly as possible in a north and south plane. Note that the needle is deflected from its north and south position in the earth's field.

The direction towards which the north-seeking pole tends is east-

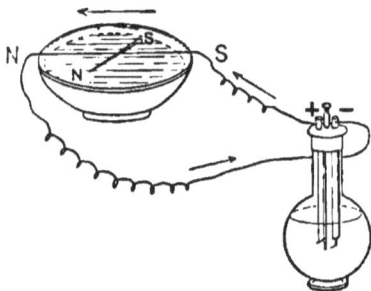

FIG. 242.

erly or westerly, according to the direction in which the current flows through the wire. If the current flows toward the north, that is to say, if the end of the wire which is towards the south is attached to the copper terminal of the battery, and the northerly end to the zinc terminal (see Fig. 242), the north-seeking end of the floating needle will tend to point towards the west. If the current is reversed, the north-seeking pole will point toward the east. But for the forces due to the earth's field, the magnetic needle would always point directly at right angles to the wire which carries the current. On account of the earth's magnetism it comes to rest, however, in the position where the forces of the earth's field are precisely balanced by those due to the action of the current. (See Fig. 243.) To test this statement, proceed as follows:

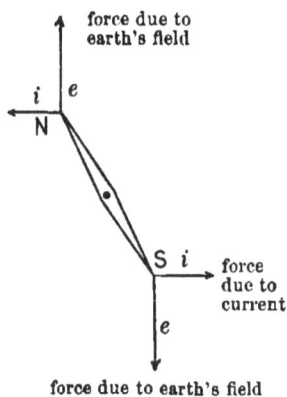

FIG. 243.

(b) Turn the ends which hold the wire in such a way as to swing the wire round into an east and west direction, and note that in this way the action upon the needle diminishes, becomes zero, and as the wire is brought around into a north and south direction with its end reversed, begins to show a deflection with its north pole towards the east. Move the wire from the neighborhood of the needle, and bring it up again from below: note that now all the above effects are reversed.

(c) Coil the wire into a large circular loop. Hold this loop in a north and south vertical plane and move it up slowly towards the needle from the east. Note the deflection which occurs, and that it increases until the wire is in the same plane with the needle, and then begins to diminish again as the wire moves westward.

243. Ampère's Rule. — The action of a wire carrying current upon a magnetic needle was first described by Oersted in 1819. The effect was later stated by the French physicist Ampère. as follows :

If one imagine himself swimming with the current and facing the needle, the north-seeking pole will always be deflected towards his left hand.

Many persons prefer the following simple rule :

Hold the right hand with the thumb extended (Fig. 244), *the fingers pointing in the direction in which the current flows and the palm towards the needle. The north-seeking pole will be deflected towards the thumb.*

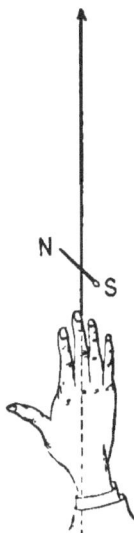

FIG. 244.

CHAPTER XXVIII

MAGNETISM

244. The Earth's Magnetic Field. — The experiments with the floating magnet described in the preceding chapter, illustrate several important points in the science of magnetism. The floating needle acts as a magnetic compass, pointing with its north-seeking pole towards the magnetic north pole of the earth. The behavior of magnet needles at various points upon the surface of the planet indicate that the earth itself may be considered as a magnet and that the surface of the earth lies within a magnetic field consisting of lines of force which issue from one of the earth's poles and return through space to the other. In the neighborhood of the equator these lines are nearly horizontal. As we approach the poles of the earth, they dip downward more and more, until finally, if one could take up his position at a point immediately over the magnetic north pole, or the magnetic south pole, he would find the lines of force vertical. In latitude 40° N. in the Western Hemisphere this dip of the lines of force is about 70°.

The directive force which causes the compass needle to point towards the earth's north pole is the horizontal component of these lines of force. If the needle were free to revolve upon a horizontal axis also, it would come to rest in a position making an angle of about 70°, as has already been indicated, with the horizon. The north-seeking pole would point downward and northward at the same time.

245. The Dipping Needle. — An instrument by means of which the amount of dip may be determined is called a *dipping needle*. It consists of a bar of steel (Fig. 245), through which is placed a transverse axle which passes accurately through the center of gravity of the bar. If the bar is not magnetized, it will be in indifferent equilibrium when mounted upon this axle. If it be magnetized, and then mounted upon the horizontal axle, the end that contains the north-seeking pole will immediately dip downward, as though mass had been added to it. The reason for this change in its behavior is that the needle is now in a condition to have its poles attracted and repelled by the magnetic force of the earth, and it tends to come to rest in the lines of force of the earth's field. Since the

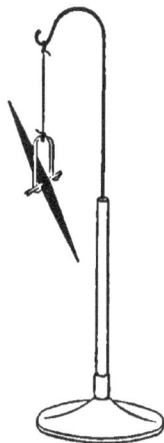

FIG. 245.

bar has freedom of rotation around a horizontal axis, it is in equilibrium only when it dips into a position such that the lines of force are parallel to the line joining its north and south seeking poles. If we measure the angle which such a needle makes with the horizon, we shall have determined the angle of dip for the locality in which the needle is mounted.

246. The Nature of a Magnetic Pole. — It is often assumed for convenience that a magnetic pole is a point lying within a magnet, generally near one end. A better definition considers the pole to be *that region of a magnet where lines of force leave the iron and enter the air, or vice versa.*

Iron affords a better path for the lines of force than the air does, so that they tend to continue within the iron of a bar magnet until the end is reached. Then they are forced outward into the air. If a piece of iron be strongly magnetized, this region throughout which lines of force are leaving the iron is quite an extended one, so that the pole must be considered as distributed throughout the entire end of the bar, instead of being concentrated at a single point. Figure 246 shows the arrangement of the lines

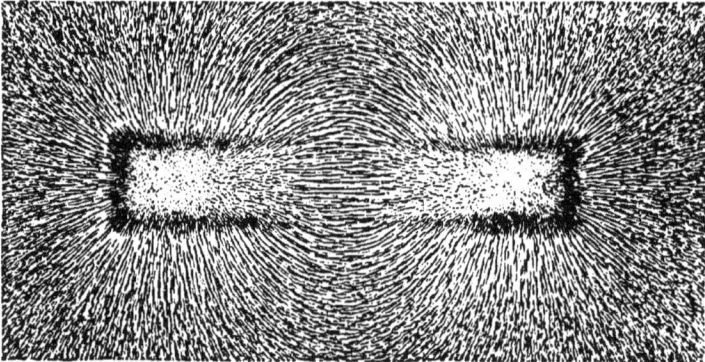

FIG. 246.

of force around a bar magnet as indicated by the pattern in which iron filings group themselves. It will be seen that everywhere near the ends of the bar lines of force are leaving the iron and entering the air, and that it would be a matter of great difficulty to fix upon any given point which could be regarded as the location of the pole. The significance of this definition of a magnetic pole may be illustrated by means of the following experiment:

247. EXPERIMENT 75. — **Making Magnets by the Breaking of a Magnetized Bar.**

Apparatus:

(1) A steel knitting needle, a pane of window glass, and some iron filings.

(2) A battery and a coil of wire.

(3) A dish of water and a magnetized sewing needle (described in Art. 240).

Procedure:

(*a*) Connect the coil of wire in circuit with three or four cells of the battery. Insert the knitting needle in the coil, so that it lies with the coil midway between its ends. Tap the needle sharply several times with a block of wood to assist in the rearrangement of the molecules, and then break circuit through the coil.

(*b*) Float the magnetized sewing needle upon the dish of water as described in Experiment 72. Bring the knitting needle near, and determine the character of its poles by the attraction or repulsion of the north-seeking pole of the floating needle. Mark the north-seeking pole of the knitting needle in any convenient way, either by attaching a bit of gummed paper to it, or by touching it with ink.

(*c*) Break the knitting needle in half, and test the magnetization of each piece by means of the floating needle. Note that the broken ends have each acquired a magnetic pole, that these poles are dissimilar, and that the new pole which is in the piece which contains the original north-seeking pole of the needle is a south-seeking pole. Each piece of the knitting needle has, in a word, become a complete magnet by itself.

The formation of the new poles may be explained simply from the consideration that the lines of force which previously traversed the needle from end to end through the iron were forced to leave the iron at the point of rupture as soon as the needle was broken.

(*d*) Lay the two pieces of the broken knitting needle in the same straight line, with an air gap of 2 cm. between the broken ends. Place over them the pane of glass, wedging the latter up into a horizontal position by means of bits of wood. Strew the glass lightly

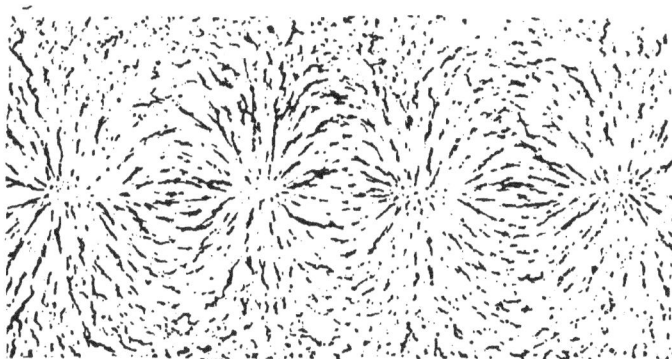

FIG. 247.

with iron filings and tap it several times. The filings will adjust themselves into a pattern (Fig. 247), which indicates quite clearly the character of the magnetic field. It will be seen from the accompanying diagram, 247 a, that there are lines of force emanating from and entering the iron in the regions where the original poles were, and likewise at the new ends formed by breaking the needles.

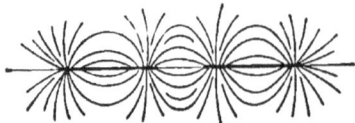

FIG. 247 a.

(e) Break each half of the knitting needle again into two pieces, and repeat the operation described in (c) and (d). It will be found that each of the new pieces is a separate magnet by itself. The character of the field obtained by mapping with filings the four pieces

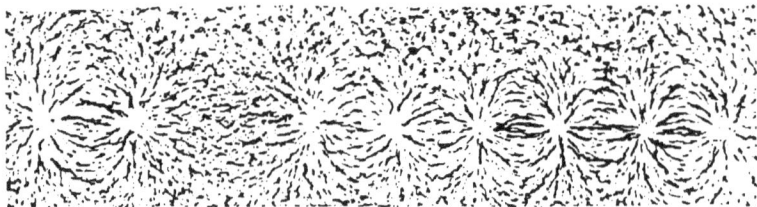

FIG. 248.

laid end to end with intervening air gaps, is indicated in Fig. 248 and in the accompanying diagram, 248 a. Each piece of the original needle will have a north and a south seeking pole. These will all

FIG. 248 a.

be of equal strength, so that, if the needle could be restored, the poles at each point of rupture being equal and opposite in character, would precisely neutralize one another, pairwise throughout.

248. Magnetization by Induction. — Any piece of iron placed in a magnetic field becomes a magnet, because lines of force enter it to find a better path than is afforded by the air, and leave it again at some other point. The region where the lines of force of the field in which the piece of iron is placed enter it becomes a north-seeking pole, and the region where they forsake the iron again becomes a south-seeking pole. Even a weak field like that of the earth is capable of producing such an effect. This may be readily shown by means of the following experiment:

249. EXPERIMENT 76. — **Magnetization of a Bar of Iron by means of the Earth's Field.**

Apparatus:
(1) A bar of soft iron 20 cm. or more in length.
(2) A dish of water, and the floating needle previously described.

Procedure:
(a) Float the magnetized needle upon a dish of water, and allow it to come to rest in its north and south direction. Hold the bar of iron as nearly as possible in the lines of force of the earth's field; namely, with one end pointing northward and downward at an angle of about 70°. Keeping the bar in its position, move it gradually towards the floating needle. It will be found that the lower end of the bar has become a north-seeking pole, and that the north-seeking pole of the floating needle is repelled by it (Fig. 249). Keeping this end of the bar as nearly as possible in its position, swing the other end round so as to bring the bar into an east and west position, where it will be at right angles to the lines of the earth's field. Note that its repellent power upon the north-seeking pole of the floating needle is thereby gradually destroyed. Continue these movements until the pole is again parallel to the earth's field, but with its other end downward; namely, in the position shown in Fig. 250.

FIG. 249.   FIG. 250.

The end which, when it was down, became a north-seeking pole has now become a south-seeking pole, by virtue of its change of position, and it attracts instead of repelling the north-seeking pole of the floating needle. Whereas the lines of force which issue from the north pole of the earth entered this end of the bar in its former position, thus constituting it a north-seeking pole, they now issue from it, so that its character is the same as the north pole of the earth. The pole which has the same character as the north pole of the earth is therefore repelled by it, or, in other words, it is a south-seeking pole.

(b) In the case of soft iron this action of the earth's field is almost entirely temporary. If the iron be not perfectly soft, however, permanent magnetization may be brought about by the action of the earth's lines. To test this point hold the bar of iron parallel to the lines of the earth's field, and tap the end of it sharply with a bit of wood several times. This operation may be performed at a distance from the floating needle. Now bring the bar into the neighborhood of the floating needle and test it for magnetization. It will probably be found that it has acquired a permanent north-seeking pole at the end where the blows were delivered, and a south-seeking pole at the other end. The difference between these permanent poles and the temporary ones produced in the previous portion of this experiment consists in the fact that the former are independent of the position of the bar in the earth's field, whereas the latter depended upon its position.

(c) These permanent poles may be destroyed by reversion of the process which produced them, and the magnetization of the pole may be reversed by a repetition of this process.

To test this statement, hold the magnetized bar near the floating needle with its south-pointing pole downwards and with the axis of the pole parallel to the lines of force of the earth's field. The floating needle will have its north-seeking pole attracted. Tap the upper end of the bar briskly while holding it in this position. Note that with each blow the attraction upon the south-seeking pole of the floating needle is diminished and that presently, instead of being attracted, the latter is repelled. The meaning of this is that the south-seeking pole has been neutralized and then converted into a north-seeking pole.

250. Permeability. — Mention has already been made of the fact that iron affords a better path for lines of force than is offered by the air. The name given to this property is *permeability*. It is measured by comparing the number of lines which pass through iron under a given magnetizing force, per square centimeter of cross-section, with those produced in air by the same force. The fact of the high permeability of iron may be demonstrated as follows:

251. EXPERIMENT 77. — Influence of an Iron Core upon the Strength of Field of a Coil.

Apparatus:

(1) A dish of water and floating magnet needle.

(2) A coil of wire similar to that used in Experiment 70 (Art. 238).

(3) A rod of iron, or a bundle of iron wires about as long as the Argand chimney upon which the coil is wound.

(4) The bichromate battery.

Procedure:

(a) Place the floating magnet at the east or west end of a laboratory table. Connect the ends of the coil to the battery by means of wires 2 or 3 m. in length. Bring up the coil, holding it with its axis in the east and west direction (Fig. 251), until a noticeable deflection of the needle is produced.

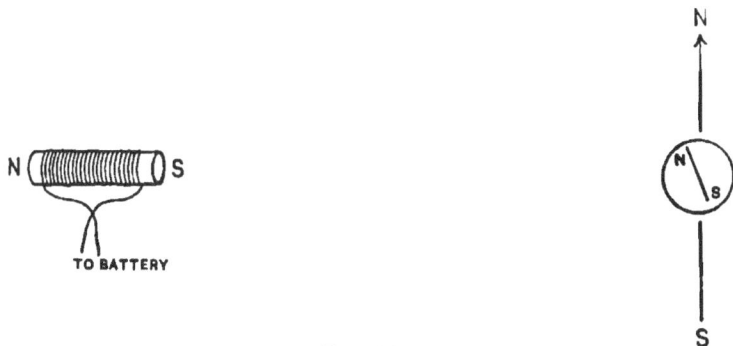

FIG. 251.

(b) Leaving the coil stationary in the position selected, insert the bar of iron from the end furthest from the needle and notice the effect. To show that the increased deflection is not due to permanent magnetization of the bar, withdraw it and insert it again reversed end for end. If iron wires are used, these may be afterwards withdrawn a few at a time and the effect noted.

The reason for the increased effect is that the number of lines of force encircling the coil is much greater when the iron core is present. The increased strength of field shows itself not only within the coil, but throughout the surrounding regions.

252. Magnetic Saturation. — Iron is a much better carrier of lines of force than air, permitting of the formation of about 3000 times as many lines for a given magnetizing force. It is not capable, however, of carrying an indefinitely great number of lines. As the magnetizing force increases, the permeability falls off, at first slowly, then very rapidly. When this rapid fall of permeability occurs, we say that the iron is *saturated*.

By using a coil of wire with many turns, and a battery of more cells, and by substituting the galvanometer for the floating needle in Experiment 76, the saturation of an iron bar may be demonstrated. The effect upon the galvanometer with and without the core is noted, with constantly increasing currents. At first these bear a nearly constant ratio, but as saturation begins, the deflection with the core does not increase so rapidly. The deflections obtained when the core is removed, measure the magnetizing forces. Those obtained when the core is in place are proportional to the magnetization of the iron.

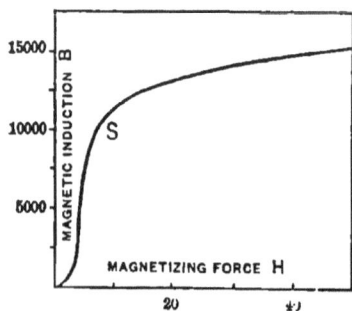

Fig. 252.

The results thus obtained, if expressed graphically, would give a curve like that shown in Fig. 252. Deflections without the core are plotted upon a scale forty times as large as those obtained with the core. The magnetization of the iron rises at first nearly in proportion to the magnetizing force. At the point *s* it bends outwards. Saturation has begun to show itself.

253. Magnetization of Other Substances. — Nickel, cobalt, manganese, and chromium show magnetic properties similar to iron, but their permeability is less. Nickel, however, is sufficiently susceptible to enable one to pick a considerable piece of it up with an ordinary horseshoe magnet. Nearly all other substances are capable of magnetization only in slight degrees. Their permeability, in other words, is very nearly unity. Some substances have less permeability than air. They are said to be *diamagnetic*. When placed in a strong magnetic field, they tend to set themselves at right angles to the lines of force. Among the most strongly diamagnetic substances are bismuth and tellurium. Oxygen, both in the form of a gas and a liquid, is quite strongly magnetic.

To test weakly magnetic or diamagnetic bodies, the material is made into a bar, which is suspended by a delicate fiber of quartz or silk in a strong magnetic field. The field employed is that of an electromagnet.

254. Electromagnets. — These are magnets in which the field is due to coils of wire carrying strong electric currents. The iron core serves simply as a carrier of the lines of force.

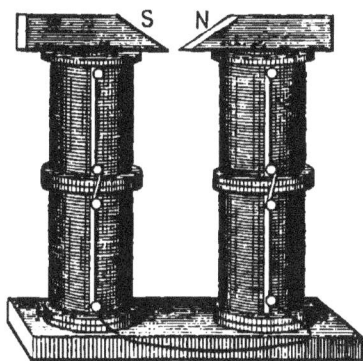

FIG. 253.

U

It is preferably of soft iron, and does not possess appreciable permanent magnetism. Figure 253 shows a favorite form. There is a bed plate of iron into which two upright cylinders of iron are screwed. Over these slip the coils of wire. Upon the cylindrical cores are placed pole pieces (S, N), also of soft iron, the shape of which varies with the experiment to be performed.

255. · Experiment 78. — Testing Bodies of Low Permeability.

Apparatus:

(1) An electromagnet. This may be of small size, like that shown in Fig. 254, the coils of which are only 5 cm. long.

(2) The bichromate battery.

(3) Some minute bars of bismuth, tellurium, zinc, lead, glass, etc., not more than 1 cm. long, and ·1 or ·2 cm. in diameter. The metals may be made into such bars by melting in a crucible, and sucking them up into the bore of glass tubes. The glass, when cold, may be broken away.

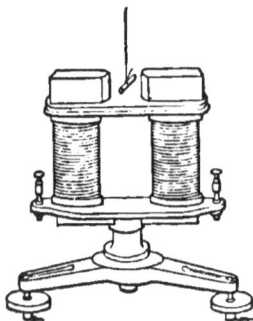

Fig. 254.

Procedure:

Hang the bars in succession between the poles of the electromagnet, using as a suspension as fine a fiber of cocoon silk as you can obtain. Note in each case whether the bar tends to set itself along the lines of force or perpendicular to them.

Bismuth and tellurium should show marked diamagnetic properties. The others, although classed as diamagnetic, more frequently are slightly magnetic from the presence of iron as an impurity.

CHAPTER XXIX

THE MEASUREMENT OF CURRENT, ELECTROMOTIVE FORCE AND RESISTANCE

256. The Galvanometer. — The arrangement described in Chapter XXVII, where the coil of wire surrounds a floating needle, contains all the essential features of the instrument known as the galvanometer. This instrument, which we have already used in certain experiments in Heat, and the construction of which is described in some detail in Appendix VI, consists of a suspended magnet which is placed in the axis of a coil of wire, and is therefore deflected from its north and south position whenever a current flows through the coil. The coil of wire may consist of one or many turns, according to the strength of the current to be measured. For the measurement of large currents, the galvanometer is frequently given the form shown in Fig. 255. In this instrument, which is

FIG. 255.

called the *tangent galvanometer*, because the tangent of the angle through which the needle is deflected is proportional

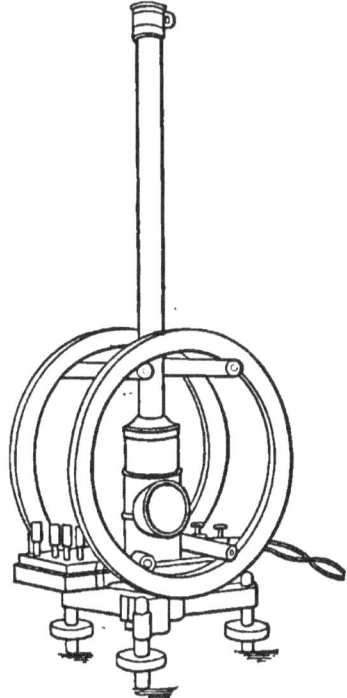

to the current, the coil is of considerable size, from 20 to
100 cm. in diameter. In many cases, as in that of the instrument selected for illustration, two coils are used, and
the needle is placed midway between them in their common
axis. The needle was originally
constructed like a mariner's compass, being pivoted upon a steel
point, and its deflection was read, in
the same way, upon a divided circle.
In modern instruments such needles
have given way to magnets which
are suspended by means of a delicate
fiber of unspun silk or of fused

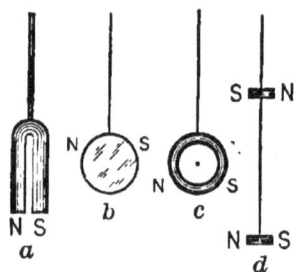

FIG. 256.

quartz. These magnets, which are given various shapes,
sometimes that of an elongated horseshoe, as shown in
Fig. 256, *a*, or of a disk or ring as shown in *b* and *c* of
that figure, or of a couple of steel strips as shown in *d*,
bear but little resemblance to a needle : the name, however, is retained. Instead of reading the deflections of
the galvanometer needle upon a divided circle,
it is customary to attach to the same a mirror,
and to observe the movements of this mirror by
means of a telescope and scale.

257. Sensitive Galvanometers. — In galvanometers for the measurement of very minute currents the coil contains a great many turns of
wire and is of small diameter. The wire is
thus brought as near as possible to the needle.

FIG. 257.

In many cases two needles are used. These
are turned north pole opposite south pole as in Fig. 257.
Such an arrangement is called an *astatic pair*. Each
needle is placed in the axis of a coil, or of a pair of coils.

The action of the earth's magnetism upon the system is very small, because the two needles oppose each other; the currents in the coil, however, flow so as to act in the same direction upon both needles. Great delicacy is thus obtained.

By means of the galvanometer numerous important electrical measurements may be made, and a great number of interesting experiments may be performed.

258. EXPERIMENT 79.— Fall of Potential in a Homogeneous Circuit.

Apparatus:

(1) The sensitive galvanometer described in Appendix VI.

(2) A cell of bichromate battery.

(3) Four meters of fine copper wire (No. 26 or 28).

(4) A resistance box of at least 1000 ohms.

Procedure:

(*a*) Strip the wire carefully of its insulation. Connect one end to the carbon pole of the cell, and stretch the wire for half its length along a laboratory table, thence around a support, and back to the battery. Attach the free end to the zinc pole.

(*b*) Connect the resistance box in series with the galvanometer,

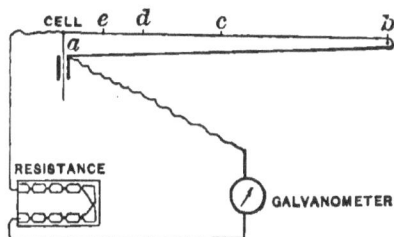

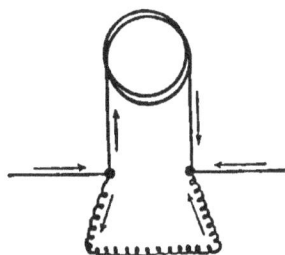

FIG. 258.   FIG. 259.

and bring wires to the terminals of the cell as in Fig. 258. Close the circuit and observe the deflection. If the deflection is too large and no more resistance is available, shunt the galvanometer. This is done by connecting a wire between the terminals to divert a portion of the current from the coils (Fig. 259). The shorter and thicker the wire, the less current will flow through the galvanometer. It is best to begin with a fine wire of considerable length and to shorten the same until a suitable deflection has been obtained. Having adjusted the shunt as above, note the deflection.

(c) Detach the galvanometer wire from the cell and make contact at *b* (in the middle of the stretched wire), *c* (at three fourths the distance along the wire), at *d* (seven eighths of the distance from *a* to *f*), and at *e* (fifteen sixteenths of that distance), successively. Note the deflections and determine the relation between the deflection and the length of wire included between the terminal *f* and the points of contact. For this purpose, plot a curve like that shown in Fig. 260. Ordinates are lengths of wire and abscissas are deflections. If the work has been carefully done this curve will be almost a straight line. The experiment will thus serve to verify the statement that *in a homogeneous circuit* (one consisting of a wire of uniform diameter and composed of the same metal throughout) *the fall of potential is everywhere proportional to the length of wire.*

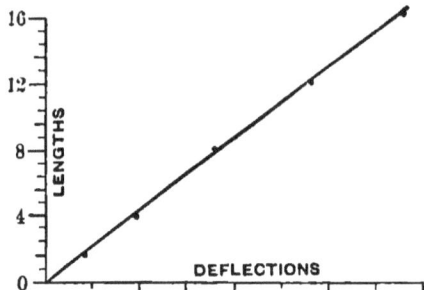

FIG. 260.

259. Resistance and Electromotive Force. — The resistance in any portion of a circuit is measured by the fall of potential in that portion. The conception of *current* leads naturally to the idea of a force needed to drive the current through the resisting portion. It is usual, therefore, to speak of *electromotive force.* This is the equivalent of *fall* of potential, although the terms are not identical in meaning in every respect. When we speak of the electromotive force of a battery, we mean its power of producing current.

260. Ohm's Law. — The current (*I*) in a given circuit is directly proportional to the electromotive force (*E*) and inversely as the resistance *R*. Thus,

$$I = \frac{E}{R}.$$

This relation is known as *Ohm's law*. It makes it possible to compute the current in any circuit in which electromotive force and resistance are known, or, indeed, to find any one of the three quantities, I, E, and R, when the other two are given.

261. EXPERIMENT 80. — **The Resistance of a Wire is inversely as its Cross-Section.**

Apparatus:
The apparatus is the same as in the foregoing determination, excepting that the circuit to be measured contains three sizes of copper wire, *ab*, *bc*, and *cd*, attached to each other as shown in Fig. 261.

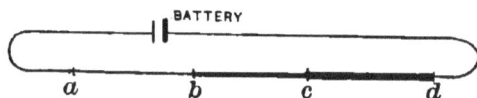
FIG. 261.

Procedure:
(*a*) Instead of connecting the wires of the galvanometer circuit to the terminals of the battery, as in the preceding determination, fasten their ends across the meter stick at a distance of 50 cm. apart, and allow about 3 cm. of naked wire to extend beyond the edge of the stick as shown in Fig. 262.

(*b*) Rub the three pieces of copper wire which constitute the part of the circuit to be tested with fine sandpaper,

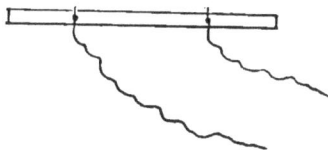
FIG. 262.

and, having cleaned the ends of the galvanometer wires in the same manner, compare the fall of potential through 50 cm. of the three wires in question by means of the deflection produced when the terminals of the galvanometer circuit are brought into contact, first with the piece of wire *ab*, then with *bc*, and finally with *cd*. Make a series of at least ten sets of readings, and average the deflections for each wire separately. The deflections indicate the fall of potential between the circuits of the galvanometer, and this fall of potential, as has been shown in the foregoing article, is proportional to the intervening resistance.

(*c*) Measure the diameters of the three pieces of wire, *ab*, *bc*, and *cd*, with the micrometer gauge, and compute the cross-section of each. Divide the average deflection obtained by contact with each wire by its cross-section. The result should be the same in the three cases. The variations in the three results will be due: *First*, to errors of observation. *Secondly*, to failure to make good contact with the galvanometer terminals; this is an avoidable error, since it depends on the proper cleaning of the surfaces. *Thirdly*, to differences in the quality of the copper of which the three wires are constructed.

262. EXPERIMENT 81. — **Specific Resistance.** — In comparing the resistance offered by different metals to the passage of the current, very great differences are found. It is customary to express resistances in terms of the resistance of a block of the substance in question, 1 cm.^{2} in cross-section, and 1 cm. long. Resistance expressed in this way is called *specific resistance*. It is, in general, impracticable to measure the resistance of so thick and so short a piece of metal; but, since resistance is always proportional to the length and inversely proportional to the cross-section, one can measure the resistance of any convenient sample, and compute from that the specific resistance of the substance. By means of the method of the fall of potential which has been employed in the two foregoing determinations, it is easy to compare the resistance of various metals. The practical unit of resistance is called an *ohm*; it is *the resistance of a column of mercury* 106·3 *cm. long, and containing* 14·4521 *g. of mercury.* This gives a cross-section of 1 mm. Mercury is selected in defining this unit instead of any solid metal, because, being a liquid, it has no structure.

263. EXPERIMENT 82. — **Comparison of the Resistance of Various Metals.**

Apparatus:

The apparatus is the same as in the foregoing experiments excepting that the circuit is made up of pieces of wire, each more than 50 cm. long, consisting of copper, iron, brass, German silver, and platinum. These are to be joined end for end by soldering. The last of the series, namely, the platinum, is connected with the battery cell by means of a copper wire. It is not necessary that the diameters be identical, but it is convenient to deal with wires of about the same size, and possessing a diameter of approximately 0·05 cm.

Procedure:

(*a*) Having cleaned the various wires with fine sandpaper, stretch them out upon a laboratory table between two supports as shown in Fig. 263.

(b) Make contact successively with the various metals, using the galvanometer terminals mounted as in the foregoing determination; note the deflection obtained in each case, and make a series of at least ten readings for each wire.

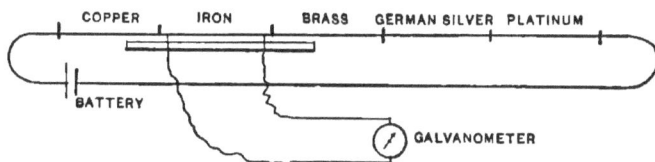

FIG. 263.

(c) Measure the diameters of the various wires with the micrometer gauge, and compute their cross-sections. Multiply the average deflection in each case by the cross-section of the wire to which it belongs. The results will be the relative resistances. The set of values thus obtained, divided by the relative resistance of copper, will give each resistance in terms of that of copper. Now, the specific resistance of copper is 1·7 millionths of an ohm. If we multiply the foregoing set of values by this quantity, we shall obtain the values for the specific resistance of the various metals.

(d) Compare your results with the following table of specific resistances :

TABLE OF SPECIFIC RESISTANCES IN MILLIONTHS OF AN OHM.

Aluminium	3·0	Iron	11·1
Brass	6·9	Lead	19·6
Copper	1·7	Platinum	13·5
Gold	2·1	Silver	1·5
German silver	23·6		

On account of varying differences of hardness and temper, samples of the same metals show a considerable difference in the resistance.

264. The Wheatstone's Bridge. — Another and more accurate method of measuring resistance is that of the Wheatstone's bridge. This method depends upon the principle, already demonstrated in foregoing articles, that the fall of potential along a circuit is everywhere proportional to the resistance. If, therefore, we have a divided circuit, consisting of two branches *abc* and *acd* (Fig. 264) of any

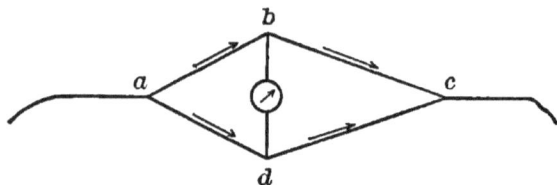

Fig. 264.

resistance whatever, and if we select points *b* and *d* so that the resistance on both sides of them in each branch are in the proportion

$$ab : bc :: ad : dc,$$

then the fall of potential through *ab* will be the same as that through *ad*, and there will be no difference of potential between *b* and *d*. A galvanometer inserted between these points will show no flow of current, however large a current may be flowing through the divided circuit. If three of the four resistances above mentioned are known, the fourth one can be computed by means of the above proportion. This arrangement is called a *Wheatstone's bridge.*

265. Resistance Boxes. — For convenience in the comparison of resistances, it is customary to place a set of coils of wire, previously adjusted to measure an exact number of ohms each, within a box. The terminals of these coils are attached to a row of brass blocks fastened to the top of the box as shown in Fig. 265, and these blocks may be connected by the insertion of plugs. The general appearance of such a box is shown in Fig. 266. The values of the coils are like those of a set of weights, *i.e.* 1, 2, 2, 5, 10, 20, 20, 50, 100, 200, 200, 500, etc.

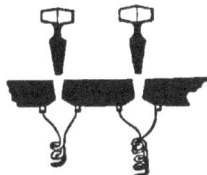

FIG. 265.

FIG. 266.

266. EXPERIMENT 83. — Measurement of the Resistance of a Coil of Copper Wire with the Wheatstone's Bridge.

Apparatus:

(1) A piece of iron or German silver wire 1 m. long; a meter stick.

(2) The resistance box used in the foregoing determination.

(3) A coil or spool of fine copper wire, the resistance of which is to be measured.

(4) A galvanometer.

Procedure:

(*a*) Stretch the piece of iron or German silver wire between two binding screws, adjusting it so that the length is as nearly as possible 1 m. Lay the meter stick alongside the wire with its ends corre-

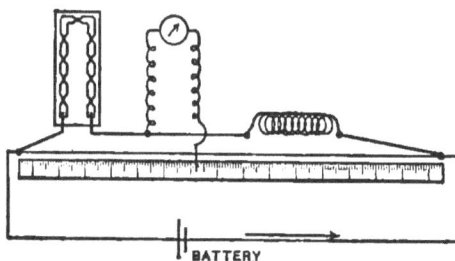

FIG. 267.

sponding to the ends of the former. Solder to the posts short pieces of copper wire. One of these leads to the resistance box, the other to the coil of wire, the resistance of which is to be determined. Con-

nect the other terminal of the resistance box and the other end of
the spool of copper wire by means of a heavy copper wire as shown in
Fig. 267.

Attach one terminal of the galvanometer at *a* and hold a wire
connected with the other terminal in the hand; connect wires from
the terminals of the cell of bichromate battery to the binding posts
at the ends of the stretched wire. This arrangement is a simple form
of what is known as the *slide-wire bridge*.

(*b*) Whatever be the resistance of the coil of wire to be tested,
which should however for convenience be of about the same as the
resistance of the box used in the other branch of the bridge, it will
be possible to find a point upon the stretched wire at which the free
terminal of the galvanometer may be brought into contact with the
wire without producing a deflection. Determine this point as nearly
as possible and note its position upon the wire. Call this point *d*.

From the principle of the Wheatstone's bridge above stated, we
have

$$ab : bc :: ad : dc.$$

Since the resistance *ad* and *dc* are proportional to the intervening
lengths of wire, we can compute the resistance of the spool of wire to
be tested in terms of that of the coils within the resistance box.

267. EXPERIMENT 84. — **Influence of Temperature upon Resistance.**

Apparatus:

(1) The slide-wire bridge described in Art. 266.

(2) Wires of iron or platinum and German silver; also a carbon
rod.

(3) Two jars and a U-shaped tube. (See Fig. 268.)   ⸱↖

Procedure:

(*a*) Substitute for the spool of copper wire used in Experiment 83,
a piece of fine iron or platinum wire wound upon a glass tube; this
wire should be uninsulated.

(*b*) Balance the Wheatstone's bridge and heat the coil of iron or
platinum wire with a Bunsen burner. Note that the galvanometer
begins to show deflection, which increases as the temperature of the
wire rises. Note that the direction of this deflection is such as to
indicate that the resistance of the wire has been increased by heating.

(*c*) Repeat the above experiments, using a piece of German silver wire, and note that the effect of temperature in this case is much smaller.

(*d*) Repeat the experiment, using a slender rod of carbon; a quarter inch arc light carbon will answer this purpose, or even the graphite of an ordinary lead pencil. Note that the effect of temperature upon the carbon is opposite to that in case of the metals, namely, that the influence of temperature is to decrease the resistance.

(*e*) Repeat the experiment, using a column of acidulated water. This column of liquid may be obtained by filling a siphon tube between two jars. (See Fig. 268.) Upon warming this tube of liquid with the flame, it will be found that its resistance is likewise diminished by heating. The facts established by means of this experiment may be stated as follows:

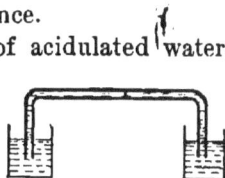

FIG. 268.

(1) The resistance of metals is increased by heating. The increase in the case of pure metals amounts to about 40 per cent for 100° rise of temperature.

(2) The influence of temperature upon the resistance of alloys is much less than in the.case of pure metals.

(3) The resistance of carbon is slightly diminished by rise of temperature.

(4) The.resistance of all electrolytes falls as the temperature rises.

CHAPTER XXX

THE HEATING EFFECT OF THE ELECTRIC CURRENT

268. Transformation of Energy by Means of the Current. — The electric current, as has already been shown, consists in the equalization of the differences of potential in different parts of the electric circuit. These differences of potential have been produced by the expenditure of energy. In the continuous discharge which constitutes the current this energy is liberated again in the form of heat. The heat thus produced shows itself wherever resistance is offered to the passage of the current and in proportion to that resistance.

269. Joule's Law. — *The amount of heat energy developed in the circuit is proportional to the product of the square of the current into the resistance.* This is known as *Joule's law.*

The development of heat at points of high resistance may be demonstrated as follows:

270. EXPERIMENT 85. — The Heating of Platinum Wire by the Current.

Apparatus:

(1) The plunge battery; a short piece of fine platinum wire about 0·025 cm. in diameter (No. 30).

(2) A similar piece of copper wire as nearly as possible of the same diameter; also 2 or 3 m. of larger copper wire.

Procedure:

(a) Stretch the platinum wire between two wooden blocks, as

shown in Fig. 269. Cut the larger wire in two equal parts and connect with the terminals of the battery.

(*b*) Take the free ends of these wires in the hands, and having scraped them clean apply them to the ends of the stretched platinum wire. Make contact between the platinum and the copper wire with the two
terminals, bringing these successively
nearer and nearer together, and note the
effect. As the strip of platinum wire in
the circuit becomes shorter, its tempera-
ture will rise until it becomes red hot and finally white hot. The platinum wire may even be melted by sufficiently reducing the distance between the copper terminals.

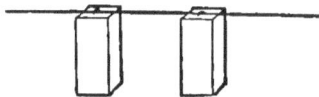

FIG. 269.

(*c*) Repeat, using the fine copper wire instead of the platinum. Note that the heating effect is much less.

(*d*) Join the copper and platinum wires end to end, and stretch between the blocks with the junction midway. Repeat the experiment, moving the contact wires inward from the blocks so as to include always equal lengths of copper and platinum between them. Note that, although the same current flows in both wires, the platinum becomes much hotter than the copper. The resistance, length for length, of the platinum wire is more than nine times as great as that of the copper, and it is in this portion of higher resistance that the heat is chiefly developed.

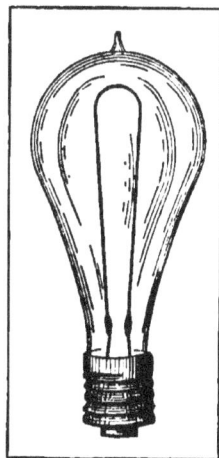

271. The Glow Lamp. — Another material which offers high resistance to the
current is carbon, and since this is capable of being heated to very high temperatures without fusion, it is used for the
purpose of obtaining light by the expenditure of electrical energy. A narrow

FIG. 270.

filament of carbon is inclosed in a glass bulb (Fig. 270) from which the air has been exhausted. The removal of the air serves a double purpose: on the one hand it prevents oxidation of the carbon, which would speedily destroy it, and on the other it reduces the loss of heat by convection. Current is conveyed to this filament by means of platinum wires sealed in the glass. Such an arrangement is called a *glow lamp* or an *incandescent lamp* (Fig. 270). •

The lamps generally used in artificial lighting require the use of a source of electromotive force much higher than those described in this book. It is possible, however, to obtain small lamps which may be brought to incandescence by the use of a very few cells of battery, or even of a single cell. The lamps used in the lighting of houses require a difference of potential at their terminals of 50 to 150 volts. The currents employed range from $\frac{1}{2}$ an ampere to $1\frac{1}{2}$ amperes. The source of current is generally a dynamo machine.

272. EXPERIMENT 86. — Determination of the Power expended in a Glow Lamp.

Apparatus:

(1) A miniature or "pea" lamp of 5 or 6 volts.

(2) A small calorimeter containing about 250 cm.² of water.

(3) A good thermometer.

(4) A timepiece.

Procedure:

(*a*) Solder to the terminals of the lamp two copper wires. Slip these through a glass tube and draw the latter up snugly to the base of the lamp. Fill the tube with paraffin, taking care that the wires do not become crossed, and that the wires from the point where they leave the lamp to the further end of the tube are well embedded in

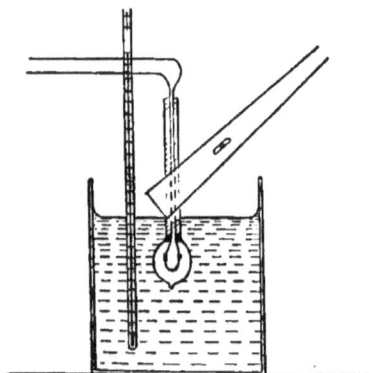

FIG. 271.

the paraffin. Mount the lamp within the calorimeter as shown in Fig. 271. Weigh out 200 g. of water, and pour the same into the calorimeter.

(b) Connect the lamp in circuit with the battery, using a number of cells sufficient to bring it to incandescence, but not enough to imperil the filament.[1] Leave the circuit open.

(c) Note the temperature of the water.

(d) At a time accurately noted, close the circuit. Stir the water and watch the thermometer until the temperature has risen about 10°. Then break circuit, noting the time. Stir the water in calorimeter and read the thermometer.

(e) Compute the gram calories of heat produced by the lamp in one second.

(f) Compute the power necessary in "horse power" (one horse power if converted into heat will produce 179·3 calories of heat per second).

273. The Arc Light. — When an electric circuit is interrupted, great heat is developed at the point of interruption during the short interval of time when bad contact exists between the terminals, which are being withdrawn from one another. The terminals become heated, and as they are withdrawn the air lying between them also becomes intensely hot. Now a hot gas is a good conductor of electricity, and if the difference of potential between the terminals is at least 25 volts, a sufficient current will flow to keep the gas permanently hot. This flow of current across a heated air gap is called an *electric arc.* The temperatures developed in such cases are very high, and if metals be employed as terminals they are rapidly volatilized. By substituting rods of carbon for the metallic conductors, we may, however, maintain the arc between them for a long time. Carbon is the only conductor

---

[1] For a lamp marked "5 volts" four cells of bichromate battery may be used.

x

which will sustain the temperatures thus created without very rapid destruction.

The carbon terminals where the current passes from the air gap to the solid conductors become heated to brilliant incandescence, and the light which they give off is the most powerful of all artificial lights. It is employed in the arc lamp.

Figure 272 gives a picture of the electric arc between carbon terminals. It is taken from a photograph. It will be seen that the most brilliant portion is the surface of the upper carbon. This is the positive terminal, that from which the current flows into the arc. The end of the positive carbon is eaten away so as to form a con-

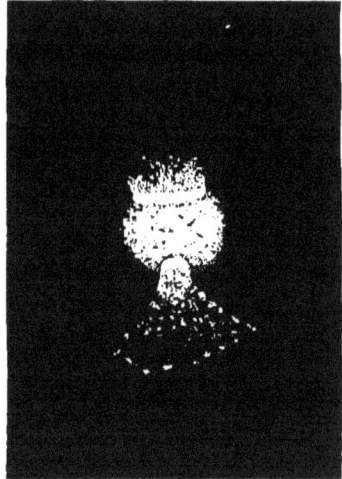

FIG. 272.

cave region called the *crater*, and it is the crater which emits the most intense light. The negative carbon is also incandescent, but does not reach so high a temperature. The light which it gives off is comparatively feeble. The temperature of the crater, according to Rossetti, is 3900° C.; that of the negative carbon, 2500° C.

CHAPTER XXXI

ELECTROLYSIS

274. The Law of Electrolysis. — The chemical effect of the electric current consists in breaking up liquids through which it passes into two parts, the metal and the acid-radical with which this is combined to form a salt. Copper sulphate ($CuSO_4$), for example, is decomposed into copper (Cu) and the acid radical (SO_4). In order that a compound may be thus decomposed, it is necessary that it be a conductor of electricity. Substances which thus conduct the current, and in conducting it are broken up, are called *electrolytes.* Nearly all liquid conductors are electrolytes; the only exceptions being molten metals. These are not chemical compounds, and consequently are not capable of electrolysis.

275. Ions. — The parts into which an electrolyte is broken up by the action of the current are called its *ions;* thus in the case of silver nitrate the ions are Ag and NO_3. These ions are separated by the action of the current, and the metallic atoms move in the direction in which the current is flowing and are deposited upon the terminal at which the current leaves the electrolytic cell. (See Fig. 273.) These metallic ions, which travel with the current, are called *kations;* the other ions, *anions*, move towards the other terminal of the cell, where they are set free. · These consist of the group of atoms known in Chemistry as the *acid radical.* Since an acid radical is an incomplete and chemically unsatisfied group, it immediately attacks the material of the terminal, or the solution itself, and enters into combination. The terminals of the electrolytic cell are called *electrodes*. The one at

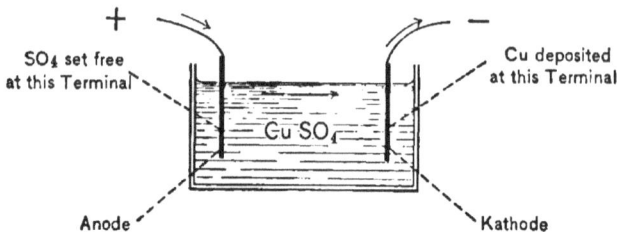

+   SO₄ set free at this Terminal   Cu deposited at this Terminal

Cu SO₄

Anode   Kathode

FIG. 273.

which the current enters the solution, and where the anions are liberated, is called the *anode*. The one where the current leaves the cell, and where the metallic iron is deposited, is called the *kathode*.

It will be seen from the above that the chemical reactions which result from the passage of a current through an electrolyte, are generally twofold. We have, first, the decomposition of the electrolyte into its ions. If a solution of copper sulphate be subjected to electrolysis the copper will be deposited upon the kathode, and the anion (SO_4) will be set free at the surface of the anode. The SO_4 anion immediately attacks the water in which the sulphate of copper has been dissolved, combining with the hydrogen atoms to form free sulphuric acid (H_2SO_4), and liberating oxygen gas which appears in bubbles at the surface of the anode. If the anode consists of copper, the SO_4 attacks it, forming copper sulphate ($CuSO_4$); if it consists of platinum or of carbon, which will not unite with the acid radical, the free acid remains in the solution uncombined.

The foregoing statement concerning electrolysis may be readily illustrated by means of the following experiments:

276. EXPERIMENT 87. — Electrolysis of Sulphate of Copper.

Apparatus:

(1) The bichromate battery previously described.

(2) A beaker; two strips of platinum foil; some small copper wire.

(3) About 100 g. of sulphate of copper.

Procedure:

(a) Dissolve the sulphate of copper in water and nearly fill the beaker with the solution. To the terminals of the battery, which should consist of three or four cells, attach pieces of copper wire each about 1 cm. in length. To the free ends of these, after removing the insulation for a distance of 4 or 5 cm., attach the strips of platinum foil. For this purpose the latter is cut, as shown in Fig. 274, and the flaps thus produced are rolled snugly around the naked wire, which should have been previously well scraped. The wire should then be bent at right angles to itself, as shown in the figure.

(b) Dip the two pieces of platinum foil thus attached into the electrolyte. Note that at one electrode gas rises, and that this occurs at the surface of the foil which is connected with the carbon pole of the battery.

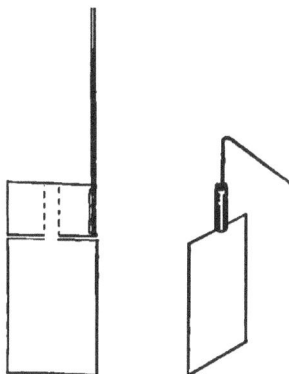

FIG. 274.

This means that the SO$_4$ radical has moved against the current, and that, at the surface of the foil, free sulphuric acid has been formed and oxygen set free as explained in previous article. The kathode, where no gas appears, will be found, if removed after the current has been passing for a few minutes, to be coated red with copper.

(c) Reverse the direction of the current by exchanging the wires at the terminals of the battery. Note that at first no gas escapes at either terminal; but that, after a period about as great as that previously has elapsed, gas begins to appear at the electrode upon which copper had been deposited and which has now become the anode. Remove the strips of platinum foil from the beaker, and note that the one which had been plated has lost its coating of copper altogether, and that the other strip is now copper plated. By means of this experiment, we verify the statement that the copper ion is carried in the direction in which the current flows and is deposited upon the kathode. The evidence with reference to the SO$_4$ ion is not so direct. The fact that free acid is produced in the neighborhood of the anode, when electrolysis occurs, can be more directly shown by means of the following experiment:

277. EXPERIMENT 88. — **Electrolysis of Sodium Sulphate.**

Apparatus:
(1) A U-shaped tube of the form shown in Fig. 275.
(2) The platinum electrodes described in the foregoing experiment.
(3) The bichromate battery.
(4) About 100 g. of the neutral sulphate sodium (Na_2SO_4); also some neutral solution of litmus.

Procedure:
(*a*) Dissolve as much of the sodium sulphate as will be readily taken up in water, and add a sufficient amount of the litmus solution to color the liquid a rich purple. Fill the tube with this solution and insert the platinum electrodes, which should dip well beneath the surface.

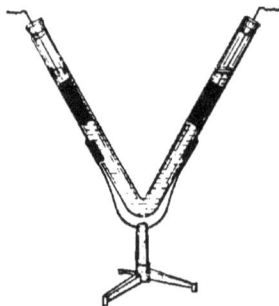

FIG. 275.

(*b*) Connect with the battery and note that gas is liberated from both electrodes; also that the solution in the neighborhood of the anode rapidly changes to a red color, while the bluish cast of the solution in the neighborhood of the kathode increases. The reddening of the solution in the neighborhood of the anode is indicative of the presence there of free acid, a point which it is one of the purposes of this experiment to establish. This fact may readily be verified by pouring a few drops of the litmus solution in a test tube, or beaker, and adding a drop of dilute sulphuric acid. The change of color thus produced will be seen to correspond with that which is going on within the electrolytic cell. A drop of ammonium added to the litmus solution will likewise be found to intensify its blueness in a manner corresponding to that which occurs in the region of the kathode.

The gases generated at the electrodes in this experiment are, respectively, hydrogen at the kathode and oxygen at the anode. The hydrogen is produced by the combination of the metallic sodium, set free at the kathode with the water thus:

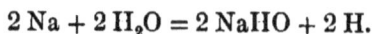

$$2\,Na + 2\,H_2O = 2\,NaHO + 2\,H.$$

At the anode the reaction is similar to that which has already been described, viz. :

$$SO_4 + H_2O = H_2SO_4 + O.$$

278. Voltameters. — *The quantity of metal deposited by a current is proportional to the strength of the current and to the time it flows.* This law, which is known as *Faraday's law*, makes it possible to measure the quantity of electricity which has flowed in the circuit. By dividing this quantity by the time, we obtain the average strength of the current. An instrument for measuring current in this way is called a *voltameter.*

279. The Water Voltameter. — An instrument designed for the measurement of current, by means of the amount of water which it decomposes, is called a *water voltameter*. In the decomposition of water by the current, hydrogen plays the part of a metal, and oxygen that of the acid radical. If, therefore, two platinum electrodes, like those described in the foregoing experiments, be inserted in a dish of water, which should be slightly acidulated to increase its conductivity, and if inverted tubes, closed at one end and filled with water, be placed over each electrode, as shown in Fig. 276, the current sent through the cell will liberate hydrogen gas at the surface of the kathode, which will rise and displace the water in the tube placed over that electrode, while oxygen will similarly be collected in the other tube. The volume of hydrogen thus produced will be twice as great as that of the oxygen, and each of them will be proportional to the time and to the average strength of

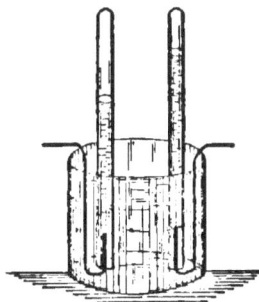

FIG. 276.

the current. If we know how many cubic centimeters of hydrogen a unit of current is capable of producing in one second of time, we can tell, from the volume of hydrogen produced in the water voltameter in a given time, how strong the current has been. The water voltameter does not give accurate results, excepting when very great precautions are taken, on account of the absorption of the gas by the liquid within which they are generated, and the occlusion, especially of hydrogen, at the surface of the platinum electrode.

Two forms of voltameters, with which it is possible to obtain much better results, are the *copper voltameter* and the *silver voltameter*.

280. EXPERIMENT 89. — **The Measurement of Current by Means of a Copper Voltameter.**

Apparatus:

(1) A copper voltameter. The form of voltameter, by means of which an accurate result may most easily be obtained, is constructed as follows: Take four pieces of naked copper wire, size No. 12. Two of these pieces should be 1 m. long, and two of them ½ m. long, each. Clean the wire with sandpaper and make it up into four spiral coils, the long piece into coils about 8 cm. in diameter, and the short pieces into coils with a diameter of 4 cm. each. Pass each coil through the flame of the Bunsen burner to remove the oil due to the hand, and plunge it into a dilute solution of sulphuric acid and then into water. Mount the coils thus prepared as shown in Fig. 277.

(2) The bichromate battery previously mentioned.

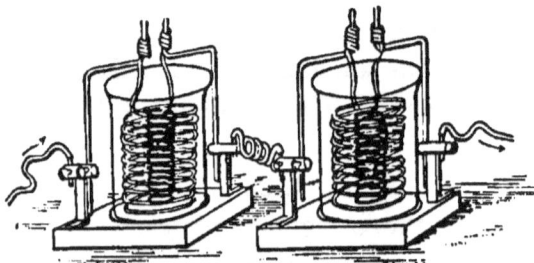

FIG. 277.

(3) Two beakers, large enough to contain the coils, when mounted as shown in the figure, and a sufficient quantity of the solution of sulphate of copper (density between 1·10 and 1·18) to immerse the coils.

Procedure:

(*a*) Weigh all four coils as carefully as possible; then mount them as described, connect them with wires so that when the current flows it will pass in each case from a large to a small coil, leaving the circuit open at one terminal of the battery. Fill the beakers with the solution until the coils are entirely submerged. Close the circuit at a time carefully noted by means of a watch or other timepiece. Allow the current to run for thirty minutes before breaking circuit, and note the time again.

(*b*) Remove the coils from the solution; dry them carefully by rolling them on pieces of white filter paper, then by dipping into strong alcohol and again rolling on filter paper. The alcohol will quickly evaporate, leaving the coils dry.

(*c*) Weigh all four coils again, and compare their weights with those previously obtained. It will be found that the small coils have gained in weight, and that the large ones have lost by almost the same amount. If the experiment has been carefully carried out, it will be found that the gaining coils increase by almost precisely the same amount. The change in the weight of the losing coils is less to be depended upon. To compute the average strength of current which has passed through the voltameter during the experiment, we divide the increase of weight of each in grams by the number of seconds of time during which the current has been flowing. This, in turn, is divided by the quantity 0·000328, which is the amount of copper in grams deposited by one ampere of current in a second of time. The result gives the average strength of the current in amperes. By means of a balance of proper delicacy current may be measured in this to within a fraction of one per cent.

281. The Silver Voltameter. — In this apparatus a silver anode is used, and a kathode consisting of a platinum bowl, which serves at the same time as the containing vessel for the solution. The solution consists of silver nitrate dissolved in water. By the electrolytic action the nitrate of silver in the solution is decomposed, and the silver is deposited in the form of minute crystals upon the surface of the platinum dish. The free nitric acid which is produced at the anode attacks the latter and eats it away. This instrument is capable of as high a degree of accuracy as the copper voltameter, but it is less convenient for laboratory purposes on account of the costly materials used and of the corrosive nature of the solution employed as an electrolyte.

282. Electrochemical Equivalents. — If voltameters employing various substances are placed in the same circuit, so that the same current passes through them all, the products of the electrolytic action will be found to differ, according to the metals deposited, in a perfectly definite manner. The amount of each metal deposited is proportional to its *atomic weight,* and is found by dividing that quantity by its valency. The amount of any metal which will be deposited by an ampere of current in a second of time is called its *electrochemical equivalent.* The electrochemical equivalent of a few of the substances most frequently employed in electrolysis are given in the following table:

TABLE OF ELECTROCHEMICAL EQUIVALENTS.

Element.	Electrochemical equivalent (grams per ampere per second).	Element.	Electrochemical equivalent (grams per ampere per second).
Hydrogen . . .	0·00001038	Oxygen . .	0·000082
Copper	0·000328	Chlorine . .	0·000367
Gold	0·000679	Iodine . . .	0·001314
Lead	0·001072	Bromine . .	0·000828
Nickel	0·000304		
Silver	0·001118		
Zinc	0·000337		

CHAPTER XXXII

THERMO-ELECTRICITY

283. The Production of Current by Means of Heat. — Whenever in an electric circuit we have two metals, and a difference of temperature exists between the places where these metals are joined, current will flow through the circuit. If, in Fig. 278, *A* and *B* are the junctions between iron and copper, and if *A* be heated by means of a flame, the current will flow in the direction indicated by the arrow. If *B* be heated instead of *A*, the flow will still be from copper to iron, but in the opposite direction. Currents gener-

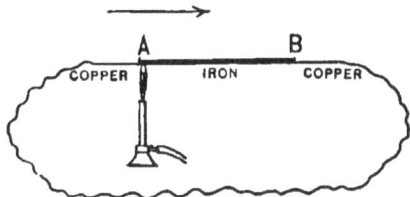

FIG. 278.

ated in this way are called *thermo-electric currents*, and the arrangement of two metals is called a *thermo-element*.

284. The Thermopile. — The difference of potential produced by heating one junction of the thermo-element depends upon the metals which are employed. The greatest effect attainable with ordinary materials occurs when bismuth and antimony form the two metals of the ele-ment. The difference of potential

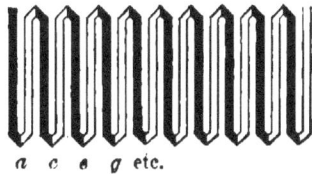

a c e g etc.

FIG. 279.

may be further augmented by placing a large number of such thermo-elements in series, as shown in Fig. 279; a bar of antimony soldered to a bar of bismuth, and this to the second bar of antimony, and so on indefinitely.

If we heat the alternate junctions *a*, *c*, *e*, *g*, etc., of such an arrangement, the difference of potential at the ends of the series is the sum of those produced in each element. Such a combination of thermo-elements is called a *thermopile*. In order to give it as compact a form as possible, the thermopile is made of parallel bars of antimony and bismuth, with an insulating layer of mica between each. These are usually arranged in the form of a cubical block, mounted as shown in Fig. 280. The bars are connected together in such a manner that alternate junctions lie together upon one face of the block. When used with a suitably sensitive galvanometer, very small differences of temperature between the faces of the thermopile may be measured.

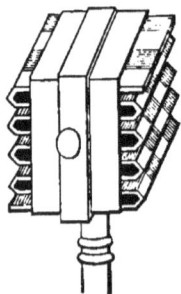

FIG. 280.

285. The Peltier Effect. — In 1834, the physicist Peltier discovered that when the electric current is sent through a circuit containing a thermo-element, one junction will be heated and the other cooled; also, that the distribution of temperatures is such as would tend to form a thermo-electric current flowing in the opposite direction from that which produces this difference of temperature. This phenomenon is called the *Peltier effect*. It may be illustrated by means of the following experiment:

286. EXPERIMENT 90. — **The Peltier Effect in a Thermopile.**

Apparatus:

(1) A thermopile and galvanometer.

(2) A bichromate battery consisting of three or four cells.

Procedure:

(*a*) Connect the thermopile with the terminals of the galvanometer. Touch one face of the pile for an instant with the finger, thus warming it slightly, and notice the direction in which the galvanometer needle is deflected.

(*b*) Disconnect the galvanometer and the thermopile. Connect the latter with the battery for a few seconds, then disconnect and reattach to the galvanometer. There will be a deflection produced this time by the relative heating and cooling of the faces of the thermopile by the passage of the current, and this will die away as the temperature distributes itself within the pile. The direction of this deflection will depend upon the way in which the battery was connected with the thermopile.

(*c*) Repeat operation (*b*), reversing the connections between the thermopile and the battery. Note that the galvanometer deflection is reversed. It appears from this experiment that a sufficient difference of temperature may be produced by sending the electric current through a thermopile to give thermo-electric currents, and that the direction of these, and consequently the nature of the heating and cooling of the junctions, depends upon the direction of the current which produced them. It is easier to establish the remaining fact of importance with reference to the Peltier effect; namely, that the thermo-electric current flows in a direction such as to oppose the current which produces it, by means of an apparatus to be described in the following experiment:

287. EXPERIMENT 91.—**Direction of the Thermo-electric Current in an Element of Antimony and Bismuth.**

Apparatus:

(1) A bar of bismuth about 10 cm. long and 1 cm. in diameter soldered at the ends to similar bars of antimony. These at their free

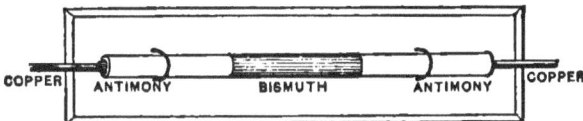

FIG. 281.

ends are also soldered to copper wires, and the whole is mounted upon a wooden block, as shown in Fig. 281.

The bars of antimony and bismuth necessary for the construction of this apparatus may be obtained by pouring the molten metals into small test tubes; after the metal has been solidified, the tubes may be broken and the bars thus obtained may be dressed with a file and soldered together. Great care must be taken in handling bismuth, on account of the brittle nature of this semicrystalline metal.

(2) The galvanometer used in the foregoing experiment.

(3) A bichromate battery.

(4) A resistance box; one containing several thousand ohms is to be preferred.

Procedure:

(*a*) Connect the zinc pole of one cell through the resistance box with one terminal of the galvanometer. Take the wire running from the carbon pole in the right hand and make momentary contact with the other terminal of the galvanometer. Note the direction in which the needle is deflected.

(*b*) Connect the galvanometer with the antimony-bismuth thermo-element above described, as shown in Fig. 282. Warm slightly first one and then the other of the junctions between bismuth and antimony, and note the deflections of the galvanometer. Having determined which one must be heated in order to send the current through the galvanometer in the same direction as when the cell was employed, trace out the direction of current in the thermo-element. It will be found that when a junction between antimony and bismuth is heated, *the current always flows from bismuth to antimony through the heated junction.*

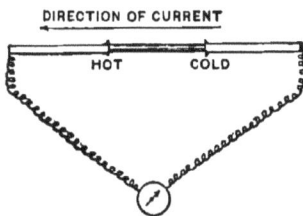

FIG. 282.

(*c*) Test the direction of the Peltier effect as follows: Disconnect the galvanometer and connect the thermo-element with three or four cells of the bichromate battery, for about thirty seconds, noting the direction in which the current flows through the element. Disconnect and make connection with the galvanometer again, taking care to arrange the circuit precisely as in operation (*b*). Note the direction of the deflection. It will be found that the deflection indicates the production of a difference of temperature such as to create a thermo-electric current flowing in the opposite direction through the element from the current which produced it.

288. The Thermo-electric Series. — By testing various metals in pairs at a given temperature, and determining the size and direction of the thermo-electric currents produced when a difference of one degree of temperature is created between the junctions, it is possible to arrange the various metals tested in a series, such that the current in the heated junction will flow from any metal through the junction into metals lower down in the series, and *vice versa.* The following is such a series, in which lead is selected as the neutral metal with which all others are to be compared. It is made out for the temperature interval 19°–20° C.

THERMO-ELECTRIC SERIES.

+	Silver.
Bismuth.	Zinc.
German silver.	Copper.
Lead.	Iron.
Platinum.	Antimony.
Gold.	−

289. The Neutral Point. — Such a series as the above is applicable only to the temperature in question. It is found, in fact, that for any given pair of metals the thermo-electric effect varies with the temperature at which the experiment is performed. Frequently the effect disappears altogether when a certain temperature is reached, and then becomes reversed. The point at which this reversal occurs is called the *neutral point* for those metals. This fact may be illustrated by means of the following experiment:

290. EXPERIMENT 92. — **The Neutral Point for Iron and Copper.**

Apparatus:

(1) A piece of iron wire about 1 m. long; copper wire and a galvanometer.

(2) A Bunsen burner.

Procedure:

(*a*) Form junctions between the iron and copper wires by twisting together the ends, which should have been previously cleaned by scraping. In this way the use of solder is avoided. Connect the free ends of the copper wire to the terminals of the galvanometer.

(*b*) Carefully heat one junction with the Bunsen burner and watch the effect upon the galvanometer. It will be seen that as the temperature of this junction rises, the deflection increases to a certain value and then diminishes again, passing through the zero point and changing sign. There is, in fact, a neutral point between copper and iron at about 275° C. If we were to measure temperature differences of the junctions, as well as deflections, we should be able to plot a curve like that in Fig. 283. Such peculiarities in the behavior of the thermo-elements make it necessary to study the thermo-electric effect very carefully before using it for the measurement of temperatures.

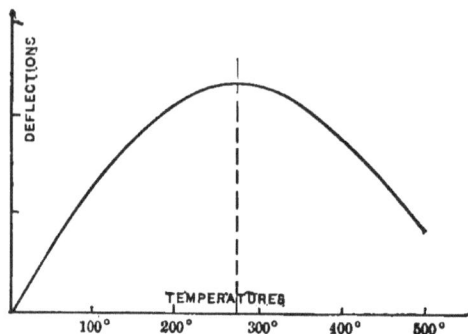

Fig. 283.

CHAPTER XXXIII

ELECTROMAGNETIC INDUCTION

291. The Production of a Current in a Wire by Cutting Lines of Force. — If a wire be moved in a magnetic field in such a direction as to cut the lines of force, an electric current will be set up in the wire. A current thus produced is called an *induced current*. Its source is the work required to move the wire through the magnetic field. The existence of such currents may be demonstrated by means of the following experiment:

292. EXPERIMENT 93. — **The Study of Currents induced by moving a Wire in the Magnetic Field.**

Apparatus:

(1) The galvanometer previously described; a piece of wire several meters in length.

(2) An electromagnet and battery.

Procedure:

(*a*) Connect the ends of the wire to the terminals of the galvanometer, forming a closed circuit. At a distance of at least 3 or 4 m. from the galvanometer, set up the electromagnet, connecting it with the poles of the bichromate battery; the latter should consist of at least 4 to 6 cells.

(*b*) Close the battery circuit, and allow the galvanometer needle to come to rest. Take the wire connected with the galvanometer, holding a piece of the same near the middle, stretched between both hands. Move this wire downward and through the field of the electromagnet from above, so as to cut the lines at right angles, as shown in Fig. 284. Observe the effect upon the galvanometer needle. Repeat this movement, but in a reverse direction, and note that the deflection is reversed. Move the wire at uniform speed

Y

through weak, and then through very strong, parts of the field, and notice that in the strong field, where the lines of force are most plentiful, the effect is greatest. Move the wire down from outside the field to a position midway between the polepiece of the magnet, and bring it to rest there. Note that, as the motion of the wire ceases, the induced current ceases to flow, although the wire remains in a very strong part of the field. Move the wire as nearly as possible in the direction of the lines of force, instead of across them, and note that the induced current can thus be diminished almost to zero. Were it possible to move the wires so as to cut no lines of force, no induced current would show itself.

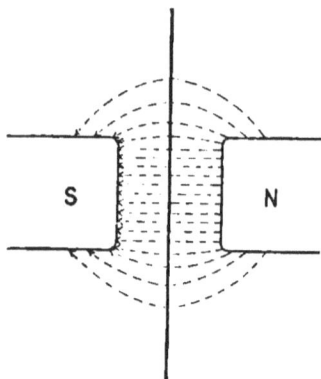

FIG. 284.

(c) The direction of the induced current is determined thus:

Lines of force which have a common direction repel each other. When we move a wire in the magnetic field, lines of force are set up in such a direction as to resist its motion. The lines of force created around the wire are circles, and the positive direction around these circles is clockwise to the person looking along the wire in the direction in which the current is flowing. (See Fig. 285.) Bearing this convention in mind, proceed to verify the statement that the direction of the induced current is such as to impede the motion of the wire through the field. For this purpose, test the poles of the electromagnet by means of a compass needle, or with the floating magnet needle used in previous experiments.

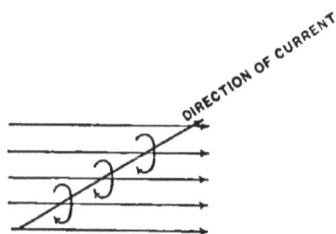

FIG. 285.

The lines of force between the pole of the electromagnet are to be considered as issuing from the north pole and entering the south pole. If, therefore, the experimenter stands facing the magnet with the north pole to his right hand, these lines will run from right to left; if the wire be moved downward through this field, the current, which will retard its motion, is one such that the circular lines around the wire are parallel to the lines of the field beneath the wire. This means that they are clockwise to the observer, and that the current is flowing through the wire in the direction in which he looks. Having moved the wire down in the manner just described, and noted the direction of the galvanometer deflection, test this point by opening the galvanometer circuit, and introducing, for an instant, a single voltaic cell. Connect this cell so that the deflection will be as above; then trace out the direction of the current.

293. The fact that a conductor, moving in the magnetic field, meets with resistance to its motion, may be demonstrated in numerous ways. One of the simplest experiments for this purpose consists in mounting a pendulum of the form shown in Fig. 286 between the poles of an electromagnet. So long as the magnet is inactive, this copper pendulum swings freely. If, however, the current be sent through the coils of the magnet, thus creating a strong field, the copper pendulum will experience such resistance to its passage across the lines of force that it will be brought to rest before it completes a single vibration. The

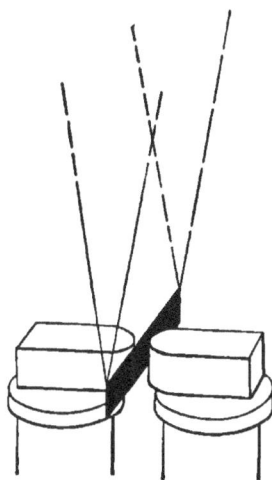

FIG. 286.

experiment is a striking one and may be easily tried; it is instructive to attempt to move the sheet of copper suddenly in the field by means of the hand. The resistance which manifests itself may then be directly felt.

If a second copper disk similar to the first, but cut nearly through with the shears in a series of vertical slits, as shown in Fig. 287, be substituted for the pendulum above described, it will

FIG. 287.

be found that it experiences comparatively little opposition when swinging in the magnetic field. The reason is that the vertical slits in the copper interfere with the circulation of the induced currents. These being less powerful, comparatively few lines of force are formed round the moving mass. Such a pendulum will make several vibrations in the field of the electromagnet before coming to rest.

294. Force exerted upon a Wire carrying Current in the Magnetic Field. — The phenomenon described in the first article of this chapter is a reversible one. Any wire placed in a magnetic field in such a position that it is not parallel to the lines of force, and which has a current flowing through it, is acted upon by a force which tends to move it at right angles to its own length, and to the direction of the lines themselves. (See Fig. 288.)

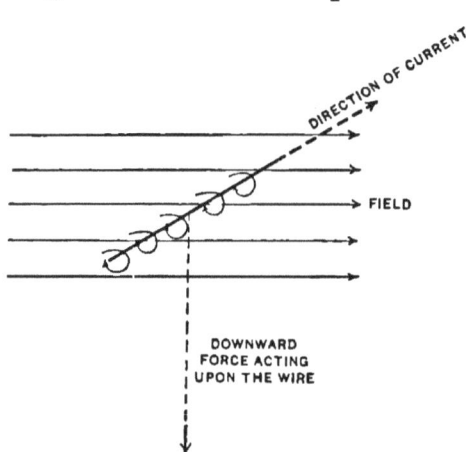

FIG. 288.

This force may be conveniently considered to be that with which the lines of force of the magnetic field, and those due to the current within the wire, repel each other upon that side of the wire where they are parallel, and in the same direction. Let *mm* and *cc* (Fig. 289) be portions of lines of force due to the magnet and to the current within the wire.

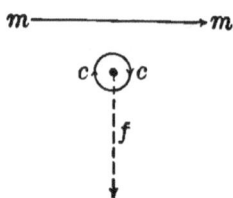

FIG. 289.

These repel each other, and the direction of the force, which may be considered as acting upon the wire, will be that of the arrow f shown in the figure. By means of the conception of lines of force and their mutual action, we can always ascertain what the forces acting upon a current in the field will be, and even their size and direction. The existence of the forces acting upon electric circuits in the magnetic field may be shown in a great variety of ways. One of the simplest devices is described below:

295. EXPERIMENT 94. — Attractive and Repellent Forces between a Magnet and a Wire carrying Current; also between Two Circuits.

Apparatus:

(1) The arrangement shown in Fig. 290. This consists of two parallel wires, each bent into the rectangular form shown in the figure.

These are hung side by side about 2 cm. apart from a wooden block by means of strips of thin copper foil.

(2) The bichromate battery.

(3) A horseshoe magnet.

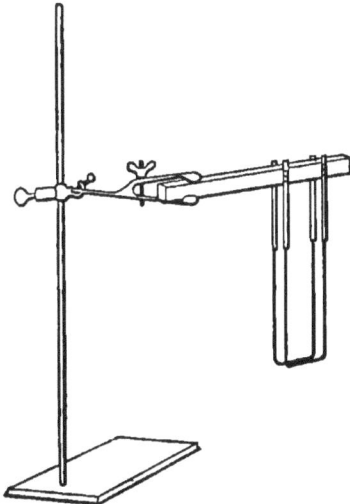

FIG. 290.   FIG. 291.

Procedure :

(a) Send current from the battery through one of the wires. Hold the magnet near, as shown in Fig. 291, and note the attraction or repulsion of the wire according to the position of the magnet poles. Trace out the course of the current, and the relations of its lines of force to those of the field of the magnet. Show that the behavior of the wire follows the principle laid down in Art. 294.

(b) Send current through both wires, and note that the wires attract each other when the currents flow in the same direction, and repel each other (Fig. 292) when they flow in opposite directions.·

The French physicist, Ampère, has stated this mutual action of electric currents as follows :

(1) *Parallel conductors carrying current in the same direction attract each other.* (2) *Parallel conductors carrying current in opposite directions repel each other.*

Fig. 292.

Fig. 293.

The effect, from the point of view of the mutual action of lines of force, is indicated in Fig. 310. Ampère pointed out likewise that two conductors carrying current which make an angle with one another will, if free to move, arrange themselves parallel to one another, and with the current flowing through them in the same direction.

296. Dynamos and Motors. — Upon the principles developed in the foregoing experiments of this chapter, two very important forms of electrical apparatus are constructed; these are the dynamo and the motor. The former is a machine for the production of electric currents

ELECTROMAGNETIC INDUCTION

by the motion of conductors across the lines of force in
the magnetic field. The latter is a machine for the
development of *power* by the movement of conductors
under the action of the magnetic field.

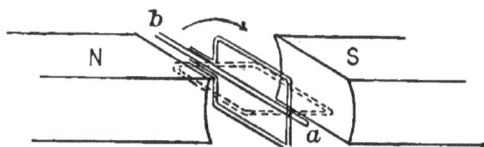

FIG. 294.

To obtain the principle of the dynamo in its simplest
form consider a loop of wire which moves in the field as
shown in Fig. 294. This is free to revolve upon *ab* as
an axis, between the poles of an electromagnet. If the
wire be made to revolve in the direction of the arrow,
it will cut the lines of force of the field at first very
slowly, and then more and more rapidly. When it has
reached the position shown in the dotted line, where
its motion is perpendicular to the lines of force, it cuts
them at the highest rate. After passing this position, it cuts fewer and fewer lines, until it reaches the
position 180° away from its starting point, where it
again moves parallel to the lines and cuts none. The
result of this motion is a surge of current through the
wire, which rises to a maximum and dies away again.
During the remainder of its revolution the wire will cut
lines of force again; at first at an increasing rate, then
more and more slowly. But these will be cut in such a
way as to send the current in the opposite direction.
Such a revolving wire, with arrangements for carrying
currents into an outer circuit, would be an alternating
current dynamo of the simplest type. To change this
alternating current into a direct current, that is to say,

into a current which is always flowing in the same direction, a device called a *commutator* is employed. This consists of two sliding contacts of metal or of carbon, called *brushes*, which rest upon a divided collar attached to the axle (Figs. 295 and 296). By this means one

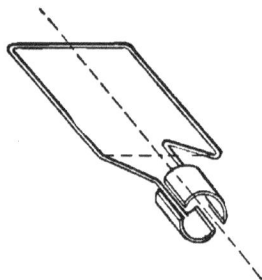

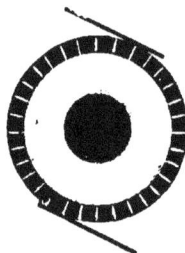

FIG. 295.

FIG. 296.

terminal of the outer circuit is connected half the time with one end of the loop of wire, the other half of the time with the other, so that the current always flows through the outer circuit in the same direction. In actual dynamo machines there is a great number of these loops of wire, which follow one another through the field continually. They are so connected together and to the commutator that each furnishes its share of current to the outer circuit in succession. This revolving combination of loops or coils is called the *armature* of the dynamo. There is no space in the present work to describe the many types of machines which are based upon the above-mentioned principles.

The electric motor is in principle of construction precisely similar to the dynamo. The distinction is that in this machine current from some outside source is sent through the wires of the armature. These then are acted upon by the lines of force of the field as described in Art. 294. The machine is so arranged that all the forces acting upon the different wires in the armature work together. The armature is thus given a rapid motion of rotation, and is capable of doing work.

297. EXPERIMENT 95. — The Production of Induced Currents by the Starting or Stopping of Currents in a Neighboring Circuit.

Apparatus:

(1) The coil of wire wound upon an Argand chimney, previously described.

(2) A coil of smaller diameter and of about the same length, wound upon a glass tube of about 2 cm. in diameter.

(3) A bundle of iron wire.

(4) The galvanometer previously described.

(5) The bichromate battery.

Procedure:

(*a*) Connect the terminals of the galvanometer with those of the larger coil, and the ends of the smaller coil with the terminals of the battery, leaving the latter circuit open. (See Fig. 297.)

(*b*) Close the battery circuit suddenly and notice the effect upon the galvanometer needle. It will be seen that the latter is suddenly deflected, but not permanently so, and that it comes to rest in its old position. Break the battery circuit again, and note that an equal and opposite deflection of the galvanometer needle is produced.

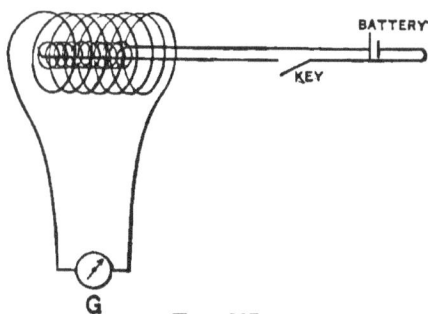

FIG. 297.

(*c*) Determine by the use of a single cell of battery the direction of current in the outer coil indicated by these two deflections, and compare this direction with that of the current due to the battery within the inner coil. It will be found that *the direction of the induced current when the circuit is closed is opposite to that of the inducing current; and that when the circuit is broken the induced current has the same direction as that which has been destroyed.* The creation of a current flowing in the inner coil is similar in its effect to the very rapid movements of a magnet, the lines of force of which correspond to those of this current, from an infinite distance to its position within the coil. The opening of circuit may be similarly compared to the sudden withdrawal of this magnet to a great distance.

(*d*) To show that this form of electromagnetic induction depends upon the action of lines of force, and increases with the number of these, the bundle of wires may be introduced into the inner coil. These act, on account of their high permeability, to enable the production of a greatly increased number of lines of force.

When the circuit is closed, all of these lines of force may be considered as having been made to cut the wires of the outer coil. The result is a greatly increased deflection. When the circuit is open again it is as though these numerous lines of force were drawn in again; cutting the coils in the opposite direction, and likewise producing a large deflection.

298. Induction Coils. — The foregoing experiment indicates the principle upon which the interesting piece of apparatus called the *induction coil* is constructed. This consists of an iron core *a* (Fig. 298) surrounded by a coil

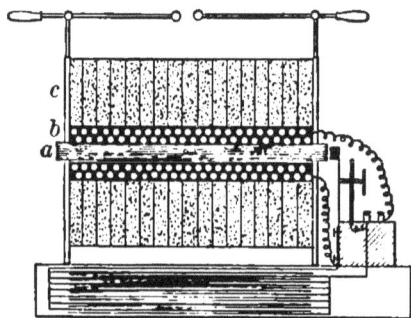

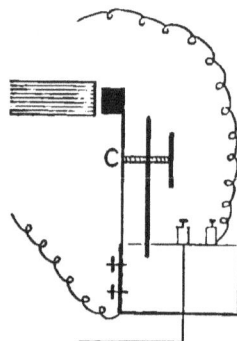

FIG. 298.   FIG. 299.

of wire *b* consisting generally of a few turns of wire. This is called the *primary coil*. It in turn is surrounded by a large coil *c* consisting of very many turns of very fine wire. The ends of the latter, which is called the *secondary coil*, are attached to a pair of terminals similar to those used upon the Holtz machine. In order to break the circuit through the primary coil, a device called an *interrupter* is employed. This is frequently given the form shown in Fig. 299. It consists of a metal spring, to one end of which is attached a disk of iron. The spring is mounted in such a position that the disk is opposite the end of the iron core of the coil. The current from the battery flows through this spring, and it makes its exit by the way of a metal screw which touches the spring at the point *C*. Thence the current passes through the primary coil back to the battery. The current through the coil magnetizes the core, and the disk at the end of the spring is attracted. Its movement toward the core draws the spring away from the screw at *C* and breaks the circuit at that point. The current then ceases to flow through the primary coil, and the core, which is of soft iron, loses its magnetism. The spring of the interrupter being no longer attracted flies back to its original position, comes into contact with the screw at *C*, and the circuit is closed again. This process goes on indefinitely so long as current is allowed to flow, interrupting the current through the primary coil many times a second.

Each time the circuit closes, a magnetic field is formed which embraces both the primary and the secondary

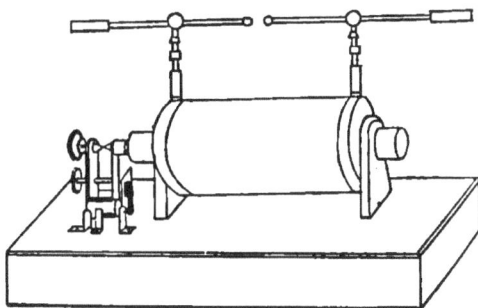

Fig. 300.

ary coils, and in each of the numerous turns of the secondary coil electromagnetic induction takes place. The number of these turns is very large, so that the difference of potential between the terminals is enormous. The result is that a spark leaps across between the balls of the discharges at each vibration of the interrupter. The sparks thus produced are similar in every respect to those produced by influence machines or by any electrostatic device. Induction coils have been constructed, giving sparks as much as a meter long. Such instruments have secondary coils containing hundreds of miles of fine wire. Induction coils giving a spark of from 1 to 20 cm. are very common. Figure 300 shows the usual form of induction coil.

299. The Telephone. — Another important instrument which depends for its action upon electromagnetic induction is the telephone. This in its simplest form consists of a permanent bar magnet (Fig. 301), around one end of which is a coil of fine wire. Opposite the end of the bar is a diaphragm of thin sheet iron against which the sound waves of the speaker's voice fall. These throw the diaphragm into vibration, thus causing periodic fluctuations in the magnetic field of the bar corresponding to the pitch of the voice. The variations in the field induce fluctuating currents in the coil of wire.

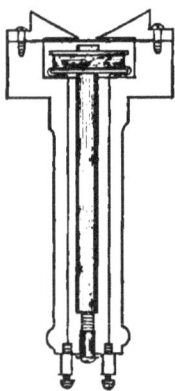

FIG. 301.

If the wires are connected to a similar instrument at a distant station (Fig. 302), the field of this is made to vary by the fluctuating currents, and the diaphragm of the latter

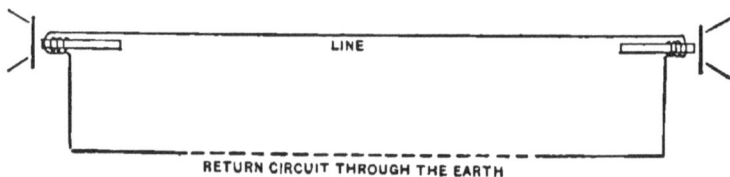

LINE

RETURN CIRCUIT THROUGH THE EARTH

FIG. 302.

instrument vibrates in unison with that of the former. Sound waves thus produced convey speech to the ear of the listener.

These telephonic currents are exceedingly weak. In order to increase them, the device known as the *carbon transmitter* is used. This consists of a diaphragm, a stylus attached to the middle of which presses against a mass of powdered carbon compressed into a disk (Fig. 303). The disk forms part of a voltaic circuit as shown in the figure. The vibrations of the disk under the influence of sound waves change the pressure of the stylus upon the carbon and, thereby, the resistance of the latter.

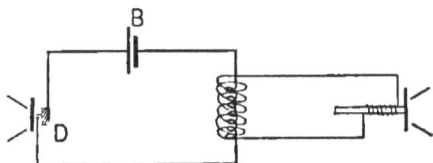

FIG. 303.

The current from the battery *B* which flows in the circuit fluctuates in response to these vibrations. It flows through the primary circuit of the induction coil *C* and induces corresponding currents in the secondary circuit. The latter is connected with the receiving telephone at the distant station, the diaphragm of which thus vibrates much more powerfully than if it were of an ordinary telephonic circuit.

PART IV — SOUND

CHAPTER XXXIV

THE PROPAGATION OF SOUND

300. Sound Waves. — Sound is conveyed to the ear by means of a wave motion of the air. The energy of sound waves is generally small; although in the case of the sounds produced by explosions — by the firing of large guns, etc. — decided mechanical effects are produced. That there is a mechanical effect produced at considerable distances, whenever even a feeble sound is made, may be shown by means of the following experiment:

301. EXPERIMENT 96. — **The Sensitive Flame.**

Apparatus:

(1) A piece of rubber tubing such as is used to convey gas to Bunsen burners, etc.

(2) Several pieces of glass tubing of such size as to fit the rubber tubing. These pieces should each be about 20 cm. long.

Procedure:

(*a*) Hold each of these glass tubes with its middle in the Bunsen burner until the glass softens, and draw it out until the diameter in

FIG. 304.

the contracted portion is reduced to about 2 or 3 mm. (See Fig. 304.) After having thus drawn out each of the tubes, cut them in two in the middle by scratching with a file and bending.

334

(*b*) Insert the large end of one of the tubes thus prepared into the free end of the rubber hose connected with the gas jet. Turn on the gas and ignite it at the contracted end of the glass tube. If the pressure of gas is sufficient for the performance of this experiment, the flame thus produced will burn with a slight roaring noise, and show a perturbed flame somewhat similar to that indicated in Fig. 305 (*a*).

1. If the flame in question, instead of taking this form, burns quietly, assuming the form shown in (*b*), the glass tube must be discarded, and another of the set just made must be tried.[1] When a tube has been found the flame from which roars in the manner just described, turn off the gas gradually until the roaring ceases. The flame will then change over into the form depicted in Fig. 305 *b*. A slight increase in the supply of gas will restore it to the perturbed form. There is an adjustment between these two conditions, such that the flame is very easily changed from the quiet to the roaring form. Such a form is called a *sensitive flame*. When this adjustment has been made it will be found that the flame changes form in

FIG. 305.

response to nearly every sound which is produced in the room. It is especially sensitive to shrill or sibilant tones. With this apparatus one can show that a mechanical disturbance sufficient to transform the gas flame from the quiet to the roaring state into which it tends to pass is almost instantaneously produced as the result of a noise in the most remote corner of the laboratory. If, for example, two wooden blocks be struck together in the further corner of the room, the flame responds. The jingling of a bunch of keys will produce the same result.

[1] In some localities, especially if the laboratory is situated at a lower level than the gas-works, the pressure is insufficient to produce a sensitive flame. Under such circumstances the experiment may be performed by filling an ordinary gas holder from which to supply the flame. The pressure may then be adjusted at will.

A shrill whistle made by blowing across the open end of a tube is also effective, and if the flame be properly adjusted the utterance of any word with a syllable containing a sibilant with moderate loudness, or even when whispered, will suffice.

Note the great promptness with which this effect follows the exciting cause. A moment's consideration will show that no air draft could traverse the intervening space with sufficient velocity to account for this promptness of response.

2. To compare the time required for a draft of air to agitate the flame, with that needed by the sound wave, turn the flame down until it is no longer sensitive, then step back for a distance of 2 or 3 m., face the flame and blow towards it as forcibly as possible. Use a short tube which may be made by rolling up a sheet of note paper or cardboard to direct the blast. It will be possible in this way to make the flame flicker; but the interval which elapses between the puffing sound when the air passes through the tube until the flame shows signs of agitation is very appreciable, whereas the response of the sensitive flame to sound waves coming from several times the distance in question is too prompt to admit of direct observation.

302. Tyndall's Experiment. — Professor Tyndall in his lectures on sound used the following apparatus for showing that it is a true wave and not a puff of air which effects the sensitive flame. Tyndall's apparatus, which is depicted in Fig. 306, con-

FIG. 306.

sists of a long tube contracted to a nozzle at one end. This is placed in a horizontal position, and a lighted candle is mounted with the flame directly in front of the nozzle. Upon clapping two wooden blocks together at the other end of the long tube, the candle flame responds by a sudden movement. To show that this is the result of a sound wave traveling through the tube, and not of a draft of air blown out from between the blocks and traveling up the flame, a portion of the air within the tube is filled with smoke. This may be done by burning within the tube a bit of paper previously soaked in a solution of sodium nitrate, or potassium nitrate, and dried. If the smoky air thus produced be observed when the blocks at the open end of the tube are brought together, it will be possible to follow its movement. It will be seen that while it is agitated for an instant by the passage of the wave, it does not travel through the tube and out, but is simply disturbed and then comes to rest again. There is some difficulty in performing this experiment, owing to the fact that the tube is apt to become a channel for feeding the flame with air, so that a constant draft moving toward the nozzle is set up and the smoke is carried along with this air current. There is little danger, however, of confusing this slow movement of the smoke toward the flame with the wave which causes the flame to jump when the blocks are clapped together.

A sound wave in air consists of a rapid to and fro motion (a longitudinal vibration) of the particles of the gas, in the direction in which the wave travels. If we could secure a picture of the condition of the air in a narrow region through which waves were moving, we should find that, instead of being everywhere equally dense, there are successive layers of great density (*d*, *d*, Fig. 307), with intervening layers (*r*) where the air is rarefied. The vibration of the air is momenta-rily in the directions of the small arrows, thus producing condensation at *d*, *d*, and rarefaction at *r*. These regions of condensation and rarefaction all move simultaneously in the direction of the large

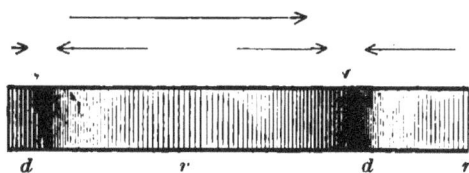
Fig. 307.

z

arrow, without changing their positions relative to one another.

303. Sound Waves produced by a Projectile. — Projectiles moving at high speeds produce air waves which are similar to those by means of which sound is transmitted and which travel at the same velocity. Instantaneous photographs of such projectiles show these. The diagram in Fig. 308 is from a photograph of a rifle bullet in flight, taken by Mr. C. V. Boys in England. The oblique lines, *a*, *a*, *b*, *b*, show the positions of the waves.

FIG. 308.

304. The Velocity of Sound. — It has been seen in the experiment of the sensitive flame that sound waves travel with a high velocity. It is only necessary to recall the following observation, which every one will have had frequent opportunity to make for himself, to show that this velocity is very much less than the velocity of light. If we watch a distant railway train when the whistle is sounded for a crossing or station, we may notice that the puff of steam is followed only after a very considerable interval, depending on our distance from the engine, by the sound of the whistle. The sight of the puff of steam has been conveyed to the eye by means of a wave of light; the sound, however, which occurs simultaneously with the setting free of the steam, by a sound wave. The interval of time which seems to elapse between the two is therefore that which the sound wave requires to traverse the space between the whistle and the ear of the observer, over that required by the wave of light to traverse the same distance. Light waves travel at such very great velocity (330,000,000 m. per second), that we may neglect the time occupied by the light altogether. By stationing one's self at a measured distance from the point at which the railway train whistles, one can therefore obtain a very good estimate of the velocity of sound. This velocity, as has been determined many times, is about 332 m. per second when the air is at 0° C. With rise of temperature the velocity rises. At 20° C., for example, it is about 343 m.

305. EXPERIMENT 97. — **Sounds will not travel in a Vacuum.**

Apparatus:

(1) An air pump.

(2) A receiver with an open neck, as shown in Fig. 309.

(3) An electric bell.

(4) A Leclanché cell, or other battery used for open circuits; also a key.

Procedure:

(a) Find a cork which will fit the neck of the receiver. Bore two holes through the cork and insert copper wires, cementing them into place. To the free ends of these wires, a

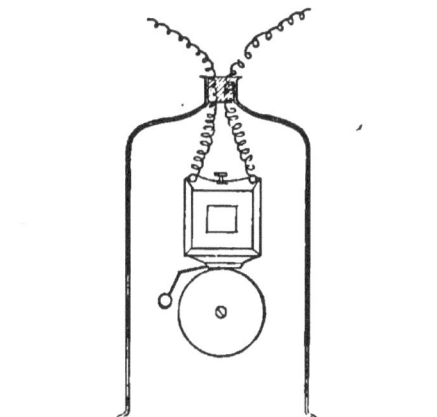

FIG. 309.

short distance within the receiver, hang the electric bell, suspending the same from its binding posts as shown in the figure. Cover the entire outer surface of the cork with beeswax and rosin cement, so as to preclude leakage.

(*b*) Attach the battery, with the key inserted, to the outer ends of the two wires which have been used in the suspension of the bell. Place the bell jar upon the plate of the air pump. Close the key and note the sound of the bell, which will be distinctly heard, although somewhat muffled by being inclosed within the heavy bell jar.

(*c*) Exhaust the air from the receiver and note the dying away of the sound of the bell.

If the air pump be in good condition, and the sealing of the neck of the receiver be really tight, the exhaustion can be carried to such a point that it is very difficult to detect the faintest sound from the bell. A certain small amount of sound will always find its way out through the body of the conducting wires. These are solids, and are capable of transmitting sound waves. Upon restoring the air to the receiver, the sound of the bell will return to its initial loudness. From such experiments as this we conclude that sound is conducted to the ear by means of waves which are transmitted through the air or through other material substances.

306. Reflection of Sound. — Sound waves, like other waves, move in straight lines and are capable of reflection. The echo is an example of reflected sound. The reflector is usually the wall of a house or a cliff. Generally speaking, we get reflection only from the surfaces of large bodies. The dimensions of the reflector must be large as compared with those of the wave. A floating block upon water, for instance, rides upon large waves without diverting them from their course, but it reflects small ripples perfectly. The laws of reflection are much more easily studied in the case of light. They will be discussed in Chapter XXXIX.

CHAPTER XXXV

VIBRATING BODIES; PITCH AND TIMBRE

307. Source of Sound Waves. — Sound is produced by the periodic vibration of bodies. To show that sounding bodies are actually in a state of movement, it is only necessary to mount a ball of wood or cork, as shown in Fig. 310, and to place the same in contact with one prong of a tuning fork, or with the lip of a bell. If we then cause the tuning fork or bell to resound by drawing a violin bow across it, the suspended ball will be driven violently from its position by the movement of the vibrating body.

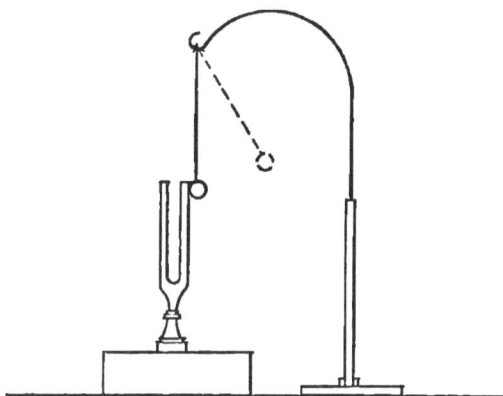

FIG. 310.

308. Classification of the Vibrating Bodies commonly used in Musical Instruments. — The sounds employed in music are commonly produced by the vibration of one of the following classes of bodies:

(1) *Vibrating plates.* The most important instruments which depend upon this class of vibrations are drums, gongs, cymbals, etc.

(2) *Vibrating rods.* This class gives us the tuning fork, the triangle, the reed, etc.

(3) *Vibrating strings.* This is the most important class of all; to it belong the stringed instruments, such as the violin, harp, guitar, zither, piano.

(4) *Vibrating columns* of air. All wind instruments, including the human voice, belong to this class.

309. Musical Pitch. — That property of sound which depends upon the rapidity of vibration is called *pitch*. Pitch is stated in terms of the number of vibrations per second. To be audible, the pitch of a vibrating body must lie between certain limits. If less than about twenty-four

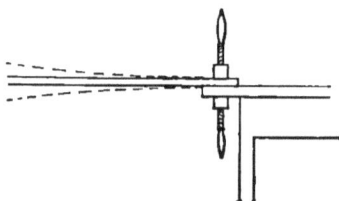
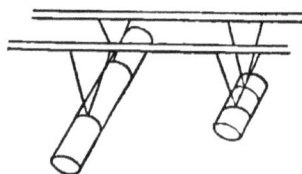

FIG. 311.   FIG. 312.

single vibrations per second take place, no sound is heard. If a lath be clamped to the table, as in Fig. 311, it may be made to vibrate in precisely the same manner as a sounding body; but it is silent. The waves which it sends forth do not affect the ear. We say that its pitch is below the audible limit. If the tip of a pencil be brought lightly into contact with the free end of the rod, it will be found possible to count the separate vibrations.

The upper limit of audibility is tested by means of small steel cylinders mounted as in Fig. 312. These, tapped with a mallet, give very rapid longitudinal vibrations. The power of hearing the tones emitted by these bars, the pitch of which rises as the bars grow shorter, varies greatly. Some persons can hear sounds of more than 32,000 vibrations; in most cases, audibility ceases between 15,000 and 25,000 vibrations.

To illustrate this point, procure a bar of tool steel, and have two cylinders turned from it, each 2 cm. in diameter. They should be 12·0 cm. and 6·10 cm. in length. They should be hardened and mounted as in Fig. 312. When tapped on the end with a wooden mallet, the longer cylinder will emit a high but audible tone. Very few persons will be able to hear that uttered by the shorter bar.

310. Musical Tones and Noises. — All bodies vibrate at a period which is natural to themselves, and depends upon their dimensions, the materials of which they are made, and their structure. Most bodies, however, when thrown into motion vibrate not as a whole, but independently, in various parts. The rates of vibration of these various parts are not necessarily related to one another in any simple manner. If they are simply related to one another in a manner which will be discussed later, the result is what we call a *musical tone*. If the vibrations of the various parts are not thus simply related, we get what we term a *noise*. The only distinction between noises and musical tones is that the former are of a complexity which the ear is unable to analyze. This difference may be illustrated by means of the sound emitted by a vibrating plate, a body which lies upon the borderland between noise and music. The method described in the following experiment is due to Chladni (1756–1827), a celebrated acoustician.

311. EXPERIMENT 98. — The Vibration of Plates; Chladni's Figures.

Apparatus:

(1) A square plate of sheet brass, which should be as nearly per-

fectly flat as possible; it should be about 30 cm. square and 0·3 cm. in thickness. Also a circular brass plate of the same diameter.

(2) A large clamp. It is best to have one of the cast-iron clamps which are made purposely for this experiment. Any clamp of size sufficient to reach from the center of the plates to the edge will do if provided with two metal buttons, or disks, to be placed between the surface of the plate and the clamp, at the center of the former.

(3) An ordinary violin or bass-viol bow.

(4) A pepper box containing dry sand.

Procedure:

(*a*) Clamp the square plate by its center. Draw the bow across its edge, and note the character of the sounds which it may be made to emit. It will be found that, while they are not altogether devoid of musical character, they are, for the most part, harsh and unpleasant. As a musical instrument, such a plate would be held distinctly inferior to any stringed instrument.

(*b*) The vibrating plate, thus set in motion, resolves itself into a considerable number of vibrating parts. The existence of these may be demonstrated as follows:

Strew the plate with sand and bow it, in the meantime pressing the tip of the left forefinger against one edge. Note the arrangement of the sand particles along symmetrically situated lines. Make free-hand sketches of the patterns thus produced. Figure 313 shows one of the numerous patterns which may be obtained. By changing the position of the bow, and placing the finger at various points upon the edge of the vibrating plate, the figures produced can be varied almost indefinitely. Chladni, in his work on "Acoustics," depicted several hundred of them. Figure 314 is reduced from one of his plates.

FIG. 313.

The lines along which the sand settles when the plate is thrown into vibration are lines of no motion; they divide the plate into vibrating parts. These lines are called *nodal lines*. The vibrating parts of the body are called *segments*.

(*c*) Repeat, using the circular plate. Note that the nodal lines upon the circular plate are equidistant radii. The number of these determines the pitch of the sound which the plate emits. In order to hold these lines in a fixed position upon such a plate, it is necessary to check the vibration at some point upon the periphery with the finger, and the plate then divides itself into an even number of vibrating segments, of which the usual number is four.

When the plate is thus divided into four parts, it utters its so-called fundamental tones. When more segments are produced, tones of higher pitch are the result.

312. Vibration of Bells. — The case of the circular plate is of interest, because its mode of vibration is identical with that of the bell. Bells, whether struck with the clapper or set into vibration by means of the bow, vibrate thus in segments separated from each other by nodal lines running from the apex to the lip of the bell. When the bell is in vibration the lip, which is circular in form when at rest, becomes distorted into the form of an ellipse. (See Fig. 315.) This ellipse changes with each vibration. That which was its major axis at one instant becomes, an instant later, the minor axis. This oscillation takes place without any movement at the nodal lines which pass through the point

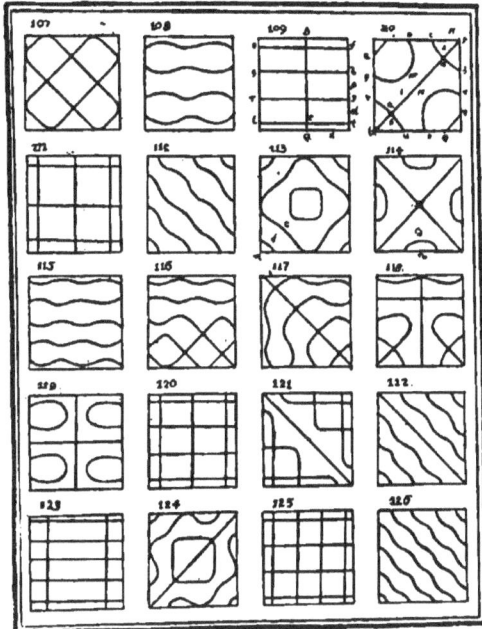

FIG. 314.

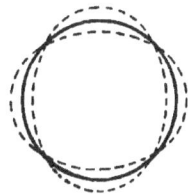

FIG. 315.

where the circle and the two ellipses in the figure cut each other.

The vibration of the bell can be illustrated by nearly filling a light glass goblet with water. If the finger be placed upon the edge and the glass be thrown into vibration with the bow, there will be four regions in which the water will remain at rest. These correspond in position to the nodal lines. Midway between these nodes, the water will be thrown into violent agitation and its surface will be broken up into a series of fine, rapidly moving ripples. (See Fig. 316.)

FIG. 316.

It was noted in the course of Experiment 97 that the sounds produced by the vibration of the square plate were decidedly harsh and unpleasant. The reason for this lies in the fact that the plate breaks up into so many dissimilar segments, all of which vibrate simultaneously, producing sounds which do not necessarily harmonize. The vibrations of bells are simpler, and they therefore fulfill more nearly the conditions necessary to the production of a musical tone.

313. Tone-Color or Timbre. — Tone color is the property which makes it possible to distinguish between sounds, even when they have the same pitch. If, for example, a cornet, a violin, and a human voice are sounding the same note of the scale, we have no difficulty in telling them apart. Each has its unmistakable character, due to the fact that it is really a composite tone produced by several independent vibrations. Such tones differ from musical chords only in the fact that one of the vibrations of which they are composed is very much louder than the others.

It is called the *fundamental*, while the others are called *overtones*. They are, as a rule, scarcely to be recognized individually; but they give the tone its color. The overtones are usually due to the vibration of smaller segments of the same body which produces the fundamental.

A good example of color tone is afforded by the vowel sounds *u* (*oo*), *o*, *ä*, *ā*, *e*. If these be sung at the same pitch their fundamentals are identical, but each possesses its own characteristic set of overtones.

CHAPTER XXXVI

EXPERIMENTS WITH TUNING FORKS; THE MEASUREMENT OF PITCH

314. The Tuning Fork. — This instrument consists of a stiff rectangular rod of steel bent upon itself, as shown in Fig. 317. To the center is attached a handle, or shank, which is frequently clamped to a sounding box. The sound of the tuning fork is caused by the simultaneous vibrations of the two prongs. This motion, which is too rapid to manifest itself directly to the unaided eye, can be made easily visible in the following manner:

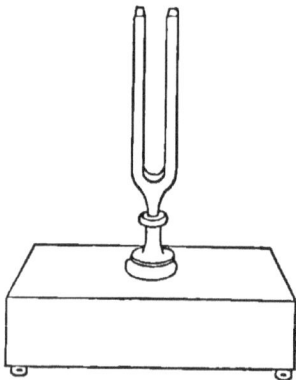

315. EXPERIMENT 99. — Observation of the Motion of a Tuning Fork by Interrupted Vision.

FIG. 317.

Apparatus:

(1) A tuning fork (the larger the better) and a bow.

(2) A disk of cardboard, or of sheet metal, about 40 cm. in diameter. It should contain four equidistant radial slots, Fig. 318, and should be mounted to revolve upon a horizontal axis at high speed. A convenient form for this apparatus is shown in the figure. It consists simply of an ordinary " whirling table " clamped in a vertical position.

The four open sectors, or slots, should be about 1 cm. wide at the periphery of the disk.

Procedure:

Set the tuning fork up in a position where it will have a bright background, or, if practicable, the sky for a background. In front of it, at a distance of about 50 cm., mount the revolving wheel. Unless the wheel is to be driven automatically, the experiment should be performed by two persons. The tuning fork is to be kept in active vibration by one of the operators; the other looks through one of the open sectors of the disk at the tuning fork, then starts the disk into motion. When the disk has reached a speed such that the sensation of flickering has vanished, the tuning fork will be seen as if by continuous vision.

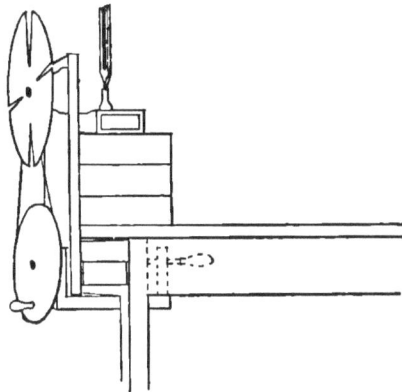

FIG. 318.

At a certain speed of the disk, however, the tuning fork will appear to vibrate very slowly. By careful adjustment of the speed it may even be made to come to rest. It is not possible to drive the disk with such regularity as to maintain this condition of affairs, but it is easy to keep the apparent motion of the tuning fork down to a rate such as to enable one to examine the character of the motion of the prongs.

The phenomena observed in this experiment depend upon persistence of vision. The observer obtains a succession of views of the tuning fork; each so short that the fork does not move appreciably before the light is cut off. Each successive, instantaneous view shows the fork in a position slightly different from the preceding one. This succession of views overlaps upon the retina and is blended by the mechanism of vision into a continuous impression. The impression produced is that of a fork vibrating with a very slow motion. Note that the amplitude of vibration of the tuning fork is very considerable; the prongs being one instant spread wide apart and at the other end of their excursion being bent strongly together as in Fig. 319.

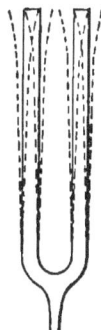

FIG. 319.

316. Experiment 100. — Tuning Fork Tracings.

Apparatus:

(1) A wooden track about 1 m. long, consisting of two parallel guides between which a block slides smoothly.

(2) A wooden support (*S*, Fig. 320) consisting of an upright piece,

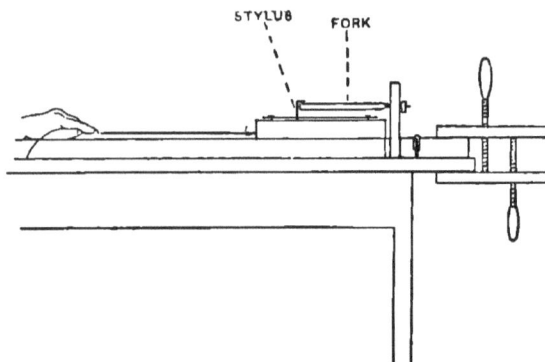

FIG. 320.

20 cm. square and 2 cm. in thickness. This is screwed to a base consisting of two blocks hinged together as shown in the figure.

(3) Two large tuning forks and a bow.

(4) Several pieces of glass (cut from window glass) about 5 cm. wide and 20 cm. long.

Procedure:

(*a*) To one end of a laboratory table clamp the support *S*. About 12 cm. above the table bore a hole through the upright to fit the shank of the tuning forks. Clamp one of the forks to the support by slipping the shank through the hole and bolting it into place with the nut used to fasten the fork to its sounding box.

(*b*) From a bit of thin sheet metal, ferrotype iron, or copper foil, cut a slender, pointed strip about 4 cm. long. Bend this strip at right angles as in Fig. 321, and fasten it with wax to one prong of the fork.

FIG. 321.

(*c*) Smoke one of the pieces of glass over a candle flame, and fasten it, smoked side uppermost, to the sliding block by means of thumb tacks. Place the block between the guides of the wooden track, and adjust the latter parallel to the prongs of the fork, and at such a height below it that when the block is at one end of the track the point of the bent strip attached to the prong will touch the smoked glass.

(*d*) Bring the fork into strong vibration by lowering it; then draw the block from under it by means of a string previously attached to the same. The motion should be as steady as possible, with a velocity of about 1 m. per second. Note that the tracing is sinusoidal, like that made by a pendulum, whence it follows that the vibration of a tuning fork is a *simple harmonic motion.* Figure 322 is from such a tracing.

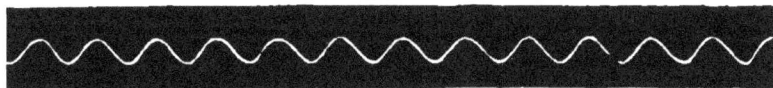

FIG. 322.

(*e*) Repeat the experiment with two forks. The second fork should be mounted by means of another hole in the support, so placed that the two forks will be side by side and about 2 cm. apart. The stylus on each should be on the inner prong. Compare the two tracings obtained, and compute from them the relative pitch of the forks.

Compare the ratio obtained from your measurements with that computed from the rates marked upon the forks by the maker.

317. Relative and Absolute Pitch. — The foregoing experiment illustrates a method of measuring *relative* pitch of tuning forks, *i.e.* of obtaining the pitch of one fork as compared with that of another. If the apparatus were so modified as to permit measurement of the velocity of the plate, the *absolute* pitch could be determined. To get the absolute pitch one must know the number of vibrations in a given space of time. If the distance traversed by the plate in $\frac{1}{10}$ second, for example, were known, the absolute pitch would be determined by multiplying the number of undulations traced within that distance by ten.

318. The Method of Beats. — When two forks or other instruments not quite in unison are sounded together, a peculiar throbbing or pulsating effect may be noticed.

These pulsations are called *beats*. The more closely the forks agree, the slower are the beats. When complete unison is attained they disappear. With increasing difference of pitch, on the other hand, the rapidity of the beats increases until they come too fast to be distinguished. The result then is discord.

Beats are due to the fact that the sound waves from the two instruments combine; at one moment to reinforce, and at the next to annul each other. If, for example, a tuning fork vibrates 255 times a second, and another 256 times a second, the waves from them will reach the ear in the same phase once every second. At those instants the sounds reinforce each other and give us *beats*. The method of beats consists in determining when two instruments come into complete unison by the slowing and disappearance of the beats.

319. The Graphical Representation of Beats. — The nature of beats may be shown graphically by an experiment simi-

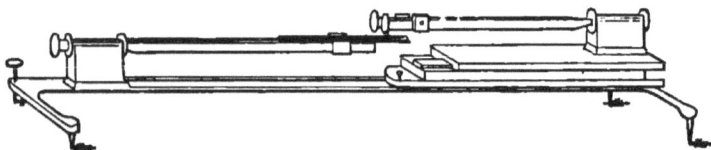

FIG. 323.

lar to that described in Art. 316. The smoked plate is clamped upon one of the forks, as shown in Fig. 323. Both forks are set in motion, and the tracing is made by moving one of them. If the forks are nearly in unison,

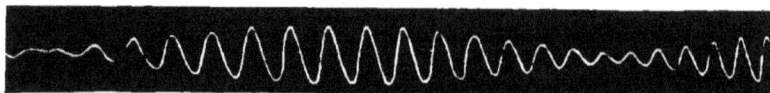

FIG. 324.

the tracing will present an appearance like that shown in
Fig. 324. The experiment should only be tried with a
fork especially constructed to carry the smoked plate.
The latter must be very securely clamped, and the other
prong must carry a counterbalancing weight.

320. **The Optical Study of Vibrations.** (*Lissajous's Method.*)
— If a mirror be attached to a tuning fork, and a beam of
light from a lantern be reflected thus to a screen (Fig.
325), an image of a small aperture may be focussed upon

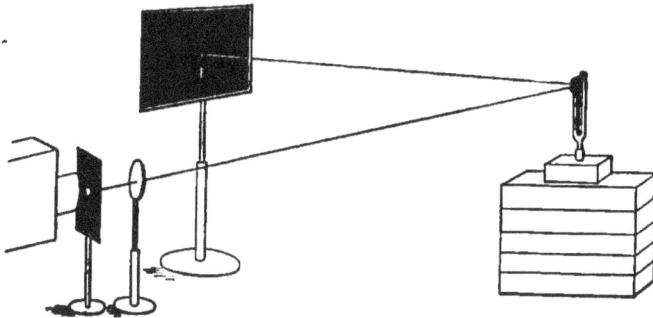

FIG. 325.

the screen. When the fork vibrates, this image will be drawn
out into a line or band. When viewed in a revolving
mirror it may be resolved into a sinusoidal curve. With

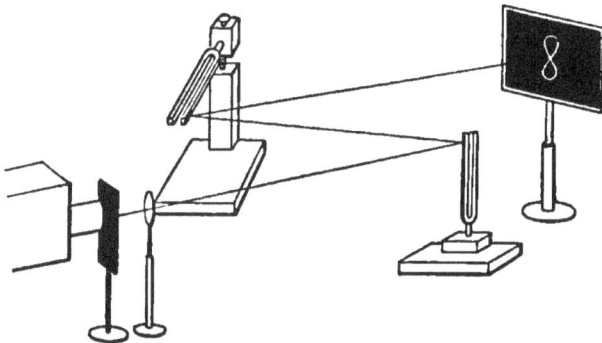

FIG. 326.

2 ▲

two forks, each provided with a mirror arranged as in
Fig. 326, movements which are characteristic of the combined vibrations may be obtained.   With parallel vibrations the foregoing experiment upon beats may be obtained.
When the vibrations are perpendicular to one another,
certain figures are obtained, the character of which depends
upon the ratio of pitch of the forks and the difference
of phase.   These are known as *Lissajous's figures*.   Figure
327 shows some of the most important.   Lissajous's figures

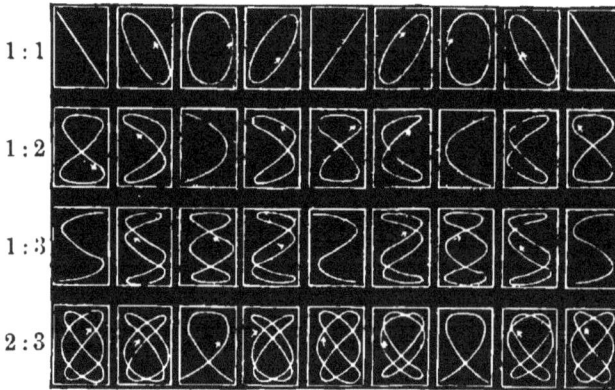

Fig. 327.

are used in the comparison and adjustment of tuning forks.
They afford one of the most delicate methods for the determination of relative pitch.

CHAPTER XXXVII

THE VIBRATION OF STRINGS

321. Modes of Vibration. — A vibrating string is always fixed at both ends, since in no other way can it be given the necessary tension. It vibrates either in a single segment, or vibrating part, or is broken up into smaller segments. These, however numerous they may be, are always simply related as to their length and rate of vibration to the whole string; and it is owing to this simplicity that the pleasant effects of vibrating strings are due. When the string is vibrating as a whole, with nodes only at the ends, it utters its fundamental tone, which is the lowest tone in pitch the string can be made to produce. If the center of the string be constrained by the finger, or by means of any stop, the string will vibrate in two segments, each of which is half the length of the entire string, and each of which gives forth a note (the octave), the vibration period of which is half of that of the fundamental.

By placing a stop at the distance of $\frac{1}{3}$ the length of the string, or at $\frac{1}{4}$, $\frac{1}{5}$, $\frac{1}{6}$, $\frac{1}{7}$, $\frac{1}{8}$, etc., of that length, the string may be broken up into the corresponding number, *i.e.* 3, 4, 5, 6, 7, or 8 vibrating segments. The pitch will rise in inverse proportion to the lengths of the segments. That the vibrating string, when thus restrained at a single point, is in fact broken up throughout its entire length into vibrating segments with nodes separating them which are at rest, can be shown as follows:

322. Experiment 101.—Location of the Nodes in a Vibrating String.

Apparatus:

(1) A bow and rosin ; a sheet of paper. (Paper which is colored upon one side is to be preferred.)

(2) A sonometer. The sonometer is a hollow box of thin spruce or other wood, about 1 m. long and 30 cm. wide. (See Fig. 328.) This

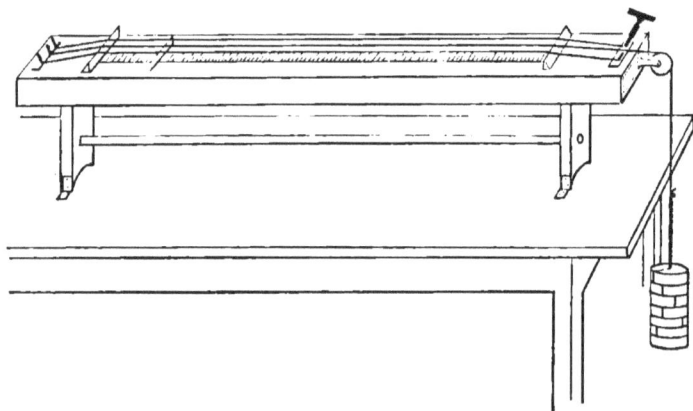

FIG. 328.

box carries keys at either end by means of which strings may be stretched. The strings pass over wedge-shaped bridges, so that they all have the same length. Adjustable bridges also are provided, by means of which the length of a single string may be varied at will. A centimeter scale runs the length of the sonometer, dividing the distance between the bridges, which should be just 1 m., into 100 parts. The construction of a good sonometer is a task requiring something of the instrument maker's skill, and also a supply of properly selected and thoroughly seasoned wood. For the following experiment a dry board of pine or spruce, rather more than 1 m. long, may be used instead of a sonometer. Across this, near the ends, 1 m. apart, V-shaped bridges as shown in the figure should be fastened. An ordinary meter stick, placed with its ends against these bridges, will serve for a scale. To one end of this improvised sonometer should be attached a pair of pulleys, over which the wires or strings to be experimented with may be stretched by means of weights.

Procedure:

(*a*) Place a steel wire upon the sonometer, and increase its tension until it gives a distinct musical tone. Place the finger upon the middle of the string, and bow it with one hand. The tone now emitted will be the octave of the fundamental. That both ends of the string are thrown into vibration may be shown by placing a small rider of paper, in the form shown in Fig. 329, upon the unbowed end of the string, anywhere between the finger and the end. This will be thrown off as soon as the bow touches the other half of the string.

FIG. 329.

(*b*) Move the finger to a position distant one third the length of the string from either end. Bow the short portion of the string, and note the change of pitch. There will now be a node, or point of rest, midway between the finger and the further end of the string. To locate this, place at the point in question one of the paper riders already referred to. It is convenient in making these riders to fold one set of them with the white and the other with the colored side out; and to use the white riders at the nodes, the colored ones upon the vibrating segment. If this rider be properly located, the short end of the string may be bowed vigorously without displacing the rider. Other riders placed anywhere else upon the intervening segments of the string will, however, be thrown off at once. We thus get direct ocular evidence that the string is vibrating in three parts, although only one of these is directly under the bow; also, that there are two nodes so situated as to divide the string into three equal parts, although only one of these is compelled to remain at rest by the presence of the finger.

(*c*) Repeat the experiment with the finger one fourth the length of the string from the end. Locate two other nodes and show that they are at divisions 50 and 75 upon the scale. Show by means of colored riders that the four intervening portions of the string are all in vibration when one of them is bowed. Note that the tone emitted by the string is two octaves above the fundamental.

323. Motion of Bowed String. — When the vibration of a string is not hindered by a stop, it is possible with the bow to produce vibrations of it in parts, together with the fundamental tone of the string vibrating as a whole. A single string, under the hands of a skilled performer, is thus made to utter sounds which are rich in tone color; and it is to these that the beauty of the music of stringed instruments is due.

324. EXPERIMENT 102.—**Study of a Bowed String.**

Apparatus:

(1) A violin string or a steel wire, tightly stretched horizontally between wooden supports.

(2) A screen with a narrow, vertical slit, so placed in front of the string that the latter will bisect the slit.

(3) The revolving mirror described in previous experiments.

(4) A bow, rosin, etc.

Procedure:

(a) Mount the apparatus so that the string will be strongly illuminated from behind. The best way to perform the experiment is to mount the string in the field of the lantern, and observe its projection upon the screen. The arrangement of the apparatus for this form of the experiment is shown in Fig. 330. In front of the middle of the string, and not more than 2 cm. from the same, set up the slit, and immediately in front of that the revolving mirror *m*, so placed that when it is at rest an image of the slit will be thrown upon the screen *S*. Upon looking at this image, one will see an element of the string in the form of a short black line bisecting the slit.

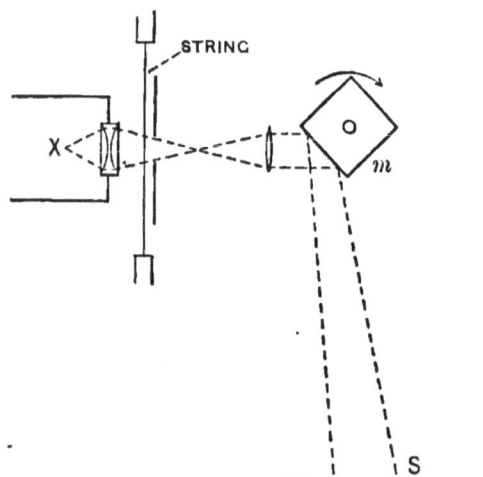

FIG. 330.

When the string is plucked or bowed, this black line is distended into a band, the width of which measures the amplitude of vibration of that portion of the string. If the revolving mirror is

turned, the image of the slit spreads across the screen, and the image
of the string forms an undulatory line, the form of which exhibits
clearly the mode of vibration. Projected upon the screen, this experiment is a very beautiful one.

(b) Vary the action of the bow, and note the changes produced in
the manner of vibration of the string. It will be found that there
are three prevailing types of motion:

1. A simple sinuous vibration which is natural to a freely vibrating string and to all vibrating bodies.

2. A sinuous vibration with more rapid motions superimposed
upon it. These, which are due to overtones of the string, *i.e.* to the
independent vibration of small segments of the string, are superimposed upon the larger curve, due to the vibration of the string as a

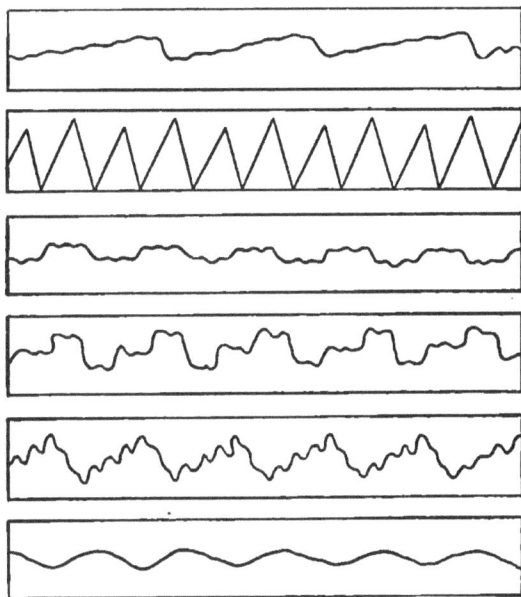

FIG. 331.

whole, like ripples upon a wave. The third type is produced by the
action of the bow when the latter clings to the string, only partially
releasing it from moment to moment and then seizing it again. This
produces a movement of the string which is forced. It is represented
by a series of oblique straight lines instead of curves. The appear-

ance is that of a succession of saw teeth. Messrs. Raps and Menzel, in Berlin, have photographed a great number of these types of vibratory motion. Figure 331 is reproduced from one of their plates. It shows the three types of vibration already mentioned.

This experiment serves to illustrate the fact that, under the action of a skilled performer, vibrating strings may be made to undergo a great variety of motions, each of which produces its characteristic effect upon the ear. It is to this variety that stringed instruments, especially those in which the bow is used, owe their expression.

325. EXPERIMENT 103. — **Laws of the Pitch of Vibrating Strings.** — It is a well-known fact, which may be verified by a moment's observation with the sonometer, that the rapidity of vibration of a string increases with the tension, and that it increases as the length of the string is diminished. It is the object of this experiment to determine the law of these changes.

Apparatus:

(1) A sonometer, some pieces of steel pianoforte wire, and a bow.

(2) A clamp, a set of iron kilogram weights, one of which should be provided with a hook. (See Appendix IV.)

(3) Three tuning forks mounted on sounding boxes; these should give the tone "ut$_2$," "mi$_2$," "sol$_2$," or the corresponding tones of the next higher octave.[1]

Procedure:

(a) Mount a fine steel wire upon the sonometer, fastening it at one end and carrying the free end over the pulley. Clamp the sonometer firmly to the table, with the end which carries the pulleys projecting over the edge. Hang weights to the free end of the wire, testing its pitch from time to time, until its sound approaches unison with that of the "ut" fork. Make careful adjustments of the weights, using the tenths of kilograms until the agreement is as good as you can get. Note the amount of weight applied, the length of the wire, and its diameter.

(b) Add more weights until the wire comes into unison with the "mi" fork. Note the amount necessary to bring this about, and repeat, using the "sol" fork. Tabulate your results as follows:

[1] The mark "ut" upon an instrument indicates pitch corresponding to "do" or C of the English notation.

TABLE.

PITCH AND TENSION OF A STEEL WIRE.

Diameter of wire, 0·025 cm.   Length, 50 cm.

Pitch	Weight applied = M	$\dfrac{\sqrt{M}}{\text{Pitch}}$
$Ut_3 = 512$ s.v.	2800 g.	0·1033
$Mi_3 = 640$ s.v.	4200 g.	0·1012
$Sol_3 = 768$ s.v.	6430 g.	0·1045
		$3)\overline{\cdot 3090}$
		Aver. = ·103

It is a well-established law of vibrating strings that the pitch will be directly proportional to the square of the weight applied. Test the accuracy of your determinations by dividing the square root of each weight in grams by the number of single vibrations made by the fork with which in each case the pitch of the string was compared as a standard: *this ratio should be constant.*

(*c*) Restore the stretching weight to the value necessary to give the tone " ut₂ " by comparison with the fork. Leaving this weight constant, insert a temporary bridge under the stretched wire, and slide it along the length of the wire until the pitch of the longer portion of the string coincides successively with that of the " mi " fork and the " sol " fork. Note the position of the bridge when the string comes into unison with each of these forks. Your readings should show that *the pitch of a string under constant tension varies inversely as the length.*

(*d*) Stretch side by side with the wire thus far used, one of greater diameter. Add weights to the latter until it also comes into unison with the " ut₂ " fork. Continue the adjustment until the strings are in unison with one another and also with the fork. Measure the diameter of the larger string, and read the weights attached to each. Compute the relative areas of cross-section of the two wires, and see in how far your observation verifies the law that strings of the same material and the same length vibrate at the same pitch when the stretching weights are proportional to the cross-section.

[If more convenient, the larger string may be stretched *in place of* the smaller one, and tuned by reference to the fork.]

CHAPTER XXXVIII

WIND INSTRUMENTS AND RESONATORS

326. The Vibration of Air Columns. — In wind instruments, the vibrating body is a column of air. The visible instrument serves simply to fix the size of the air column. The vibrations of the latter give the pitch, those of the wooden or metal walls affect only the timbre.

The vibration of an inclosed air column may be excited in a variety of ways. If, for example, the end of an argand lamp chimney, or any open tube, be struck with the palm of the hand, it will utter a musical tone due to the vibration of the inclosed column of air. If the hand be pressed against the end of the tube and suddenly removed, a faint tone will likewise be heard which is the octave above that obtained by closing the end. The chimney when closed with the hand forms what is known as a *closed pipe*. When both ends are open it is an *open pipe*. *The fundamental tone of a closed pipe is an octave below that of an open pipe of the same size.*

To maintain the sound of a pipe, what is called a *standing wave* must be produced. In other words, the air column must be made to vibrate longitudinally with a motion like the longitudinal vibration of a solid bar. This condition may be produced by the motion of a solid, the pitch of which corresponds with that of the air column, as shown in the following experiment:

327. EXPERIMENT 104. — **Resonance of an Air Column excited by a Tuning Fork.**

Apparatus:

(1) A tuning fork and bow.

(2) A tall cylindrical vessel of glass.

(3) A beaker or flask filled with water.

Procedure:

(*a*) Put the tuning fork into vibration and hold it over the mouth of the cylinder.

Pour water into the latter slowly from the flask. Note that when the water reaches a certain level, the cylinder emits a strong musical tone, agreeing in pitch with the tuning fork; and that this tone dies away again as the level rises. This observation indicates that the tuning fork is capable of exciting only a column of certain length, and that the fundamental tone of the latter has the same pitch as the fork.

(*b*) Repeat the above operation, ceasing to add water when the tone of the cylinder is loudest. Measure the length of the air column from the mouth of the cylinder to the surface of the liquid. Note the pitch (in single vibrations) marked upon the shank of the fork.

Since a tuning fork sends forth one wave in the air for each complete (or double) vibration, we can compute the distance between successive waves, and compare this wave length with the length of the vibrating air column. For example, with a fork marked 512 s.v., 256 waves will be started every second. The velocity of sound is 33300 cm., and these waves are therefore $\frac{33300}{256} = 130+$ cm. apart.

The wave length 130 cm. is that of the fundamental tone of the fork, and likewise of the air column which responds to it. Measurement of the length of the latter, however, will give about 32·5 cm. $= \frac{130}{4}$. ∴ *The fundamental wave of a closed pipe is four times the length of the pipe.*

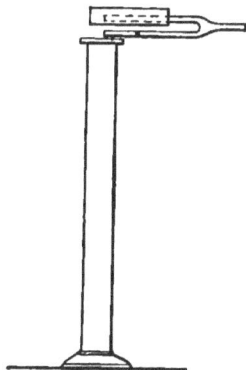

FIG. 332.

(*c*) The sound waves produced by the two prongs of the fork interfere, and tend to a certain extent to neutralize each other. The resonance of the cylinder will therefore be more intense when the wave from one prong is suppressed. To verify this statement, roll a piece of paper into the form of a tube large enough to inclose one prong of the fork without touching it. Hold the vibrating fork above the mouth of the cylinder as in Fig. 332. Slip the paper tube over the upper prong of the fork, and withdraw it several times, noting the successive accessions and diminutions of loudness thus produced.

328. Resonators. — An air column adjusted so as to vibrate in unison with a sounding body, and to reinforce the tone of the latter, is called a *resonator*. The glass cylinder in the foregoing experiment, for example, is a resonator; the sounding boxes, upon which tuning forks are mounted, also the sounding boards of pianos and the bodies of violins, etc., are resonators. Every inclosed body of air acts as a resonator to some particular wave length. Thus it is that a seashell, or other hollow body, held to the ear, murmurs continually. The outer air is crowded with sound waves, which constitute the volume of ever-present noise. The resonator responds only to the single wave length with which it is in tune.

By means of a set of resonators corresponding to the various tones used in music, it is possible to detect the various vibrations contained in sounds far too complex to be analyzed by the unaided ear and to express them in musical notation.

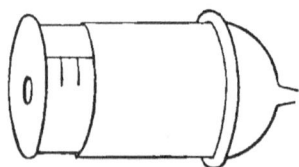

FIG. 333.

Such resonators are usually given the form shown in Fig. 333. Two brass cylinders, one partly closed, the other drawn out conically to a small open tip which fits the ear, slide one within the other. The length is thus adjustable, and the resonator will respond, according to its length, to any single tone within a range of several whole tones of the scale.

By means of sets of such resonators, many interesting analyses of complex sounds have been made. The various tones which make up the roar of Niagara, for example, have been isolated, and written down in the form of a complex chord.

329. Essential Parts of a Musical Instrument. — It is obvious, from the preceding articles, that a musical instrument has two essential parts: (1) a vibrating solid, the motion of which determines the pitch; (2) a resonator which reinforces the sound, and gives volume and character to the instrument. In wind instruments, the vibrating solid is a less conspicuous part than in stringed instruments. In the voice, the vocal chords play this essential part; in reed instruments, a small tongue of sheet metal (the reed) in the mouthpiece; while, in some instruments, the lips of the performer are the vibrating solids.

330. The Organ Pipe. — This is the simplest type of wind instrument. It consists of a long, hollow box or tube (Fig. 334), generally of wood, and of rectangular cross-section. It is sometimes open at both ends (an open pipe); sometimes one end is closed (a closed pipe). The draught of air blown through the mouthpiece is projected against the sharp, wedge-shaped lip l, and is thrown into violent

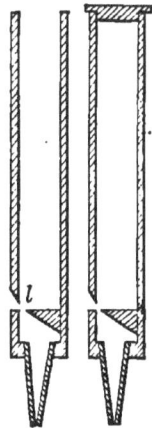

FIG. 334.

oscillations. The effect is similar to that produced by blowing sharply against the edge of a piece of cardboard. Of the numerous waves thus produced, those which are of the proper pitch are reinforced by the air column within the pipe, which acts as a resonator.

When the air within the organ pipe is thrown into vibration by the action of the mouthpiece, a sound wave travels the length of the pipe, to the further end, where it is reflected and returns upon its course. If a second wave follows the first one after the proper interval, a standing wave is formed, and the tone is maintained. If this wave has a wave length equal to four times the length of the pipe, for a closed pipe, the tone which is produced is called the *fundamental tone.* If twice as many waves per second be started at the mouthpiece, the octave of the fundamental will be produced; and if waves be started still more rapidly, other overtones corresponding to the overtones or harmonics of a vibrating string will be formed. In an open pipe, the fundamental tone has a wave length which is only twice as long as the pipe itself, and which is, therefore, an octave higher than that of the closed pipe.

The pitch of an organ pipe depends upon the length of time required for a sound wave to travel from the mouthpiece to the further end of the pipe and back again. If the pipe be filled with some gas in which sound travels more rapidly than in air, the pitch will be raised in proportion to the increase of velocity. This statement may be verified as follows:

331. EXPERIMENT 105. — Pitch of an Organ Pipe when filled with Illuminating Gas.

Apparatus:
An organ pipe and a retort stand.

Procedure:
Support a closed pipe in a vertical position, as shown in Fig. 335. By means of a rubber tube connected with the gas mains, fill the pipe with illuminating gas, and withdraw the tube. Sound the pipe at

once, and notice that the pitch grows lower as the gas is supplanted by atmospheric air. By the use of hydrogen, in which sound travels much more rapidly than in the mixture used for illuminating purposes, the effect is still more marked.

332. Influence of Temperature upon the Pitch of an Organ Pipe. — Since the velocity of sound in air rises with the temperature, it follows that a pipe filled with hot air will have a higher pitch than if the air were at a lower temperature. This may be shown by the introduction of a lighted candle into an organ pipe which is mounted vertically. The change of pitch of the pipe can be most easily detected in this by comparing it with a second instrument, with which, when cold, it is in unison.

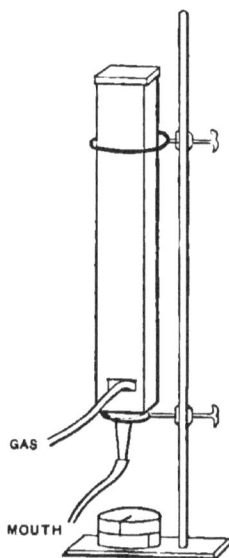

GAS

MOUTH

FIG. 335.

333. Overtones in Wind Instruments. — The shrill overtones which are produced by a powerful blowing of wind instruments are due to the breaking up of the air column into numerous short standing waves. The presence of these may be illustrated by adjusting to any ordinary whistle, the end of which is left open, a glass tube about 1 m. in length, and 2 or 3 cm. in diameter. If a small

FIG. 336.

quantity of lycopodium powder be introduced into this tube, and distributed throughout its length, and if the tube be then placed horizontally, and be strongly blown for an instant, it will be found that the dust tends to arrange itself in the form of equidistant transverse ridges. These ridges lie in the nodes of the wave. (See Fig. 336.)

334. The Manometric Flame. — This is an ingenious device for studying the vibrations in organ pipes and other wind instruments. An opening in the wall of the pipe (Fig. 337) is closed with a flexible diaphragm over which is a hollow chamber *R*. This is connected with the gas main through a tube *g*, and with a small circular gas jet through another tube *t*.

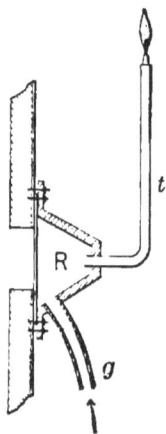

Fig. 337.

Changes of pressure within the organ pipe are transmitted through the diaphragm to the gas in *R*, and thence to the jet. The flame dances in time with the vibration within the pipe, and its image, viewed in a revolving mirror, appears as in Fig. 338.

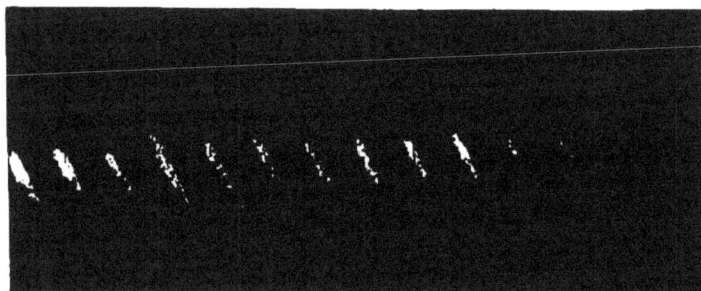

Fig. 338.

335. Experiment 106. — Analysis of Speech by Means of the Manometric Flame.

Apparatus:

(1) A manometric flame apparatus of the form shown in Fig. 339.

This consists-of an argand chimney, to one end of which a cork is fitted. Through this cork two glass tubes pass, one of which is connected with the gas, the other with a jet made by drawing a glass tube down to about 0·1 cm. diameter. Over the other end of the chimney is stretched a piece of the thinnest obtainable sheet rubber. A conical mouthpiece of pasteboard completes the instrument.

(2) A revolving mirror.

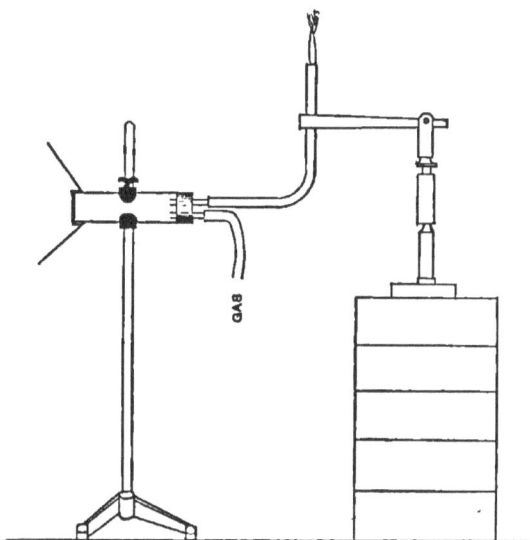

FIG. 339.

Procedure :

(*a*) Turn on the gas, light the jet, and adjust to a flame height of about 2 cm.

(*b*) Speak or sing into the mouthpiece, watching the flame meantime in the revolving mirror. Observe the forms of the serrated image corresponding to various articulate sounds, particularly to syllables containing hissing or explosive elements, *p*, *pr*, *t*, *s*, etc., and to the vowels, *ä*, *ā*, *e*, *o*, *ou*, etc.

Figure 340 shows the image produced by the word INK.

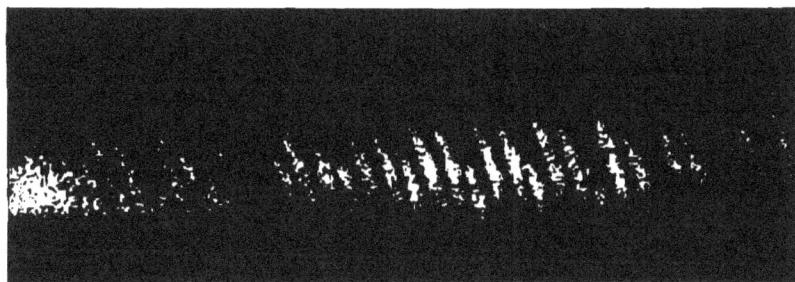

FIG. 340.

2 B

PART V — LIGHT

CHAPTER XXXIX

REFLECTION AND REFRACTION

336. The Propagation of Light. — Light, like sound, is propagated by means of a wave motion, but light waves differ from sound waves in many respects. The velocity is much greater, as has been already pointed out in Chapter XXXIII. A light wave travels from the sun to the earth, for example, in eight minutes, which requires a velocity of over 300,000,000 of meters per second. A

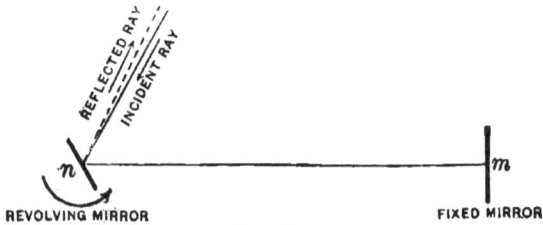

Fig. 341.

sound wave, which travels in air 332 m. per second, would take nearly fourteen years to traverse the same space.

Enormous as the velocity of light is, it has been found possible to measure it experimentally. A small mirror n (Fig. 341) is made to revolve several hundred times per second. A ray of sunlight reflected from it to a second mirror m (situated at a considerable distance), and back again, is found to take a slightly different course when reflected the second time. The mirror has had time to revolve through a small angle while the ray was traveling to the fixed mirror and back again. By measuring this minute angle and the distance between the mirrors, the velocity of light has been found to be 300,574,000 m. (186,680 miles) per second.

Light waves differ from sound waves also as to wave length and frequency. The sound waves used in music have wave lengths lying between 16 feet and 3 inches. The number of vibrations varies between about 32 and 4000 vibrations per second. The light waves to which the eye is sensitive have wave lengths lying between 0·000076 and 0·000039 cm.; they oscillate between 392,000,000,000,000 and about 757,000,000,000,000 times per second. Sound waves in air consist of longitudinal vibrations. Light is due to the transverse vibration of an imponderable medium which is supposed to fill all space, and which is called the *luminiferous ether.*

337. Reflection of Light. — The law of reflection of light is usually stated thus: *The angle of incidence is equal to the angle of reflection.* These angles, which are marked *i* and *r* in Fig. 342, are the angles between a line drawn normal to the reflecting surface, and the directions of the incident and the reflected ray respectively. This law may be verified as follows :

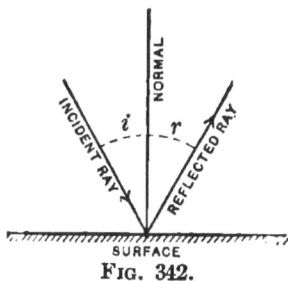

Fig. 342.

338. EXPERIMENT 107. — Equality of the Angles of Incidence and Reflection.

Apparatus :

An ordinary plane mirror, to the frame of which a wooden pointer has been attached at right angles to the reflecting surface.

Procedure:

(*a*) Mount the mirror in a darkened room with freedom of rotation upon a horizontal axis. (See Fig. 343.) Send a beam of horizontal light to the mirror from a projecting lantern or from a porte lumière[1] placed outside the room and reflecting the sun's rays in through a hole in the shutter.

(*b*) Rock the mirror to and fro upon its axis, and notice that the pointer always bisects the angle between the incident and the reflected ray. The path of these rays may be rendered more easily visible by filling the air in front of the mirror with chalk dust.

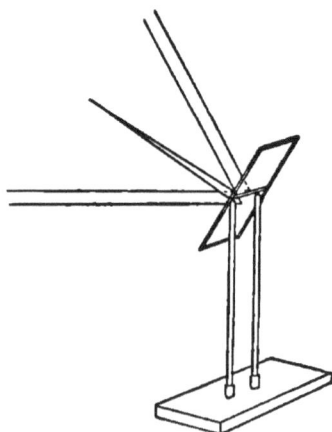

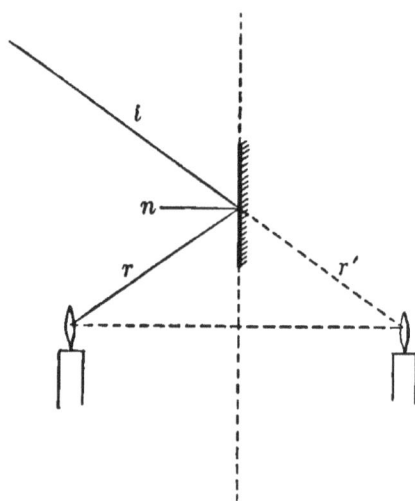

FIG. 343.   .   FIG. 344.

339. Position of the Image behind a Plane Mirror. — The image formed by a plane mirror is of the same size as the object, and its distance behind the mirror is the same as that of the object in front of the latter. The line joining

[1] Porte lumière : an arrangement for introducing a horizontal ray of sunlight into a darkened room. It consists of an adjustable mirror placed outside of the window shutter. Such mirrors are mounted so as to have a freedom of motion upon two axes, one of which should be parallel to the polar axis of the earth. ·When driven by clockwork, so as to maintain the reflected ray constantly in a fixed position in spite of the motion of the earth, the instrument is called a *heliostat*.

a given point in the object and the image of that point is bisected by the mirror, and is perpendicular to the surface of the same. To an observer looking towards a plane mirror the image appears in the direction of the ray from the face of the mirror to his eye. The position of the image is found by turning the reflected ray r (Fig. 344) until it coincides in direction with the incident ray i; *i.e.* into the position r' behind the mirror. The sum of the distances $i + r$ will be the same as $i + r'$. The image will therefore be as far behind the mirror as the object is in front of it.

340. Concave and Convex Mirrors. — In the case of curved surfaces, although the law of reflection is the same *for each point of the surface*, as in the case of plane mirrors, the results produced are very different. The most important case is that of the *concave mirror*. This has a surface which is a portion of the inner surface of a sphere.

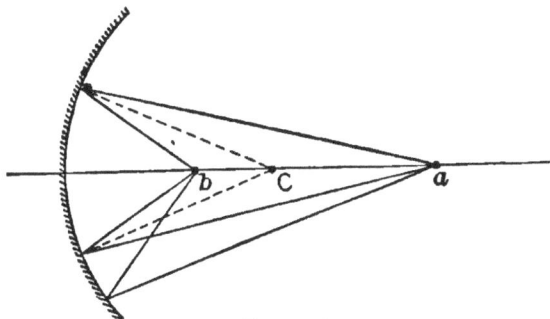

FIG. 345.

(1) Consider the rays of light from a point a in front of the mirror. Those which reach the surface of the mirror are reflected, each one in accordance with the law of reflection given in the foregoing article. As will be seen from Fig. 345, all these rays come together at a

single point b in front of the mirror. If, on the other hand, the source of light were situated at b, rays emanating from it, and reaching the mirror, would all be reflected to a. The points a and b are called *conjugate foci.*

(2) If the source of light be situated at the center of the mirror C, the angles of incidence and reflection are reduced to zero. Since all rays strike the surface of the mirror normally, they will be reflected directly back upon their course, and will be brought to a focus at C.

(3) Consider the rays from a point F halfway between the center of the mirror and its surface (Fig. 346). The reflected rays will be every-where parallel to the axis of the mirror. They can be considered as meeting only in infinity. If parallel rays, namely, those from some very distant source, such as the sun or a fixed star, fall upon the mirror, the latter being placed so that its axis is parallel to the rays, they will come together at the point F which, as above stated, is halfway between the surface of the mirror and its center of curvature. This point at which parallel rays are brought to a focus is called the *principal focus* of the mirror.

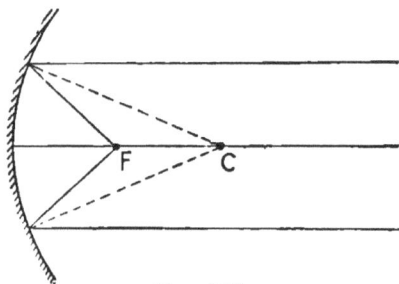

341. Real and Virtual Images. — Consider a candle $C_1 C_2$ in front of a concave mirror. The rays from each portion of it, after reflection from the face of the mirror, will be brought to a focus at some given point. Thus the rays from the tip of the candle at C_1 (Fig. 347) will come to a focus at c_1; those from the base of the candle C_2 at c_2.

Intermediate points in the object will come to a focus at points lying between these two. These various foci of light reflected from the surface of the candle produce an

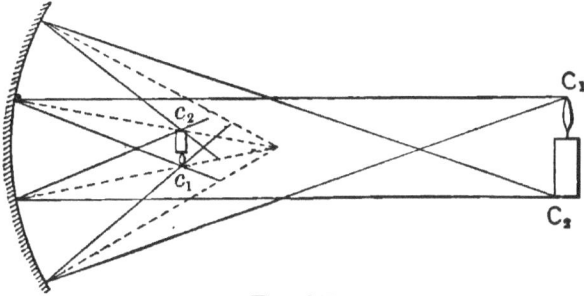

FIG. 347.

image of the candle. This image, unlike the image in the plane mirror, is in front of the mirror. It is not, generally speaking, of the same size as the candle itself. It is inverted instead of being erect. Images thus formed by means of a concave mirror are called *real images*. They differ from images behind the mirror, which are called, to distinguish them, *virtual images*, in one important particular. The rays are *actually present* where the real image is formed; they only appear to pass through the position of the virtual image.

In order that a real image may be formed by means of a concave mirror, the object must be placed beyond the principal focus. The rays from an object situated between the principal focus and the mirror itself will not converge to any focus in front of the mirror after reflection, but will diverge, as shown in Fig. 348. These divergent rays, if we imagine them produced behind the mirror, will, however, come to focus so as to form a *virtual image* at *i*. This image, like the image in the plane mirror, will be erect, but its size is not the same as that of the object, nor is it situated the same distance behind the mirror that the object is in front.   Its size and position, which vary,

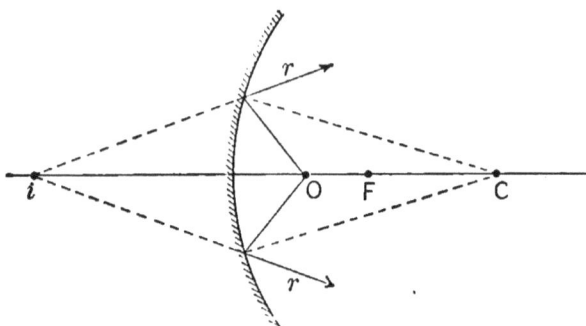

FIG. 348.

may be determined from the direction of the reflected rays.   The intersection of the paths of these behind the mirror indicate the position of the image.

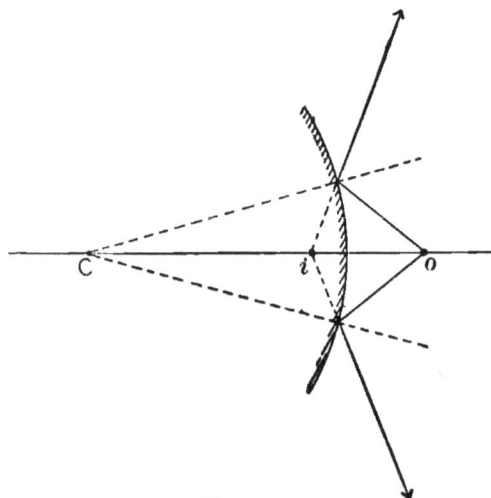

FIG. 349.

The images produced in *convex* mirrors are always virtual, since the reflected rays from any object in front of the mirror will diverge.   (See Fig. 349.)

The convex mirror has no very important application in optics, but concave mirrors are much used. In reflecting telescopes, real images of stars and other distant objects, formed by the use of such a mirror, are magnified by observing it through an eyepiece. The largest telescopes ever constructed have been of this kind. The great reflecting telescope of Lord Rosse, the mirror of which was six feet in diameter, is shown in Fig. 350.

FIG. 350.

342. Refraction. — When a ray of light falls upon any surface, a portion of the ray penetrates the body. If the latter be transparent, this portion of the ray will be transmitted; if the body be opaque, the wave motion will be destroyed and its energy converted into heat. We say in such a case that the ray has been absorbed. If a ray of light passes through the surface of the transparent body, normally, its direction is not changed. If, however, it meets the surface at an oblique angle, it will be bent from its path. If the medium into which the light thus passes is *denser* than that from which it comes, as, for example, when a ray passes from air into water, or glass

(Fig. 351), it will be bent *toward the normal* to the surface. When it passes through a surface between a denser

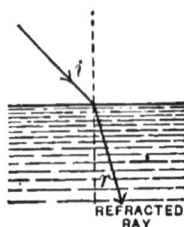

FIG. 351.

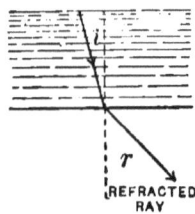

FIG. 352.

and a rarer medium, as when the ray passes from glass into air (Fig. 352), it is bent *away from the normal* to the surface. A ray thus bent from its course is said to be *refracted*. The law of its change of direction follows a fixed law; viz.:

343. THE LAW OF REFRACTION. — *The ratio of the sine of the angle of incidence to the sine of the angle of refraction is constant.*

The law may be expressed thus,

$$\frac{\sin i}{\sin r} = n.$$

The quantity n in the above equation is called *the index of refraction*.

344. EXPERIMENT 108. — **Displacement of a Ray by passing through a Sheet of Plate Glass.**

Apparatus:

(1) A projecting lantern, or porte lumière.

(2) A piece of plate glass.

(3) A small revolving circular stand.

Procedure:

(a) In front of the condenser set up a diaphragm containing a vertical slit, and focus the image of the latter upon the screen.

(*b*) Mount the glass vertically upon the revolving stand, as in Fig. 353, securing it in place with wax.

Place the glass in front of the objective of the lantern in the path of the ray, and turn the stand upon its vertical support.

Note the displacement of the image, and that when the light traverses the glass at right angles the displacement is zero.

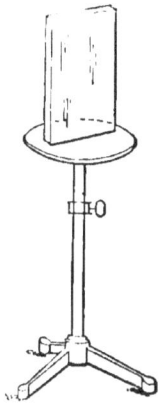

If the experiment were performed with a piece of homogeneous glass, with absolutely flat and parallel surfaces, it would be possible to show from the relation between the angle through which the glass is turned and the displacement of the image, that a ray of light in passing through a refracting medium, the faces of which are parallel, suffers lateral displacement but no permanent change of direction. In other words, the ray is bent away from the normal to the surface (Fig. 354), upon emerging from the denser medium, just as strongly as it had been bent towards it upon entering that medium.

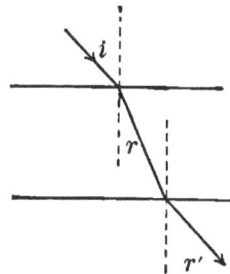

Fig. 353.

Fig. 354.

345. The Passage of Light through a Prism. — If, in the foregoing experiment, a prism of glass be interposed in the path of the ray instead of the plate glass, the path of the ray, after traversing the prism, will no longer be parallel to its original direction. If we map out the path of the ray by means of chalk dust, as described in Experiment 106, we find the refracted ray bent toward

Fig. 355.

the normal, and the emerging ray bent away from the normal, as in the case of the plate glass. The angle at which the emerging ray meets the second surface, how-

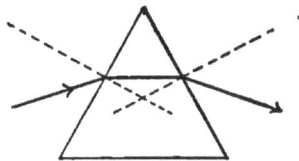

ever, is such that the ray acquires a new direction. (See Fig. 355.) It will be noted, on trying the experiment, that the image is broadly fringed with color. This effect is due to what is called *dispersion*, a phenomenon which will be fully considered in Chapter XL.

346. Total Reflection. — When a ray emerges from a denser into a rarer medium, the angle r (Fig. 356), which the emergent ray makes with the normal to the surface, is, as we have seen, always greater than the angle i of the ray in the denser medium. If the angle i be increased, the emergent ray will finally take a direction parallel to the surface. Any further increase in the angle a will make the angle greater than 90°. The ray, therefore, will not emerge at all, but will take a new path within the denser medium. This phenomenon is called *total reflection*. It may be conveniently studied as follows:

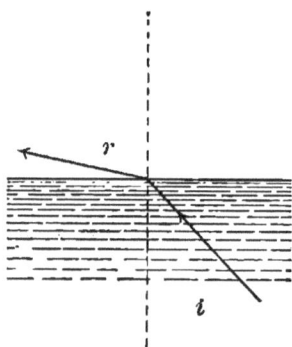

Fig. 356.

347. Experiment 109. — Total Reflection in Water.

Apparatus:

(1) A large beaker, or other glass vessel. The tanks with glass sides used for aquariums are well suited to this experiment.

(2) A projecting lantern or porte lumière.

(3) A plane mirror.

Procedure:

(*a*) Cover one side of the beaker, or tank, with black paper, leaving a horizontal slit about 5 cm. long, and 1 cm. wide. Fill the beaker with water in which a few drops of milk, or a pinch of starch, sufficient to give it a slightly cloudy appearance, has been added. Mount the beaker upon a high stand, and throw a beam of sunlight obliquely upwards upon the slit, as shown in Fig. 357.

(*b*) Observe the path of the ray within the liquid, and trace out the emerging ray in the air above by means of chalk dust. Note carefully also the existence of a reflected ray within the liquid. This ray must not be confused with that produced by total reflection. It

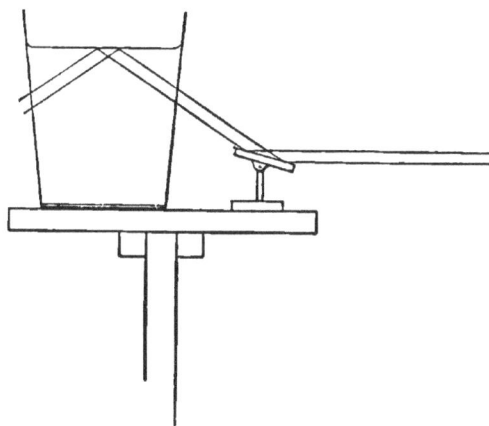

FIG. 357.

is a case of ordinary reflection from the inner surface of the denser medium. So long as there is an emergent ray in the air, total reflection does not take place.

(*c*) Increase the angle which the ray makes with the surface of the water, by movement of the mirror, until the critical angle is reached at which the emergent ray becomes parallel to the surface. Upon further increasing the angle, note the disappearance of the refracted ray, and the sudden appearance of the totally reflected ray within the liquid.

(*d*) To the observer looking at the surface from below at an angle greater than the critical angle, the surface appears to be an opaque mirror. Objects lying in the air beyond are entirely hidden. To verify this statement, float a short lighted candle upon a cork on the surface of the liquid, and look up through the liquid at various angles.

CHAPTER XL

DISPERSION

348. The Composition of White Light. — It has already been shown that, when an image has been thrown upon the screen, and a prism is interposed in the path of the ray, the image, which is displaced from its position by refraction, exhibits colors. This show of color is due to the fact that white light is composed of various rays differing in wave length. Each wave length has its distinctive color. The amount of refraction which a ray undergoes, however, depends upon its wave length. Short waves are more bent from their course than long ones. The result is that the violet rays, which have the shortest wave lengths, are most displaced by the action of the prism, while the red wave lengths, which are longest, are least displaced. This phenomenon of varying refraction according to the wave of length is called *dispersion*. When a beam of white light passes through a prism, each of the countless wave lengths of which it is made up is displaced, and forms a colored image upon the screen. These images overlap, but they do not coincide. In the center of the composite image thus formed, the rays from the overlapping images recombine to give the appearance of white; at the ends, however, red shading into yellow, and violet into blue, respectively, show themselves. If, in place of the circular aperture, a vertical slit be used, these colored images will be reduced almost to a line. They will then overlap scarcely at all, and each color will stand out unmixed.

349. The Spectrum. — The number of different wave lengths which are present in a beam of white light is very great, perhaps infinite; consequently, when we disperse such light by means of a prism, we do not find a series of colored images separated from each other by a dark space, but a continuous series merged into a band of gradually changing colors. This is called the *spectrum*. Its color changes by insensible gradations from red to violet. At least one hundred and fifty intervening tints may be distinguished, but comparatively few of these have received names. Newton, who was one of the earliest students of this subject, named seven. These, in the order of decreasing wave lengths, are : red, orange, yellow, green, blue, indigo, violet.

350. EXPERIMENT 110. — **The Spectrum is made up of a Series of Overlapping Colored Images.**

Apparatus:
(1) A lantern, with a diaphragm, containing a circular opening.
(2) An equiangular prism of flint glass.
(3) A piece of ruby glass.[1]

Procedure:
(*a*) Set up the diaphragm in front of the lantern with the beam of light, making an angle of about 70° with the wall. Adjust the lens so that an image will be formed upon a screen at a distance equal to that of the lens from the wall. (See Fig. 358.) In front of the lens, mount the prism, as shown in the figure. The rays falling upon the prism will be refracted and dispersed, and will form a very impure spectrum upon the screen.

(*b*) Hold the piece of ruby glass between the diaphragm and lens, or, indeed, anywhere in the path of the ray, and note the effect upon

[1] Ruby glass is a variety of red glass selected for use in photography because it transmits only red light, with a small amount of orange and a trace of yellow. These are the colors to which the plates used in photography are least sensitive.

the spectrum. It will be seen that, owing to the opacity of the glass to all rays except the red and orange, the spectrum is reduced to a red and nearly circular image of the aperture. This image, which is somewhat elongated horizontally, is not of one color throughout. On one side it is of a deep red, and on the other it is orange or yellow. Ruby glass really transmits more than one wave length of light, and the image consists of the overlapping series of images which go to form the red and orange ends of the spectrum. If we could take away all of these images but one, we should find the spectrum reduced to an image of the aperture uniform in color and of the same appearance as though the light had not been passed through the prism.

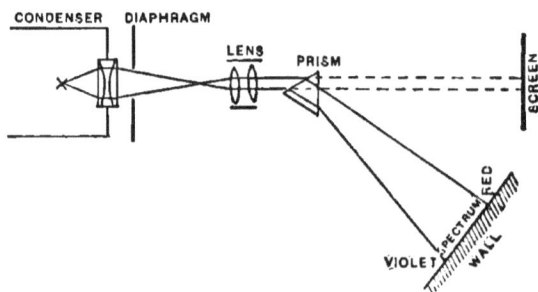

FIG. 358.

(c) In order to produce a spectrum which contains but one color, we must use as a source of light some incandescent vapor. The most convenient vapor for this purpose is that of sodium. Remove the lamp from the lantern and put in its place a Bunsen burner. Introduce into the flame, by means of a pair of tweezers improvised from iron wire, a small piece of metallic sodium. The yellow flame which results from the burning of this metal contains light of one wave length only (monochromatic light). The spectrum formed upon the screen will, therefore, be reduced to a single yellow image of the aperture in the diaphragm. Were the sodium flame reduced almost to a point like the arc light or a lime light, this image would be well defined and circular in form. Owing to the great size of the flame, the image is not very distinct in its outlines.

351. EXPERIMENT 111. — **The Formation of a Pure Spectrum.**

Apparatus:

(1) In order to reduce the width of the overlapping images which form the spectrum, and thus to prevent them from mixing, it is neces-

sary, as has already been pointed out, to use a slit in place of the circular aperture of the last experiment. A slit suitable for use in the present experiment may be constructed as follows :

Take two wooden blocks 10 cm. × 10 cm. × 5 cm. each. Bevel the end of one, as shown in Fig. 359. The edges which are in contact, when placed as shown in the figure, should match perfectly. Lay the blocks together, as above, upon a flat surface, and attach a pair of brass hinges as shown. When the beveled block swings upon these hinges, the matched edges will move apart, forming an opening with parallel sides (a slit). A spring clip screwed to the square block and rubbing against the surface of the beveled block will suffice to hold the arrangement in any desired position. Screw the square block to the face of a larger block (20 cm. square) so that the slit will bisect a round hole, 2 cm. in diameter, in the center of the latter.

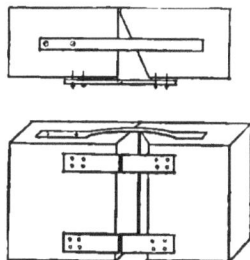

FIG. 359.

(2) The lantern, lens, and prism described in the foregoing experiment.

Procedure:

(a) Open the slit to a considerable width (about 1 cm.). Adjust the apparatus so as to give a well-defined spectrum. The prism should be turned into such a position that the deviation of the ray from its course is as small as possible. This is called the position of minimum deviation.

(b) Close the slit gradually and note the effect upon the appearance of the spectrum. It will be seen that while the brightness diminishes, the intensity of color throughout deepens. The central portions of the spectrum, which were greatly mixed by the overlapping of the broad images, now begin to show the rich colors of the pure spectrum. These changes continue until the slit, the edges of which are necessarily imperfect, come into contact at certain points. These points bar the passage of the light and produce dark, horizontal striations through the spectrum. The more perfectly straight and smooth the edges of the slit and the more nearly parallel its two jaws, the narrower it can be made without introducing these striations.

(c) Repeat the above operation, using a sodium light in place of the light of the lantern. Note that the yellow image to which the

2 c

spectrum is reduced grows narrower as the slit is diminished in width, until it becomes a mere vertical yellow line.

352. Classes of Spectra. — Spectra are usually classified as follows :

(1) *Continuous spectra.* (The spectra of glowing solids or liquids.)

(2) *Bright-line spectra.* (The spectra of glowing gases or vapors.)

(3) *Absorption spectra.* (Spectra produced by the dispersion of light which has passed through some medium capable of absorbing certain wave lengths and transmitting others.)

In the foregoing experiments we have seen examples of continuous spectra, of the simplest sort of an absorption spectrum (when ruby glass was interposed and the entire spectrum was cut off, excepting the rays at the red end). Also the simplest possible form of a bright-line spectrum (that of sodium vapor).

353. Dark-line Spectra. — It is a universal property of matter that all substances absorb precisely those kinds of light which they are capable of emitting when rendered incandescent, and no others. It follows, therefore, that gases, the spectra of which consist of bright lines, will absorb light only of the wave lengths which they emit, and will produce spectra with black lines corresponding in position with the bright lines which the gas in question radiates. This principle, which is of the utmost importance in the science of spectroscopy, may be demonstrated as follows :

354. EXPERIMENT 112. — **Reversal of the Sodium Line.**
Apparatus :
(1) The lantern with slit, lens, and prism.
(2) A sodium flame.

Procedure :

(*a*) Mount the apparatus so as to produce a fairly pure spectrum upon the screen.

(*b*) In the path of the ray, between the slit and lens, introduce a large sodium flame. This may be conveniently produced as follows:

Take a piece of asbestos wicking or, if this cannot be obtained, some loosely woven cotton or hemp yarn. Roll into a loosely wound ball at the end of a wire holder; dip the ball into alcohol and lay upon it two or three small bits of metallic sodium. Mount the wicking just below the path of the ray, midway between the lens and slit, and ignite. The result will be a large flame with the characteristic yellow color of burning sodium. The rays passing through this mass of sodium vapor, which is opaque to the yellow light, will suffer absorption. Note that the spectrum is otherwise continuous and possesses a black line in the region corresponding to that where the bright line of sodium in the previous experiment appeared. The identity in the position of this black line and the bright line of sodium may be shown by removing the sodium flame from between the lens and the slit and introducing one in the place of the light within the lantern, as described in Experiment 109.

355. The Fraunhofer Lines. — When, in 1819, the sun's spectrum was observed, using a narrow slit instead of the wide apertures which had been employed by Newton and other earlier students of this subject, it was found that the

FIG. 360.

sun's spectrum was filled with black lines instead of being, as had been previously supposed, continuous. The first to notice these lines and to describe them was a German physicist by the name of Fraunhofer, and they are still known by his name. He observed the presence of many hundreds of these lines, and designated the most prominent of them by the letters A, B, C, D, E, F, G, and H. The location of these lettered lines is shown in Fig. 360. Many years later it was shown by Kirchhoff and Bunsen that the bright line of sodium, the position of which corresponded exactly with the black line D of Fraunhofer, was capable of reversal, that is to say, of being changed into a black line by the interposition of sodium vapor. Experiment 112 is a reproduction in the simplest form of their experiment. They concluded that the lines of Fraunhofer, and the innumerable other black lines of the solar spectrum, were produced by the passage of the light of the sun through incandescent vapors in the sun's atmosphere; and they showed in support of their view that hundreds of bright lines, which constitute the bright-line spectrum of iron, correspond exactly in position to dark lines in the solar spectrum, and that other of these dark lines have their counterparts in the bright lines due to glowing vapors of various materials found in the earth's crust.

356. EXPERIMENT 113. — **The Absorption Spectra of Chlorophyl and of Potassium Permanganate.**

Apparatus:

(1) The lantern with slit, lens, and prism as previously described.

(2) About 200 c.c. of chlorophyl solution. This is readily prepared by placing a handful of green clover leaves or of fresh grass in alcohol. Warm slightly, bruise the leaves by stirring, and shake well for several minutes, then pour off the liquid, which will be of a rich green color. This solution, if bottled, will hold its color for a considerable time.

(3) A solution of potassium permanganate made by dissolving a few crystals of that substance in water.

Procedure:

(*a*) Having adjusted the apparatus so as to project a well-defined spectrum upon the screen, interpose a flat glass cell containing the chlorophyl solution in the path of the ray. Note the formation of dark bands produced by the absorption of the colors corresponding to those regions of the spectrum which are absorbed in passing through the solution. One of these bands lies in the red, two in the yellow and green. The extreme red end of the spectrum likewise suffers absorption, as does to some extent the entire blue end.

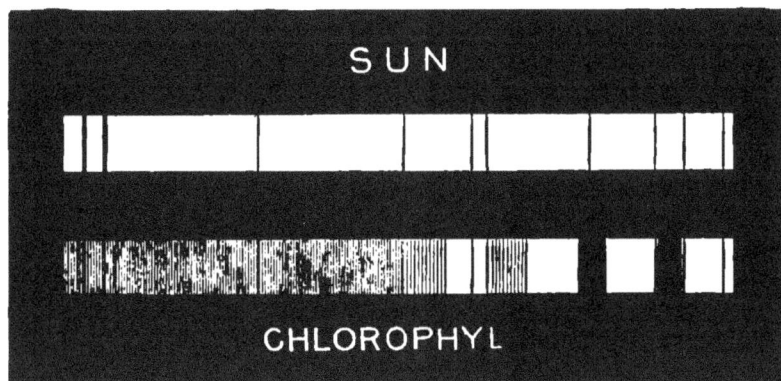

FIG. 361.

(*b*) Substitute for the chlorophyl a very dilute solution of the potassium permanganate, and note that an entirely different set of bands are produced. Strengthen the solution from time to time, and note the increased blackness of the bands as the absorption becomes more marked; also the reappearance of new ones. Note the distinction between the absorption bands produced by the passage of light through a liquid and the black lines which are obtained when a vapor is interposed in the path of the ray. Figure 361 shows the characteristic absorption spectrum of chlorophyl.

CHAPTER XLI

LENSES

357. Lenses defined. — A piece of glass or other transparent material, the surfaces of which are curved in such a manner that a beam of transmitted light is brought to a focus in consequence of refraction, is called a *lens*. The only types of lenses which it will be necessary to

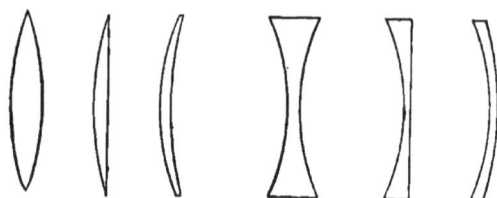

Fig. 362.

consider here are those in which the faces are portions of a spherical surface (spherical lenses). Sometimes the radius of curvature of one face of a lens is infinite, in which case that face is a plane. Sometimes the center of curvature is on one side and sometimes on the other, giving, respectively, convex and concave faces. All spherical lenses belong to one of the six forms shown in Fig. 362. Of these the first three are thicker in the middle than at the edges. All such lenses bring the ray to a focus beyond the lens, as shown in Fig. 363. The

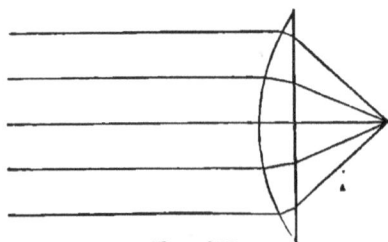

Fig. 363.

other three forms are thinner in the middle than at the edges. Such lenses cause parallel rays to diverge. (See Fig. 364.) The focus of such a lens lies in an imaginary

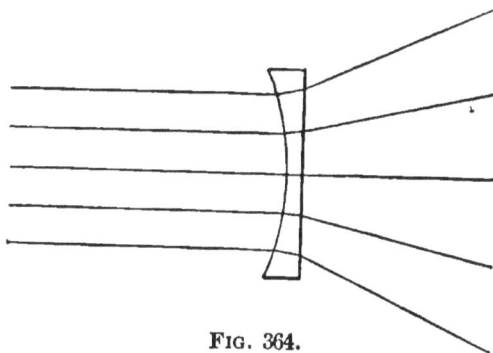

Fig. 364.

point on the side of the lens from which the rays have come. The first class is called *converging*, the second *diverging* lenses.

358. Formation of Images by Means of Lenses. — If we have a source of light at a point f on one side of a converging lens, rays reaching the lens from this point will be refracted, and will come to a focus at a point f' on the other side. In the same manner a source of light at f' will have its rays brought to a focus at f. The points f and f' are called *conjugate foci of the lens*.

If the point f be moved to an infinite distance so that the rays from it are parallel, f' will approach the lens, reaching finally a point F'. This is called the *principal focus of the lens*. If, in the same way, a source of light at f' be moved off to an infinite distance, its rays will come to a focus at F (Fig. 365). The distance from the lens to the principal focus is called the *focal length of the lens*. The position of f and f' is defined by means of the equation

$$\frac{1}{u} + \frac{1}{v} = \frac{1}{L},$$

in which u and v are the distances of f and f' from the
lens and L is the focal length. If instead of a single

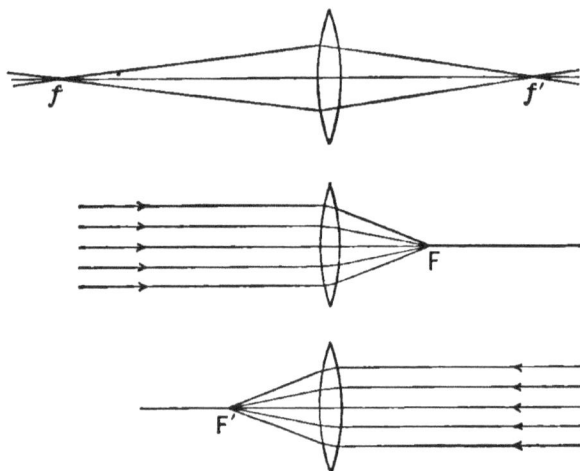

FIG. 365.

point we have an object on one side of the lens, as in
Fig. 366, light from all points of the object will be brought
to a focus at the corresponding point behind the lens, and
the result will be an image. This image, like the image

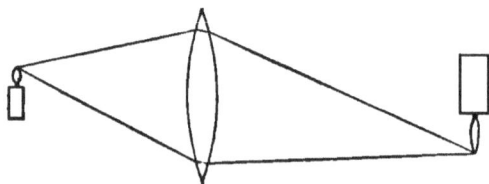

FIG. 366.

formed by a concave mirror, will be inverted. Its size and
its distance from the lens will depend upon the size and
distance of the object in accordance with the principles
stated above. If the object be brought nearer the lens
than the principal focus F, its image will vanish at infinity
on the other side. The rays after refraction now become

divergent instead of convergent. Their only focus is on the other side of the lens, and is imaginary. The image ceases to be real and becomes virtual. It is erect and on

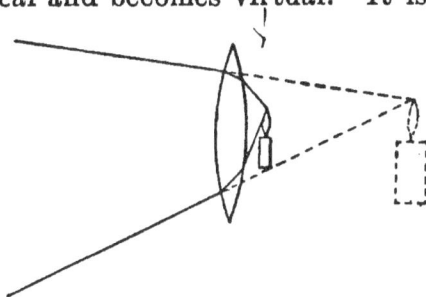

FIG. 367.

the same side of the lens as the object. (See Fig. 367.) Diverging lenses are capable of forming only virtual images.

359. The Telescope. — This is one of the most important of optical instruments. In the case of the telescope a real image of a distant object is produced by means of a lens or a concave mirror. This real image, which

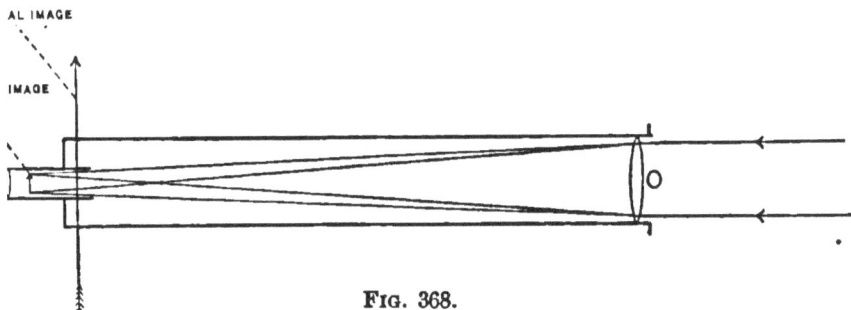

FIG. 368.

is necessarily much smaller than the object, but near at hand, is magnified by means of another lens or combination of lenses called the *eyepiece*. Figure 368 shows the simplest possible arrangement of a telescope. The lens *O*, by means of which the real image is formed, is called

the *object lens*, or the *objective*. The method of magnifying the real image at the eyepiece consists in bringing
a small lens so near to the image as to produce a virtual
image of it. As a rule, a combination of two lenses is
used both in the objective and eyepiece instead of one;
but the object of the lenses, whatever the construction
may be, is that stated above.

The size of the image produced by a lens is proportional;
first, to the size of the object, and, secondly, to the focal
length of the lens. It is independent of the aperture
of the lens. The brightness of the image, which is a
matter of great importance in telescopes used in astronomy, where the objects to be studied are very dim, is
directly proportional to the aperture of the lens. The
statement that the size of the image is independent of
the aperture of the lens may be verified by means of the
following experiment:

360. EXPERIMENT 114. — **Measurement of the Image formed by a
Lens of Varying Aperture.**

Apparatus:

(1) A converging lens; the longer the focus and the greater the
diameter of this lens, the better.

(2) A candle; a screen.

(3) A sheet of paper.

Procedure:

(*a*) Mount the lens as shown in Fig. 369, so as to throw an inverted
image of the lighted candle upon the screen.

(*b*) Measure the diameter of the candle itself just below the wick,
and the diameter of the corresponding portion of the image. The
ratio of this diameter gives the magnifying power of the arrangement.
The image may be larger or smaller than the object, according to their
relative distances from the lens.

(*c*) Cut a hole in the piece of paper as nearly circular as convenient,
and about 0·5 cm. in diameter. Insert the paper just behind the lens

until the hole corresponds with the center of the latter; the aperture of the lens is now reduced to the size of the hole. Fasten the paper in this position and note, first, that the image of the candle is still

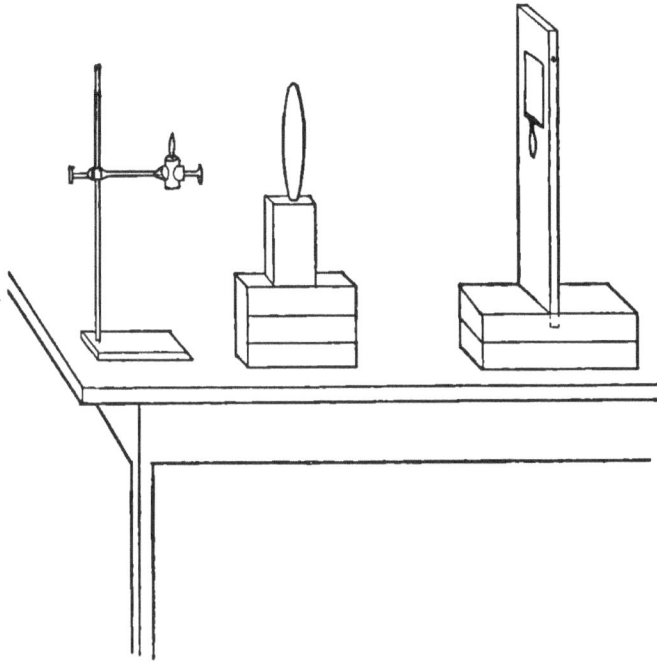

FIG. 369.

present; second, that it is much less brilliant than before. Measure the diameter of the candle as in operation (*b*). It will be found that the image is of the same size as when the entire aperture of the lens was utilized.

361. EXPERIMENT 115. — **Relation between the Size of the Image and the Focal Length of the Lens.**

Apparatus:

(1) Two converging lenses which differ considerably in focal length.

(2) A candle.

(3) Two small cardboard screens.

Procedure:

(*a*) Mount the lenses 2 m. apart, as shown in Fig. 370, with the lighted candle midway between them. Shift the two cardboard screens back and forth behind the lenses, until the images of the candle are in focus upon them. Measure the diameter of the candle in each image, and also the distance from the lens to the screen. It will be found that these diameters are proportional to the respective distances.

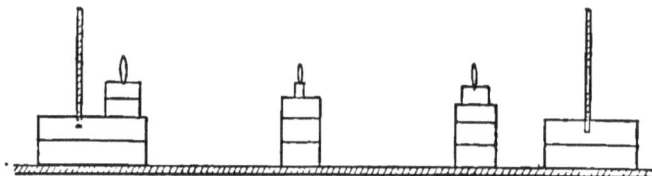

FIG. 370.

362. Magnifying Power of a Telescope. — The magnifying power of a telescope depends not only upon the size of the real image formed, but upon the size of the virtual image obtained by the use of the eyepiece. Magnifying power is expressed in terms of the size of this virtual image as it appears to the eye, compared with the size of the object itself, when viewed without the aid of the telescope. It is sometimes possible to make this comparison directly, as in the example afforded by the following experiment:

363. EXPERIMENT 116. — **The Measurement of Magnifying Power.**

Apparatus:

A spyglass or reading telescope similar to that used with the galvanometer. (See Appendix VIII.)

Procedure:

(*a*) Set up the telescope at a distance of several meters from a vertical scale (a white board divided to decimeters is useful for this purpose), and get the latter carefully in focus. Close the left eye and get the right eye well fixed upon the image of the scale in the eyepiece; then open the left eye and look with it at the scale itself. It will be found possible with a little practice to see the image and the scale simultaneously, and with a slight movement of the eyes to bring these side by side.

(*b*) Having acquired the power of seeing with the two eyes independently as above, note the number of scale divisions which the image of one scale division, as viewed in the telescope, corresponds to. This number gives directly the magnifying power of the instrument. If several eyepieces differing in magnifying power are available, — and it is often possible to adapt the eyepieces of microscopes to this purpose, — measurements may be made with each.

When it is desired to determine roughly the magnifying power of a spyglass or field glass of any kind, there are many familiar objects out of doors which may be made to serve as a scale. The tiers of brick in the wall of a house or chimney, for example, afford a very good scale of comparison, provided the distance is not so great as to make them undistinguishable to the naked eye. The equidistant pickets of a fence may often be made to serve as a horizontal scale for such a purpose.

364. Spherical Aberration. — We may regard a lens as made up of a great number of elements or parts, each one of which is capable by itself of forming a complete image. If, for example, in Experiment 114 the sheet of paper with the small aperture had been moved around so as to admit light from different portions of the lens successively, the image would continue to exist whatever the position of the aperture might be. The image obtained by the use of the whole lens is formed by the overlapping of these images. The images formed by different portions of a spherical lens do not, however, coincide accurately in position. There is therefore a blurring of the resultant image. This is due to what is called *spherical aberration*. The fact that the images formed by the various parts of the lens do not perfectly coincide may be shown by the method just indicated; namely, by mounting a diaphragm with a small aperture close to the lens, and then moving it around. It will be seen that the image shifts its position slightly as the aperture moves across the lens from side to side.

A lens with spherical surfaces, in point of fact, brings

a beam of light which passes through it near the edge, to a focus (a) at a point slightly nearer the lens than the focus (b) of the beam which passes through its center. (See Fig. 371.)

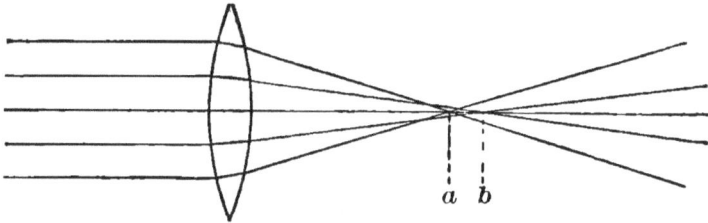

FIG. 371.

Spherical aberration is reduced by using parabolic instead of spherical surfaces, and by the use of a diaphragm or stop which reduces the aperture of the lens to as small an area as the dimness of the image will allow.

365. Chromatic Aberration. — Another error of lenses is due to the fact that different colors are differently refracted. On account of dispersion of light within the lens, violet, which is the most strongly refracted, comes to a focus at a point nearer the lens than do the other colors; and red, which has the longest wave length, comes to a focus at the greatest distance. Between these two are the foci of all intermediate colors. The consequence is that a series of colored images are formed one behind the other. For the most part these colored images overlap, and their colors mingle and neutralize each other. Around the edge of the beam of light from a lens, however, there is always a fringe of color. This can be observed by closely examining the image of a candle formed by one of the lenses used in the foregoing experiments. The fact that violet comes to a focus nearest the lens, and red farthest away, can be demonstrated by the following experiment:

366. EXPERIMENT 117. — Chromatic Aberration of a Lens.

Apparatus:

(1) A lantern.

(2) A diaphragm with a round hole, to be used in front of the condenser.

(3) A simple converging lens.

Procedure:

(*a*) Remove the objective of the lantern and set up the ordinary lens in its place, so as to bring the opening in the diaphragm to a focus at a distance of 2 or 3 m. in front of the lantern.

(*b*) Take one of the cardboard screens used in Experiment 115, and move it along the path of the ray beyond the lens. It will be seen that the cone of light lying between the lens and the focus is surrounded by a ruddy fringe or border. This is due to the fact that red light is not so greatly bent from its course as are the other colors.

(*c*) Move the screen away from the lens until it passes through the focus and intercepts the diverging cone beyond that point. It will now be found that the red fringe has been supplanted by a bluish one.

367. Achromatic Lenses. — Achromatic aberration in lenses is corrected by making use of a combination of two lenses differing greatly in dispersive power. If we take a bicon-

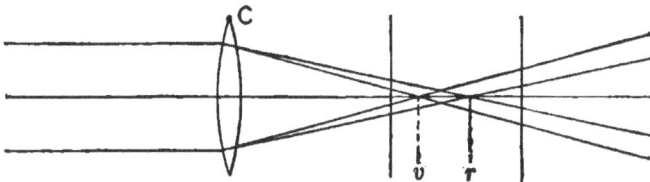

FIG. 372.

vex lens (*C*, Fig. 372) of crown glass, the dispersive power of which is slight, the violet light will be brought to a focus at *v* and red light at *r*. By placing behind this a diverging lens of flint glass (*F*, Fig. 373), which is much more highly dispersive, the violet rays, which tend to a focus too soon, will be bent outward more strongly than will the red, and it is possible to construct these lenses so that both sets of rays will come to a common focus at *vr* (Fig. 374). Such a combination is called an *achromatic lens*. Achromatic lenses are always used in the objectives of telescopes, microscopes, and other optical instruments where chromatic aberration would be detrimental.

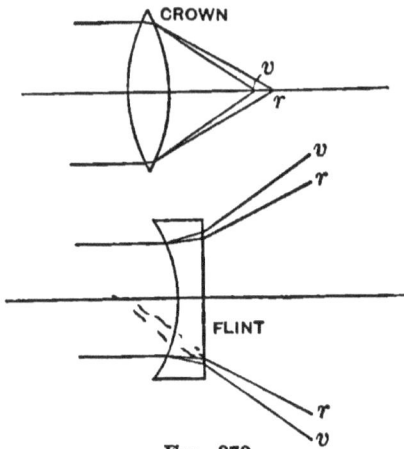

FIG. 373.

368. The Microscope. — The essential parts of this instrument, namely, the objective and the eyepiece, are the same as in the telescope; but the objective is modified to adapt it to the work for which the microscope is intended. This consists in producing an enlarged real image of a small object near at hand, and further magnifying this image by

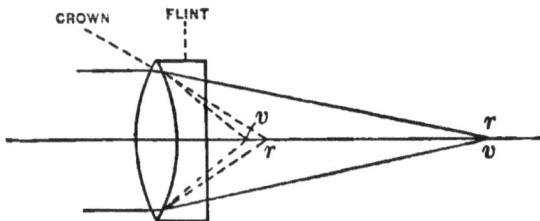

FIG. 374.

means of the eyepiece. The objective is therefore a very small, short-focused combination of lenses so constructed as to be achromatic. When this is brought near the object to be observed, a real image is formed in the eyepiece, the diameter of which is great, as compared with that of the object, in proportion to its distance from the lens. A virtual image of this is formed by means of an eyepiece, as has already been explained in the case of the telescope. In Fig. 375, the essential parts of a microscope are shown.

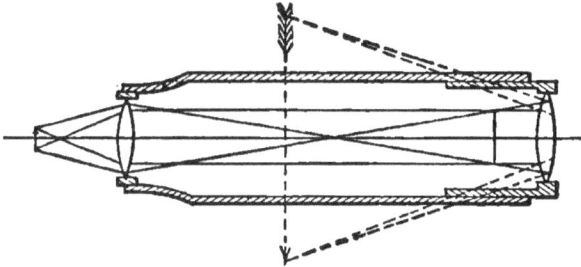

Fig. 375.

For simplicity, the objective and eyepiece are represented as simple lenses.[1]

As in the former case, the magnifying power is expressed by means of the increased apparent diameter of the virtual image, not in terms of its area. The magnifying powers of microscopes sometimes exceed one thousand diameters.

In physics, the microscope is chiefly used for purposes of measurement. Three different methods are employed:

(1) The stage upon which the object under observation is mounted is moved along under the objective by means of a micrometer screw.

(2) The cross-hair of the eyepiece is moved through the field of view by means of a micrometer screw.

(3) A glass scale, divided to tenths of a millimeter, is placed in the focus of the eyepiece so that the image of objects under observation will be superimposed upon it. In order to reduce these readings to centimeters, it is necessary to place under the objective another scale ruled

---

[1] For further information concerning the microscope, see Appendix IV; also Gage's *Microscopical Methods.*

2 D

to such fine divisions that the whole of at least one division will come within the field. The size of the image of this division upon the eyepiece micrometer having been noted, the value of a scale division of the eyepiece micrometer can be computed. This, which is the simplest method of micrometric measurement, can be illustrated by means of the following experiment (see Appendix IV):

369. EXPERIMENT 118. — Measurement of the Diameter of a Hair.

Apparatus:

(1) A microscope with a low-power objective (1 or 2 cm.), and an eyepiece provided with a glass micrometer.

(2) A scale ruled upon glass, the divisions of which do not exceed tenths of millimeters.

Procedure:

(*a*) Place the glass ruling upon the stage, and focus the microscope upon it. Determine the value of one division of the eyepiece micrometer in terms of the glass ruling, the value of which must be known.

·(*b*) Lay the hair upon the stage, holding it taut by means of wax, and adjust so that it will lie parallel to the lines of the eyepiece micrometer. Read its diameter upon the scale of the latter, and reduce those readings to centimeters by means of the foregoing determination of the value of the scale division of the micrometer. The diameter of small wires, and of any objects minute enough to come entirely within the field of the microscope, can be measured in this way.

(For a full discussion of the systems of lenses used in various optical instruments, see *Elements of Physics*, Vol. III, Chaps. I to VI.)

CHAPTER XLII

POLARIZATION, DOUBLE REFRACTION AND INTERFERENCE

370. Polarization by Reflection. — We may consider a ray of ordinary light, which may be said to contain vibrations in every possible plane, to be made up of two component vibrations at right angles to each other. Imagine a plane mirror placed obliquely to this ray, and that one of the components of the ray vibrates in a direction parallel to the surface of the mirror, while the other makes an angle with the same. It is natural to suppose that these two components would be affected differently upon reaching the surface. In point of fact, it is found that one of them tends to penetrate the glass, and the other to be reflected from it. At a certain angle, called the angle of complete polarization, this tendency produces a complete separation of the two components of the ray. One component, as

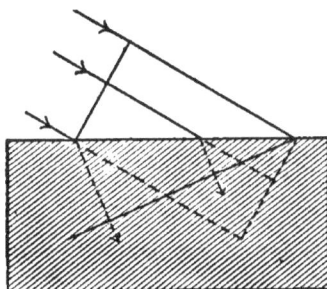

FIG. 376.

shown in Fig. 376, penetrates the material of the mirror, and is transmitted as a refracted ray; the other is reflected. This reflected ray contains only vibrations in one plane. It constitutes what is called a beam of *plane polarized light*. To show that it consists of vibrations in a single plane only, a second mirror may be interposed in its path at the angle of polarization. If this mirror be turned, as shown in Fig. 377, it will reflect the polarized ray because the vibrations meet its surface in the same manner as in the case of the first mirror. If, however, it be turned through 90°, the ray will enter the mirror, and

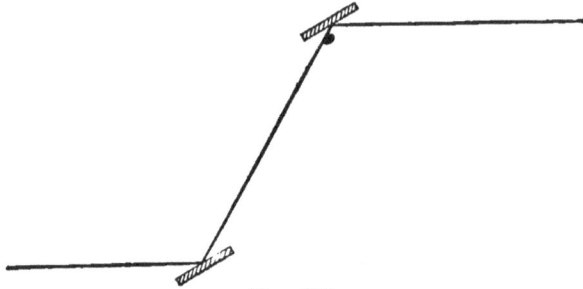

FIG. 377.

there will be no reflected ray. For this experiment, it is necessary to use mirrors in which the reflection occurs from the front surface only. The most suitable material is black glass, in which the refracted ray is entirely absorbed. A metallic reflecting surface does not polarize light as above described.

371. Polarization by Means of Double Refraction. — Certain crystals are so constituted that the velocity of light transmitted by them depends upon the plane of its vibration. If a beam of ordinary light enter such a crystal, it will be resolved into two components, which will vibrate at right angles to one another. These two rays will travel through the crystal at different velocities. One of them, therefore, will be relatively retarded, and will fall behind the other. If the beam of light enter the crystal obliquely, the ray which is most retarded will be bent further from its path than the other, and the two beams, on emerging, will follow different paths. (See Fig. 378.) Every such crystal divides the light which it transmits into two op-

positely polarized rays, one of which is called the *ordinary*,
and the other, to distinguish it, the *extraordinary*, ray.
The phenomenon is called *double refraction*.

FIG. 378.

372. The Nicol Prism. — In order to obtain a single
plane polarized ray by double refraction, it is necessary to
suppress one of the rays within the crystal. This is done
by means of a device known as the *Nicol prism*.

A rhomb of Iceland spar, which is a doubly refracting
substance, is cut obliquely through, and cemented by
means of Canada balsam. The position of the surface thus
formed is such that the ordinary ray is totally reflected,
while the extraordinary ray is transmitted. (See Fig. 379.)
In this way, a single beam of polar-
ized light is obtained, the vibration
of which depends upon the position
of the prism, and turns with it.
Polarized light entering a Nicol

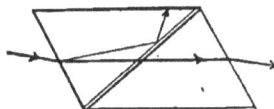

FIG. 379.

prism, if vibrating in the plane of the extraordinary
ray, will be transmitted; if vibrating in the plane of the
ordinary ray, it will be extinguished. Light which vibrates
in any intermediate plane is resolved into two components;
one of which takes the plane of vibration of the extraordinary ray, the other that of the ordinary ray. The
former component only is transmitted. Ordinary light is
likewise resolved into the two above-mentioned components, and is converted into polarized light.

In the study of polarization, two Nicol prisms are com-

monly used; the first of these is called a *polarizer;* the
second, an *analyzer.* (See Fig. 380.) They are mounted
with a common axis about which each is capable of revolution. Such an arrangement is called a *polariscope.*

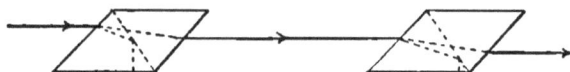

FIG. 380.

373. EXPERIMENT 119. — Double Refraction in Calcite.

Apparatus:
(1) The lantern.
(2) A pair of Nicol prisms.
(3) A rhomb of calcite.

Procedure:

(a) ·Place before the condenser of the lantern the diaphragm with
circular opening previously described, and focus the image of the
same upon the screen. Mount between the diaphragm and the objective the rhomb of calcite r, as shown
in Fig. 381. Note that, owing to
double refraction in the crystal, two
images appear.

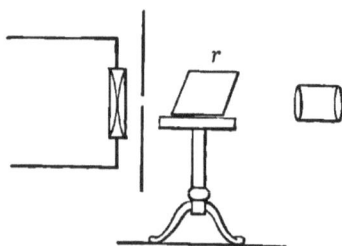

FIG. 381.

(b) Hold one of the Nicol prisms
between the calcite rhomb and the
objective, and turn it in the hand, the
axis of rotation in the path of the ray.
Note that in certain positions of the
Nicol prism one of the images is
entirely extinguished, while the other appears undimmed, and that
by turning the Nicol prism further through 45° both images appear
of equal brightness. Note that a further revolution through 45°
extinguishes the other image, and so on until the circle is completed.

The interpretation of this result follows directly from the foregoing
discussion of double refraction. The two images are formed by rays
of light which are polarized, one at right angles to the other. When
the plane of polarization of the Nicol prism corresponds with the
plane of vibration of one of these rays, that ray is transmitted while
the other is extinguished. In the position 45° from this, both rays

are resolved into components, one of which is transmitted and the other extinguished; thus two rays appear, and of equal brightness.

(c) Hold the Nicol prism between the diaphragm and the calcite rhomb and turn as before. In this case we have a single polarized ray passing through the crystal. This is resolved into components vibrating in the planes of the ordinary or extraordinary rays in calcite. There are, however, two positions in which one of these components becomes equal to zero, and two in which the other component is destroyed. In these four positions a single image only appears; in intermediate positions both images are present, their relative brightness depending upon the size of the respective components.

(d) Remove the calcite prism and place the two Nicol prisms in the path of the ray. Turn one of these until their planes of polarization form a right angle with one another. The polarized ray produced by the polarizer is now incapable of passing through the analyzer, and no image of the aperture appears. Nicol prisms in these positions are said to be *crossed*. As we turn the analyzer slowly from this position, the image reappears, increasing in brightness to a maximum when the planes of polarization of the two prisms coincide.[1] Having crossed the Nicol prisms, insert the calcite crystal between and note the restoration of light to the field of view.

Anything which changes the plane of vibration of the light entering the analyzer, so that it does not vibrate at right angles to the plane of polarization of that prism, will cause light to be transmitted to the screen. Any double refracting substance, of which there are many, interposed between the Nicol prisms, will have this effect. Plates of mica, selenite, quartz, calcite, etc., etc., if held between the Nicol prisms, will answer for this purpose. There is, indeed, no simpler method of determining whether a substance is doubly refracting than to place a piece of it between crossed Nicol prisms. Many liquids, such as solutions of sugar and of tartaric acid, produce a similar effect, not because they are capable of double refraction, but because they have power of rotating the plane of light transmitted through them. In the case of sugar this property is utilized for determining the strength of the solution. A polarizer arranged for this purpose is called a *saccharimeter*.

---

[1] The brightness of the image is proportional to $\cos^2 a$, where a is the angle between the planes of polarization of the Nicol prisms.

374. EXPERIMENT 120. — Double Refraction produced by Stress.

Apparatus:
(1) The lantern.
(2) A pair of Nicol prisms; a block of plate glass, about 4 cm. square; an iron clamp.

Procedure:
(*a*) Set up the Nicol prisms between the condenser and the objective lens of the lantern and cross them. Hold the block of plate glass between the Nicol prisms, and note that no effect is produced thereby.

(*b*) Place the block of glass in the jaws of the clamp so that pressure may be applied at the center of opposite edges; hold it thus clamped in the field and increase the pressure by turning the screw of the clamp. Note that light appears in the field of the crossed Nicol prism, with a beautiful colored pattern due to interference of the transmitted rays. Relieve the pressure, and note that the glass returns to its former condition. Similar effects may be produced by holding a strip of glass in the field and bending it. Also by clamping a glass rod or tube horizontally so that one end passes through the field of the lantern. If this rod be rubbed longitudinally with the moistened hand, so as to cause it to vibrate, it will be strained in such a way as to produce total refraction and to cause light to pass the polarizer.

375. The Interference of Light Waves. — Two rays of light of the same wave length, moving along the same path in the same direction, produce a motion of the ether

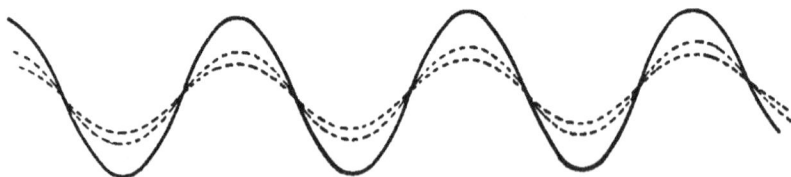

FIG. 382.

which is the resultant of the two separate motions due to the two waves. If the two waves are vibrating in the same plane and in the same phase (see Fig. 382), the resultant motion will be the sum of the two. If the two waves are of opposite phase (see Fig. 383) and are equally bright, their motions will annul each other, and the

FIG. 383.

resultant motion of the ether will be their difference. In the latter case, when the amplitudes are equal, darkness is produced.

This combination of the motions of two light waves to produce darkness is called *interference*. It forms the basis of many very beautiful optical phenomena, of which only a few of the simplest can be described here.

376. EXPERIMENT 121. — Colors of the Soap Film, produced by Interference.

Apparatus:
(1) A flat dish containing a soap solution.
(2) A lamp chimney of the form shown in Fig. 384.

Procedure:
(a) Dip the base of the chimney into the soap solution and remove it carefully. The film which stretches itself across the base of the chimney, and which can be removed with it, immediately finds its way to the narrowest part of the base, where it forms a plane diaphragm.

FIG. 384.

(b) Set up the chimney so that the soap film will be in a vertical position, and observe the changes which it undergoes. The film, at first colorless, grows thinner by the evaporation of its particles and by a process of drainage, in consequence of which the liquid flows downward under the action of gravity and escapes. Very soon colors begin to show themselves in the upper portion of the film, and, if the latter remain undisturbed, these colors will arrange themselves in parallel horizontal bands which move slowly downward until the entire surface is covered with them. As time goes on, new ones appear continually at the top of the film and join in this downward movement. Finally, if the film is not broken, the upper portion changes color again to a neutral gray, sometimes almost black, and no more colored bands are formed. Note that the bands follow the order of the colors of the spectrum, running from red to violet. The explanation of these phenomena is as follows:

The rays which fall upon the film are reflected from the front and from the back surface. A ray a, b (Fig. 385), which passes through the film and is reflected from the back surface, afterwards traveling along a common path d, e, with a ray c, d, reflected from the front surface, has had a longer distance to travel, and differs from the latter ray in phase. If the distance through the film and back again is such that the two reflected rays are in opposite phase, they destroy each other. Any very thin film that reflects light from its two surfaces will produce complete interference between light of some given wave length. Other wave lengths will not interfere completely. If the ray which falls upon the film consists of white light, that which is returned from it will be white light minus some color which has been destroyed by interference. The result, therefore, of interference in the case of the soap bubble and of other thin films is to produce variations of color instead of mere gradations of light and shade. Owing to the vertical position of the film and the action of gravity upon it, the liquid tends to the bottom, giving the film a wedge-shaped cross-section. It is, therefore, always thinnest at the top. As we go from the top downward with the increasing thickness, the different colors of the spectrum are caused to interfere, and the result is the arrangement of the colors in horizontal bands as above described. Anything which stirs up the liquid within the film, which may be done by blowing upon it gently, will destroy this arrangement, mixing the colors up temporarily. Upon being allowed to rest, the old arrangement of bands re-establishes itself. The neutral color, which the film finally assumes before breaking, is due to the fact that it has become so thin, as compared with the wave length of light, that there is not a sufficient difference of phase between the waves reflected at its two surfaces to cause interference.

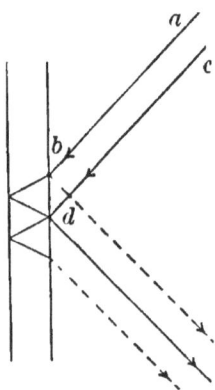

FIG. 385.

(c) Repeat the above-described observations in a darkened room, using as the sole source of illumination a sodium flame. Note that the interference bands are now alternately black and yellow. Those regions in the film, the thickness of which is such as to produce opposite phases in the waves of sodium light, appear black; those regions where the thickness is such as to give the same phase in the waves reflected from the two surfaces possess the greatest brightness.

377. Interference Colors in Nature. — Many of the most beautiful color effects in nature are produced by interference of the light reflected from thin films. The pearl, the opal, and other gems owe their colors to interference. The coloring of many insects and of the plumage of some birds, also the coloring of the linings of shells, is ascribable to the same cause. All the effects included under the term *iridescence* are due to interference.

378. Newton's Rings. — Another case of the interference of light is that produced when a lens is laid upon a plane glass plate, or when two lenses differing in curvature are brought together. If the surfaces be thoroughly cleaned so that they come well together, interference occurs between the light reflected from the inner surface of the upper lens and the upper surface of the

FIG. 386.

glass beneath. The intervening path through the air increases as we pass from the center where the glass surfaces are in actual contact outwards (Fig. 386). The result is a series of circular interference bands following the order of the colors of the spectrum. These are known as *Newton's rings.*

Whenever two glass surfaces, which do not fit each other precisely, are brought together, interference bands show themselves. If the surfaces be irregular, these take on irregular forms. If two surfaces could be made which fitted each other precisely, so that the distance between them was everywhere the same, the interference color would be uniform for the entire surface. This fact affords an exceedingly delicate means of testing the precision with which glass surfaces are ground. If two plates which fit each other very perfectly are laid one upon the other, the interference bands will run across nearly in straight lines, and will be very broad. If now the finger be placed upon the upper plate with slight pressure for a moment, this pressure will be sufficient to distort the upper plate and throw these bands out of shape and position. Upon removing the finger, the effect is found to remain after the pressure is relieved, owing to the distortion of the upper plate on account of unequal heating. The interference fringes require several minutes to return to their original position again. The observation should be made by sodium light, because the number and sharpness of the bands is thus greatly increased.

379. Interference Phenomena in Polarized Light. — When a double refracting crystal is placed between crossed Nicol prisms, as described in the foregoing chapter, and the light is thus restored, colors frequently show themselves. These are due to the interference of the ordinary with the extraordinary ray. These have difference of phase owing to the fact that their velocities in the crystal differ from one another. Crystals of different thicknesses bring different colors into interference. Thus a bit of mica which is made up of overlapping layers or plates, and which is not uniform in thickness, will frequently show, when placed between the Nicol prisms, a complicated field of various colors.

CHAPTER XLIII

VISION AND THE SENSE OF COLOR

380. Essential Parts of the Eye. — The eye consists of a nearly circular chamber, the eyeball, within a round opening, in front of which is a lens (Fig. 387). By means of this lens real images of objects before the eyes are thrown upon a sort of screen or background, which is called the *retina*. The retina consists of a network of nerve fibers of very complicated structure which form the ends of the optic nerve. This bundle of nerves extends from the brain to the eye, and enters the latter through an opening at the back. Thence the nerve fibers spread out over the interior of the eyeball, forming the retina.

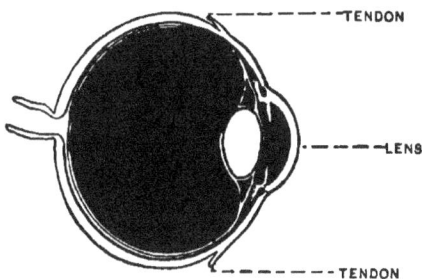

FIG. 387.

In order to bring near and distant objects to an equally good focus upon the retina without changing the distance between the lens and the image, the lens itself is made to change its form by means of a ring of muscles surrounding it. This process is called *accommodation*. When distant objects are to be brought to a focus, these muscles contract, stretching the lens radially, and thus making it thinner and increasing its focal length. To see objects near at hand, which, in this stretched condition of the lens would be brought to a focus too far back, the focal length of the lens must be diminished. This is done simply by relaxing the muscles of accommodation. In the normal eye in youth the process of accommodation makes it possible to see with equal sharpness objects 15 or 20 cm. distant from the eye and those miles away. With advancing age the power of accommodation diminishes, and as a rule the focus is better for distant than for near objects. This lack of accommodation is made good by the use of spectacles.

381. Near and Far Sightedness. — In many eyes the focal length of the lens is somewhat too great. Persons whose eyes have this peculiarity are said to be far-sighted. They can see distant objects sharply defined, but the power of accommodation does not bring objects near at hand into focus.

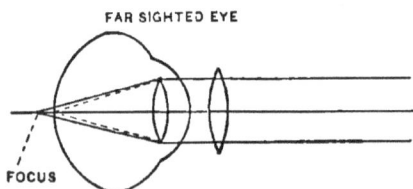

FAR SIGHTED EYE

FOCUS

FIG. 388.

This difficulty is remedied by the use of a convex lens in front of the eye. (See Fig. 388.) In other persons the focal length of the lens is too short, and only objects quite close at hand can be brought to a focus by the use of the muscles of accommodation. The attempt to adjust the eye continually to objects farther away produces serious nervous strain. The use of glasses is therefore very important in cases of near-sightedness. The lens to be used is one which will increase the focal length, namely, a divergent lens (Fig. 389).

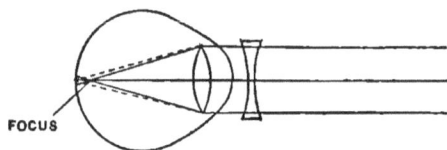

FOCUS

FIG. 389.

In order that the lens of the eye may be capable of changing form under the action of the muscles of accommodation, it is kept moist by the liquids which fill the eyeball. These are called the *humors* of the eye. That between the retina and the lens is called the *vitreous humor*. In front of the lens is the *aqueous humor*. It is held in place by the cornea, a convex window not unlike a watch glass in shape. The cornea is a transparent continuation of the coat of the eyeball called the *sclerotic coat*. Between the lens and the cornea is a colored diaphragm with a central circular aperture. This diaphragm is the *iris* and the opening is the *pupil of the eye*. The iris serves the same purpose as the adjustable diaphragm in the lens of a camera. It regulates the amount of light falling upon the retina, and by "stopping down" the lens greatly increases the sharpness of vision. It contains a set of radial muscle fibers, which by their involuntary action continually adjust the opening to suit the conditions of illumination.

The eye is moved chiefly by means of four muscles attached to the sclerotic coat. Two of these, which are attached above and below by means of the tendons shown in Fig. 387, give the eye motion in a vertical plane. The other pair enable us to move the eye in a horizontal plane.

382. The Region of Distinct Vision. — Although the retina covers a large portion of the interior of the eyeball, it is only a small region directly in the axis of the lens which gives distinct vision. The process of seeing consists in moving the eye continually, by means of its muscles, so as to bring different portions of the image of the object at which we are looking into this region of distinct vision. Were the eye incapable of such movement, we should get no distinct impressions concerning portions of the image which lay outside this region of distinct vision.

In order to test this point, it is only necessary to fix the

eye upon one end of a printed line in the middle of a page
of a book held about 20 cm. away. It will be found impossible to read half across the line, although we have an
indefinite impression of the presence of the remaining
words. It will be found in the same way that if we
look at the upper line of a page, and are successful in
preventing the tendency of the eye to fall in the direction to which our attention
is directed, the third line
below cannot be read.

The movements of the eye,
above described, which are
purely automatic, constitute
a process of measuring the
angular distances between
different parts of the image.
From these measurements we
get our estimate of shape
and size, and from them indirectly, in the case of familiar objects, the size of which
is known to us, we estimate
distance. From our judg-
ment as to the distance of
the object on the other hand,
we estimate its size.

FIG. 390.

The influence of our estimate of distance upon our judg-
ment concerning the relative sizes of things is illustrated
in Fig. 390, in which the three human figures are all of
the same height. The perspective is so arranged, however, as to make them seem at different distances, and the
observer unconsciously assigns to each a size proportionate
to its apparent distance.

383. Binocular Vision. — In the matter of estimating size and distance, we are greatly aided by having two eyes situated some little distance apart. The consequence is that we get two slightly different views of the object, and our best means of judging distances is by comparison of these. The process of viewing objects simultaneously with both eyes is called *binocular* or *stereoscopic vision*.

The extent to which we are dependent upon stereoscopic vision, in our estimate of the position of things, may be illustrated by means of the following very simple experiment:

Close one eye, and approach, from some distance, an object like the edge of a table. When within easy arm's reach, attempt to place the point of a pencil, held in the hand, in contact with the object in question. It will be found impossible to do so until, after successive trials, the distance has been gauged by means of the motion of the arm. With two eyes, however, the point may be touched at the first attempt.

384. Color. — Color is produced by any process which modifies the composition of white light. When light is dispersed, forming a pure spectrum, each ray possesses its individual color. Colors thus produced are called *spectral tints*. In nearly all cases, however, bodies owe their color to the removal of some portion of white light, either by interference or by absorption. The production of color by interference has already been considered. (See Chapter XLII.) Nearly all cases of the production of color fall under the second head, namely, absorption. When light falls upon a colored object, such as a pigment, the rays which penetrate the body are reflected from within. These rays suffer absorption, but some wave lengths are more

2 E

diminished in intensity than others. The remaining light produces an effect upon the eye other than white, and gives the body its characteristic color. Thus a pigment like red lead absorbs nearly all the shorter wave lengths, and only red light, with an admixture of yellow, is transmitted. Green pigments are produced by the absorption of both red and violet, blue by the absorption of the red and green.

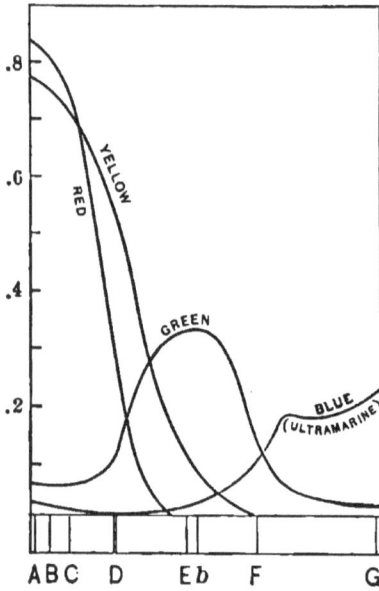

Measurements of the different wave lengths reflected by three typical pigments, and expressed graphically by means of curves, are shown in Fig. 391. None of the colors produced in these cases are pure; that is to say, there is, in all cases, a considerable admixture of light other than that which gives the prevailing color tone. *Purity of color* denotes absence of these admixtures. The only perfectly pure colors are those of the spectrum formed by a narrow slit. Colors differ also in *saturation*. This term denotes the absence of admixture of *white light*. In proportion as the white light is done away with, and the tint becomes more and more intense, it is said to be more and more saturated. There is, in all ordinary cases, a considerable amount of white light present in the rays reflected by pigments, because not all the light penetrates the substance, and thus a considerable portion is reflected from the surface, and this does not suffer absorption.

385. Dependence of Color upon the Character of the Incident Light. — Colored objects are capable of modifying the light which falls upon them only by process of absorption. They can therefore reflect back no rays which do not reach them. The colors which they present contain no light which has not been thrown upon them from without. In proportion as the source of illumination is rich, the effect will be varied upon the bodies which it illuminates. Daylight, for example, is richer in violet and blue rays than any artificial light. Colors seen by daylight, therefore, send the eye a larger proportion of these tints. If, for daylight, we substitute gaslight, which is comparatively weak in violet rays, objects of a bluish cast tend towards green. Differences in the color of bodies, which are perfectly obvious by daylight, are scarcely to be distinguished when the bodies are seen by lamplight. If, for lamplight, we substitute, in turn, the sodium flame, which possesses only one color, yellow, we find all pigments, whatever may be their colors by daylight, reduced to a mixture of yellow and black. This may be strikingly illustrated by taking a pile of colored worsteds containing a large number of different tints (the collection used in the Holmgren test for color blindness [Art. 390] is well adapted for this experiment), and viewing them, first by daylight, and then in a perfectly darkened room, by the light of the sodium flame. The matter of the dependence of color upon the source of illumination may be more definitely brought out by means of the following experiment:

386. EXPERIMENT 122. — Observation of Pigment Colors by Means of the Spectrum.

Apparatus:

(1) The lantern with lens and prism.

(2) Holmgren worsteds, or any other collection of brightly colored objects.

Procedure:

(*a*) Adjust the lens and prism so as to project a spectrum upon the screen. Darken the room completely.

(*b*) Select a skein of scarlet worsted, and move it slowly through the spectrum from red to violet. Note that, while illuminated by the red rays of the spectrum, it appears in its natural tint and very bright. As it passes out of the red into the orange, it grows rapidly darker in color. When the green is reached, the worsted has become black, nor does it resume its natural lively tint again in passing through any of the shorter wave lengths of the spectrum. This worsted absorbs nearly all the wave lengths of the spectrum excepting those at the red end. It appears black, except when illuminated by rays which it is capable of transmitting.

(*c*) Repeat, using a piece of green worsted, and note that, at the red end of the spectrum, this appears very dark, and that it grows rapidly brighter when moved through the yellow into the green. It retains some portion of its brightness throughout the blue. Specimens of blue worsted, or of any artificially blue body, tested thus will show, as a rule, some power of reflecting red light. They will appear brighter when illuminated by the red than by the yellow or green. It is only in the blue and violet light, however, that they take on their usual appearance. There are scarcely any pure blues among pigments. Nearly all are purples; that is to say, they reflect a considerable percentage of red light as well as of blue.

(*d*) Repeat the above-described observations, using a fresh blade of grass. The coloring matter of this is chlorophyl, the spectrum of which has already been described in Experiment 113. Although the dominant color tone of grass is green, not all the wave lengths of red are absorbed by it. As we move the blade of grass through the spectrum, we find that it is capable of reflecting quite strongly in certain regions of the red, and becomes suddenly black in passing into others; the latter are the regions of the absorption bands of chlorophyl.

387. The Theory of Color Sensations. — To understand the different sensations produced upon the mind by different wave lengths of light, it is necessary to consider the process by which impressions of color are acquired. · The ray entering the eye falls upon the retina, and excites the ends of the fibers of the optic nerve, which are spread out in the form of a network within the eye. By means of the excitation of these nerves, messages are sent to the brain.

Various theories have been proposed to account for the different sensations produced by the action of different wave lengths of the spectrum. The simplest of these, and the one usually adopted by physicists as a working hypothesis for the explanation of color phenomena, is known as the *Young-Helmholtz Theory of Color.*

According to this theory, there are in the retina three distinct sets of nerves : one of which conveys the sensation of red, another the sensation of green, and the third a sensation of violet to the brain. We may call these nerves the *red nerve*, the *green nerve*, and the *violet nerve*. The sensations which they produce, *i.e. red*, *green*, and *violet*, are called the *three primary color sensations.* By the combination of them, all known color sensations may be produced. It has been found that a trained observer can distinguish about ten thousand differences of color, comparatively few of which have received names. Every one of these may be formed by an admixture of the three primary colors in proper proportions. The *red nerve* is chiefly affected by the longer wave lengths of the spectrum. Objects, then, send such rays to the eye give us the imp we call red. All wave lengths have so ever, upon these nerves. If we could is

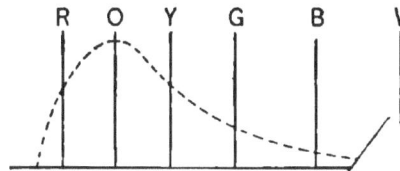

FIG. 392.

of red, and test our color sense by subjecting the *red nerve* successively to the different wave lengths of the spectrum, we should be able to express the intensity of the effect by means of a curve, as shown in Fig. 392. In the same way the *green nerve* is sensitive to all the wave lengths of the spectrum, but chiefly to those which lie in the region which we call green. The effect of different wave lengths upon it would be expressed by the corresponding curve (Fig. 393). The effect of the various wave lengths of the spectrum upon the *violet nerve* would give us still another curve, as shown in Fig. 394.

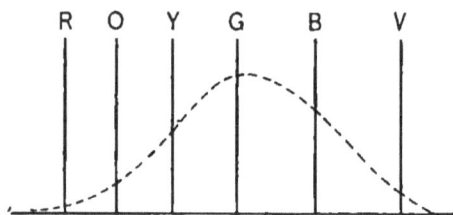

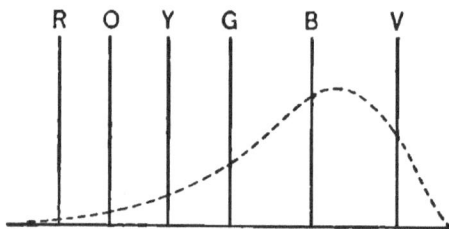

FIG. 393.

When a ray of light falls upon the retina, we may suppose these three nerves to send messages simultaneously to the brain, where they are combined in proportion to the intensity of the effects produced. This combination gives us the sensation of color. Nearly all the phenomena of color are satisfactorily accounted for by the use of this theory.

FIG. 394.

388. EXPERIMENT 123.— Contrast Effects.

Apparatus:

(1) A sheet of red and also a sheet of green paper. The colors of these should be as nearly pure as it is possible to obtain them.[1]

---

[1] A sample book of colored papers, which may be procured from various manufacturers, will be found useful in the performance of these experiments.

(2) A small piece of red paper differing slightly from the fore-going in color. A corresponding piece of green paper also differing slightly from the larger sheet. These pieces should be about 5 cm. × 10 cm.

Procedure :

(*a*) Cut the small pieces of colored paper in half, making two squares of each, 5 cm. in diameter.

(*b*) Place one of the red squares upon the red sheet, and the other upon the green. Note that the latter immediately takes on a much livelier tint than the former, although it has been cut from the same piece. This difference is due to a difference in the condition of the nerves of the eye. When we look at the red paper superimposed upon the background of red, the *red nerve* of the retina becomes fatigued by continued excitation. The impression carried to the brain is therefore weakened. When we view the same piece of paper upon the green background, it is the *green nerve* in the eye which becomes fatigued. In both cases the impression may be regarded as a com-bination of the primary sensations of red and green, because, as has already been stated, all wave lengths affect these two nerves to some extent. When the *red nerve* is fatigued, the percentage of green is greater than the normal, and the result is to deaden the intensity of the color effect. When the *green nerve* is fatigued, the relative strength of this impression is weakened, and since this is an admixture which interferes with the brilliancy of the colors, the disappearance of it heightens the impressions of red. These effects are called *contrast effects.*

(*c*) To illustrate this point further, attach to one of the small green squares a thread about 50 cm. in length. Place the square upon the red background, fixing the eye upon it intently, and taking care not to allow it to wander for an interval of about thirty seconds ; then withdraw the green square suddenly by means of the thread, and do not follow it with the eye. It will be found that that portion of the sheet of red paper which had been covered with the green now appears of a much livelier tint than the surrounding surfaces. The effect lasts for several seconds. It will be found upon moving the eye that the brilliant patch is not a fixture upon the red surface, but moves as the eye moves. The explanation is as follows :

In the small region of the retina upon which an image of the green patch had been formed, the *green nerves* become greatly fatigued.

When the patch is removed, this portion of the retina sends to the brain messages of red and green, in which the latter element is therefore much weakened. The result is a livelier impression of red than would be obtained were the green nerve of its ordinary sensitiveness.

(*d*) Repeat this observation, using a red square with thread attached, upon the green background. It will be found that the result is a greatly increased sensitiveness to green in the part of the retina previously exposed to the red square.

389. Color Blindness. — Most people possess the three primary color sensations, above described, but about four per cent of the male population lacks either the sensation for red or the sensation for green. Such people are said to be *color blind*, and they are spoken of as *red blind* or *green blind* according to the character of the defect. This peculiarity of vision is almost unknown among women. It is an organic deficiency which cannot be overcome or in any way modified by training. Although the color sense of those individuals who have two instead of three primary colors is entirely different from that of the remainder of the race, they learn in childhood to use the same color names as those who have normal vision, and oftentimes they go through life without knowing of the difference which exists between their sense of color and that of their neighbors. Color blindness can be detected with certainty by means of the very simple method of Professor Holmgren of Upsala, Sweden. This method was devised shortly after a terrible railway accident in that country, which had been caused by the color blindness of an employee. It is as follows:

390. The Holmgren Test. — The outfit for this test, which may readily be performed by any intelligent person, consists of a large number of small skeins of colored worsted. The colors are reds, greens, blues, purples, and their mixtures, together with a variety of neutral grays and browns. Two skeins, one a pale *apple green*, the other a light reddish purple or magenta, are termed the *confusion* samples. The person to be tested is shown the confusion samples in turn, and is requested to select from the other worsteds all those which appear to him most nearly related to the former.

Color-blind observers select worsteds which no one with normal vision could be brought to consider as in any way similar to the confusion samples. In the selection made by a red-blind subject we may find, for example,

With the Apple Green.	*With the Magenta.*
(1) Light yellows and straw colors tending towards gray.	(1) Purples and violets, both light and dark (in none of which red is predominant).
(2) Pale pinks.	(2) Blues, both light and dark.
(3) Light browns or fawn colors.	

The same observer, when requested to pick out those samples which resembled most nearly a skein of scarlet worsted, will include in his selection all the dark browns and olive greens.

The importance of knowing the character of the color vision of those who have to make use of colored signals is now so well recognized that mariners, pilots, soldiers, railway employees, etc., and in many countries the whole school population are tested for color blindness.

APPENDIX I

TABLE OF THE RELATION OF BRITISH MEASURES TO THE METRIC SYSTEM

(1) LENGTH.

The unit is 1 centimeter (cm.) = 0·393704 inch.

$\frac{1}{10}$ centimeter = 1 millimeter = 0·0393704 inch
100 centimeters = 1 meter   = .39·3704 inches
1000 meters   = 1 kilometer = 39370·4 inches
                           = 1093·6 yards
                           = 0·62137 mile

(2) AREA.

The unit is 1 sq. centimeter ($\overline{cm.}^2$) = 0·155000 sq. inch.

100 sq. millimeters = 1 sq. centimeter = 0·155000 sq. inch
10,000 sq. centimeters = 1 sq. meter   = 1550·00 sq. inches
                                  = 1·19599 sq. yards

(3) VOLUME.

The unit is 1 cu. centimeter ($\overline{cm.}^3$) = 0·06102 cubic inch.

1000 cu. centimeters = 1 liter   = 0·26417 U. S. gallon
                             = 0·22008 Imperial gallon
1,000,000 cu. centimeters = 1 cu. meter = 35·314 cu. feet
                             = 1·3079 cu. yard

(4) MASS.

The unit is 1 gram = 15·432 grains = 0·03215 oz. (Troy)
                                  = 0·03527 oz. (Avoirdupois)

$\frac{1}{1000}$ gram = 1 milligram = 0·015432 grain
1000 grams = 1 kilogram = 2·2046 pounds (Avoirdupois)

These tables give only the metric values which are used in physical measurements. In the complete series of quantities of the metric system, parts and multiples of each fundamental unit are given names by means of the prefixes, milli- ($\frac{1}{1000}$), centi- ($\frac{1}{100}$), deci- ($\frac{1}{10}$), deca- (10), hecto- (100), kilo- (1000), myria (10,000).

APPENDIX II

THE USE OF CROSS-SECTION PAPER

The results of very many physical measurements are capable of being expressed graphically by means of a curve. Such curves show the relation between two quantities, which it has been the object of the measurement to compare, in a much more definite manner than can be done by means of tables. One readily acquires the power of interpreting such curves.

There are many systems of co-ordination by means of which curves can be plotted, but the simplest and most useful of them all is the system of rectilinear or Cartesian co-ordinates. The paper used for the plotting of curves with Cartesian co-ordinates is divided into equal squares. It is a matter of convenience to have the fifth and tenth lines heavier than the intervening ones. The distance between lines should be large enough so that the position of a point anywhere upon the paper, with reference to the intervening lines and to the nearest heavy lines, can be seen at a glance without straining the eyes. It should, indeed, be possible to estimate a tenth of the smallest divisions with ease.

The smallest squares which can be readily used with the naked eye are square millimeters. There is a certain convenience in having the paper divided by lines the distances of which are in accordance with the metric system, but for many purposes, especially where the results obtained are not accurate to more than two places of decimals, a paper with larger squares is to be preferred. Half centimeters would not be too coarse a division for many purposes.

It is of much more importance that the lines shall be sharply defined, strictly parallel, and precisely at right angles to one another than it is to have the squares of a given size. Paper ruled at right angles by means of ordinary ruling pens is not sufficiently accurate. To be satisfactory, it must be printed from an engraved plate. Such plates are quite expensive, and it is therefore not feasible to get paper of any desired size. One has to choose between those which are

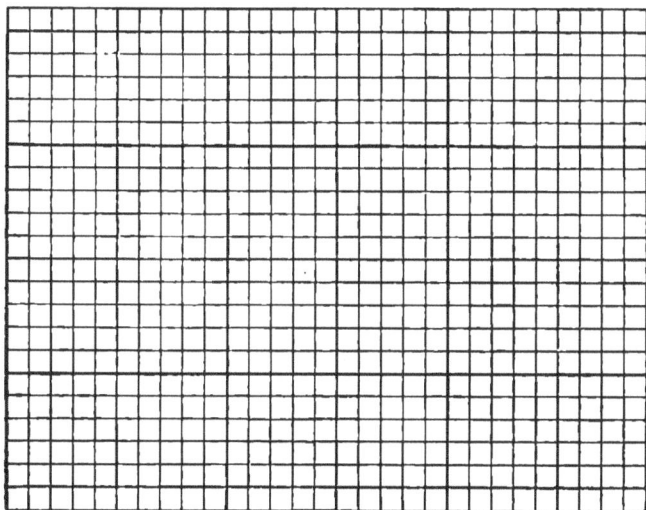

FIG. 395.

already on the market. Figure 395 shows a portion of a sheet of cross-section paper, the squares of which are large and open. Such paper is well adapted for the plotting of many of the curves obtained in the experiments described in this book.

APPENDIX III

BALANCES AND WEIGHTS

The measurement of mass is one of the chief occupations of the experimenter in physics, and a laboratory, however simple in its equipment, must therefore be provided with balances and weights. Every laboratory should contain at least two balances; one for the weighing of considerable masses, namely, from 5 to 500 g., and a finer balance for the weighing of masses less than 100 g. The latter should be a simple form of analytical balance such as is used in chemistry. It is well to place the very finest grade of such instruments in the hands of beginners. Both balances should be of good construction in their essential features and both should be provided with a pointer and scale. It is desirable that the fine balance should be inclosed in a glass case to obviate the disturbances from blasts of air. The balance for heavy masses should have broad pans and a stiff beam.

Should an analytical balance prove too expensive, for the finer weighings, a substitute may be provided by purchasing a portable hand balance with pans, and mounting the same upon a homemade wooden stand. This balance can be purchased for a small sum of any dealer in chemical supplies. It is a useful instrument, but should be used as an accessory to the better form and not as a substitute for it, if it be possible to obtain the latter.

There should be at least two sets of brass weights, the first running from 1 kg. to 1 g. for use with the heavy balance, the second from 100 g. to 1 mg. for finer weighings. It will be found advantageous likewise to have on hand a considerable number of extra weights below 1 g., since these on account of their small size are easily lost. For all ordinary practice work, what are technically called *second-quality weights* are sufficiently accurate. They cost much less than the weights of the first quality; upon the final adjustment of which much labor has to be expended. A set of the latter is valuable in any laboratory as a reference standard by means of which the accuracy of the second-quality weights may be tested. They should be reserved for such purposes, and not put into general use.

In addition to these weights, the laboratory should be furnished with a large number of iron weights. These cannot be found in the stock of dealers in scientific apparatus, but they may be obtained to order at any iron foundry. It is convenient to have three sizes, 5 kg., 1 kg., and $\frac{1}{10}$ kg. Of each size there are two distinct forms. One is a disk of

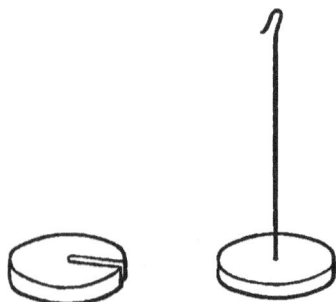

FIG. 396.

cast iron into which an iron rod bent into a hook at one end and flattened at the other is inserted, as shown in Fig. 396. The other form is a disk with a slot extending in past the center, of a sufficient width to enable the weight to slip over the weight to which the hook is attached. These two forms should be provided in all the sizes just mentioned. If patterns be provided in the following dimensions, the castings will be very nearly of the proper weight.

DIMENSIONS OF PATTERNS FOR WEIGHTS.

5 kilograms
$\begin{cases} \text{Diameter} & \text{16·2 cm.} \\ \text{Thickness} & \text{3·6 cm.} \\ \text{Width of slot} & \text{1·2 cm.} \end{cases}$

1 kilogram
$\begin{cases} \text{Diameter} & \text{10·0 cm.} \\ \text{Thickness} & \text{1·0 cm.} \\ \text{Width of slot} & \text{1·3 cm.} \end{cases}$

$\frac{1}{10}$ kilogram
$\begin{cases} \text{Diameter} & \text{4·4 cm.} \\ \text{Thickness} & \text{0·9 cm.} \\ \text{Width of slot} & \text{0·7 cm.} \end{cases}$

For the hooks (1 and 5 kg.) iron rods 1 cm. ($\frac{3}{8}''$) in diameter and 26 cm. long may be used. The disks are cast without slots and about 0·4 cm. thinner than the slotted weights. For the $\frac{1}{10}$ kg. hook weights, rods 0·5 cm. ($\frac{3}{16}''$) and 26 cm. long may be used. The disks need be only 0·4 cm. in thickness.

Cast-iron weights of the forms just described are not sufficiently accurate just as they come from the foundry, even for ordinary practice experiments. They should therefore be adjusted. If the patterns are of the sizes just described, the weights will nearly all be slightly in excess of the desired value. Only such castings as contain openings and blow holes of considerable size will fall below the normal weight. The weights should be taken in turn and placed upon the pan of the large balance. It having been found by about how much they are in excess, a hole or a set of shallow holes may be drilled in the lower side. When a lathe is not available, it is necessary to have the adjustment made in a machine shop.

The hook weights, of which there should be one for every ten of the slotted weights, are most readily adjusted by cutting off the end of the iron rod before bending the same. The size of the disk given for such weights in the preceding table is such as to allow for rods of the length indicated in the table. After adjustment the weights should be painted with asphalt varnish.

APPENDIX IV

READING MICROSCOPES

For the measurement of very small distances the instrument employed is the reading microscope. Measurements with this instrument are made in one of the following three ways:

1. *Method of the Eyepiece Scale.* — The eyepiece scale is the simplest attachment by means of which micrometric measurements can be made. It consists of a set of lines ruled upon a glass slide. This is slipped into the eyepiece of the microscope through openings cut opposite to one another in the tube. These openings are so situated that the scale upon the glass will be precisely in the focus of the eyepiece. The real image formed by the objective will then be seen superimposed upon this glass scale, and its dimensions can be read in

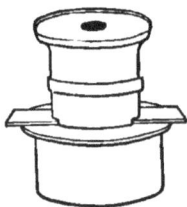

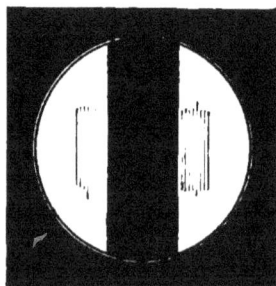

FIG. 397.   FIG. 398.

terms of the scale divisions. For eyepieces of ordinary power, the lines of the glass scale should be one one-hundredth of a centimeter apart. It is customary to rule fifty such lines upon the eyepiece scale. In order to reduce the reading obtained by means of this device to centimeters, it is necessary to calibrate the eyepiece scale by observations upon some object of known size placed under the microscope. This object may be another scale ruled upon glass or metal, or where that is not available, a fair degree of accuracy may be obtained by placing in the field of the microscope some object, such as a wire, the diameter of which has been carefully determined with a micrometer gauge. The diameter of this object is observed in scale divisions, and the value of one scale division in centimeters is computed from the observation. Figure 397 shows an eyepiece provided with an ordinary micrometer scale, and Fig. 398 the appearance of the scale as it is seen through the eye lens.

2. *Method of the Eyepiece Micrometer.* — The eyepiece micrometer is a device by means of which a cross hair in the focus of an eyepiece is moved across the field of view by turning a micrometer screw. The position of the cross hair is read by means of a scale upon the periphery of a drum-shaped head attached to the screw. Figure 399 shows a microscope provided with the usual form of an eyepiece micrometer. There is

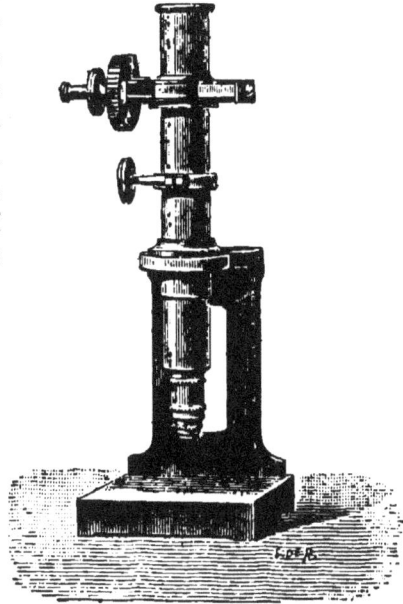

FIG. 399.

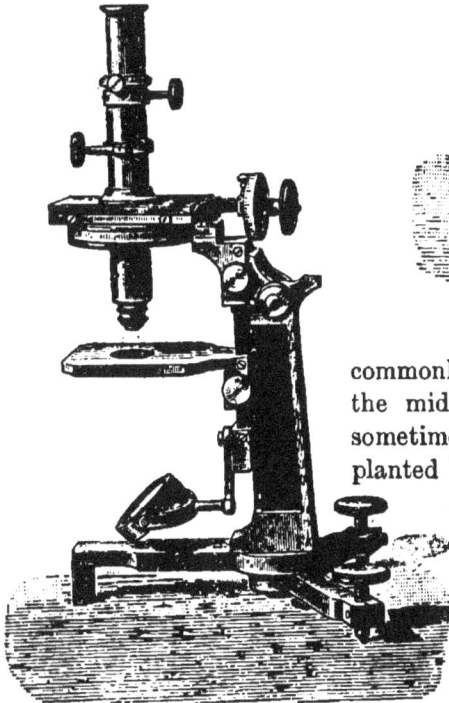

commonly a fixed cross hair to mark the middle of the field of view, and sometimes the moving hair is supplanted by two crossing each other at an oblique angle. In the latter

FIG. 401.

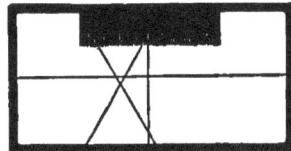

FIG. 400.

case the appearance of the field is that represented in Fig. 400.

This is a more elaborate device than the eyepiece scale, and admits

2 F

of greater accuracy of observation. Like the former, however, it must be calibrated by reference to a scale or an object of known size placed under the microscope.

3. *Method of the Micrometer Stage.* — In this method the micrometer screw is attached to the stage of the microscope, and the object is moved through the field of view by turning the screw. There is in the eyepiece a fixed cross hair. This method has one advantage over the foregoing, namely, that it is the movement of the object itself which is measured. The pitch of the screw being known, therefore, it is not necessary to calibrate the instrument. Sometimes the micrometer screw is so arranged as to move the body of the microscope across a fixed stage. Figure 401 shows a microscope provided with the latter adjustment.

APPENDIX V

FILTERING PUMPS

The filtering pump, or aspirator, was originally designed for the production of the partial vacuum employed by chemists to hasten

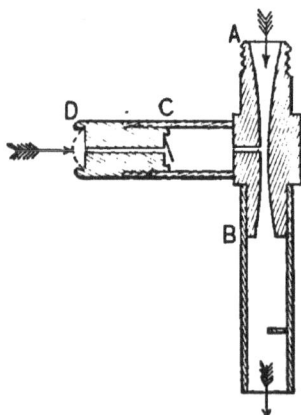

FIG. 402.

the process of filtering; it is, however, a very useful instrument in the physical laboratory. The principle of these pumps may be seen from a consideration of Fig. 402. *AB* in this figure is a metal tube attached to the water faucet. This is contracted to narrow aperture, and opposite the contraction a horizontal tube, *CD*, also of small bore, enters. Upon opening the faucet, the stream of water flows through the vertical tube. If the pressure be considerable, this is broken in falling, and upon passing the opening of the horizontal tube, it entraps air which it carries down with it and discharges below. If the horizontal tube be attached to a closed receiver, the air from this receiver will be rapidly removed by the process just described, and a partial vacuum will be formed.

It is possible to make an aspirator out of metal or even glass after a few trials, which will work fairly well. The precise relation between the parts is, however, an essential matter, and it is better to buy the instrument, which may be had at a low price from any dealer in chemical apparatus. The best forms of aspirator are capable of maintaining a vacuum corresponding to a pressure of less than 5 cm. of mercury.

APPENDIX VI

·THE CONSTRUCTION OF A SENSITIVE GALVANOMETER

It is possible by paying attention to the essential parts of the instrument to construct a galvanometer which, although seemingly rude, will be sufficiently sensitive not only to perform the various experiments described in this book, but for work of much greater delicacy. Such an instrument, made in the very simplest manner, is shown in Fig. 403.

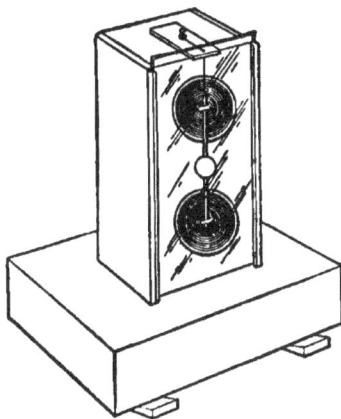

FIG. 403.

To construct it take a pine block 5 cm. × 10 cm. × 20 cm. From the same material cut out a wooden base 20 cm. square and 5 cm. in thickness. For the coils, of which in the form of instrument under consideration there are two, one above the other, take silk-covered copper wire. The size of the wire will depend upon the uses to which the galvanometer is to be put. For the experiments described in this book rather coarse wire should be selected, namely, No. 20 to 24.

To wind the coils, take a large empty spool, of the kind used for the coarser varieties of linen thread. From some thick sheet brass or copper, cut out a disk corresponding in size to the head of the spool, namely, about 4 cm. in diameter, and to the center of this disk solder a brass rod, which must be small enough to pass freely through the hole in the spool. This rod should be about 10 cm. long. Saw the spool in two transversely at a distance of 3 cm. from one end. Insert the brass rod through the hole in that part of the spool which has the longest shank, as shown in Fig. 404, and having drawn it snugly up into place, clamp it there by means of a hand vise. The result of this combination is a reel upon which the coil is to be wound. One head of this reel is the metal disk, the other the wooden head of the spool. Wind the reel full of the wire selected for the galvanometer, bringing both ends over the edge of the spool head.

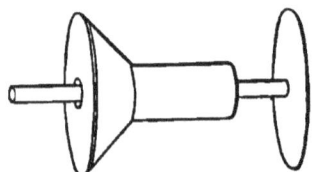

FIG. 404.

Dip the coil, reel and all, into a bath of melted paraffin, and allow the liquid to work its way well into the spaces between the wires. After cooling, heat the metal head of the reel carefully until it is possible to detach it from the coil of wire. We now have the coil still mounted upon the spool with one flat face exposed. Make a second precisely similar coil, using for the purpose another empty spool. These coils are to be mounted in the block of wood described in the first paragraph of this appendix. With an adjustable auger bore two holes through the block with a diameter just sufficient to admit the heads of the spools. The distance between the centers of these holes should be about 6 cm. Insert the coils into these holes, pushing them through until the exposed faces left by the removal of the metal disk are flush with one face of the block, and cement them into that position by the application of paraffin. The four terminal wires from these coils are to be brought out to the back of the block and fastened. The block is then mounted in an upright position upon the center of the wooden base and is fastened by means of two brass screws from below. To the front edges of the block attach strips of wood, grooved so as to admit a window of plate glass as shown in Fig. 403. This window is so placed as to leave a clear space of 1 cm. between the block and the glass.

It now remains to provide a suitable suspension for the galvanometer. For this purpose remove one of the thin bamboo splints from an ordinary Japanese paper fan and cut from it a strip about 10 cm. long. At points equidistant from the two ends of this strip mark transverse lines. The distance between these lines must equal that between the centers of the two coils of wire in the block. These marks indicate the positions of the needles. The needles may be constructed of bits of steel, cut from a watch spring, about 6 cm. in length. The two pieces should be as nearly as possible of the same size, and they should be cut from the same strip of metal. They may be magnetized simultaneously; either by placing them in the axis of a coil of wire through which a strong current is flowing, or between the poles of an electromagnet. After magnetization they are to be mounted upon the bamboo strip over the lines already drawn. The north-pointing pole of one must correspond in position to the south-pointing pole of the other, as indicated in Fig. 405. Midway between the needles the mirror is to be mounted. The simplest plan is to purchase a plane galvanometer mirror, 1 cm. in diameter. It is possible to make good mirrors by silvering a considerable number of microscope cover glasses by the method described in Kohlrausch's *Physical Measurements.* Unless one desires to make a considerable number of mirrors, however, it is better to purchase them of a dealer in physical apparatus.

FIG. 405.

The bamboo strip is to be suspended by means of a fiber of unspun silk from a cocoon. One end of this is attached to the middle of one end of the strip by means of a drop of shellac. The free end is then passed up through a hole in a strip of sheet zinc and is fastened to the top of the block. The height must be such as to bring the needles opposite the centers of the two coils. A bit of beeswax will suffice to fasten the fiber. The zinc strip may be given the form shown in Fig. 403, or it may, preferably, be bent upward into a sort of goose neck and then down as in Fig. 406. In this form it is readily adjustable and the suspension can be brought closer to the face of the block or moved from the same by the use of the pliers. In setting up the galvanometer the suspension should be brought as close to the face of the block as it can without striking. It may be found necessary to protect the suspension from draughts by filling in the opening at the top of the plate glass window with a layer of cotton batting.

FIG. 406.

To set up the instrument thus constructed, three wooden wedges may be made to take the place of leveling screws. The galvanometer must be placed upon a firm support free from tremor and at a distance from large masses of iron, such as iron columns and girders and from iron steam pipes, etc. It should be turned with the axis of its coils approximately perpendicular to the magnetic meridian and should be left so that the suspended parts swing freely.

To adjust the direction in which the astatic pair of magnets are to place themselves, a *controlling magnet* consisting of an ordinary bar magnet should be provided. This is to be laid upon the table or shelf where the galvanometer is mounted or upon any convenient stand in the neighborhood, and it must be shifted until the galvanometer needles come into the desired position. By placing this magnet so that its lines of force *oppose* the field of the earth, the sensitiveness of the instrument can be increased. By placing it with its lines of force *parallel* to the field, the sensitiveness of the galvanometer will be reduced.

The Reading Telescope. — In order to make measurements with a galvanometer like that just described, it is necessary to have a reading telescope. The simplest and one of the best forms for this purpose consists of an ordinary spyglass, which may be purchased of any optician for a small sum. The objective of this spyglass should be at least 3 cm. in diameter. The draw tube of such telescopes is not long enough to admit of focusing them upon objects near at hand; but by unscrewing them at the joint nearest the eyepiece and removing the "erecting lens," the instrument is readily adapted to use as a reading telescope. The image which it now produces is inverted, but it comes to a focus much nearer the objective and within the range of the draw tube. The spyglass thus converted into a reading telescope should be mounted on a wooden stand of the form shown in Fig. 407. A

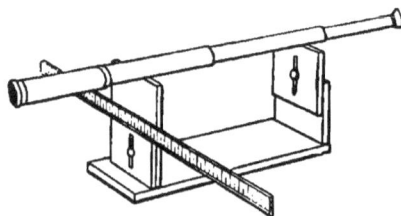

Fig. 407.

paper scale at least 50 cm. long, and divided either into millimeters or tenths of inches, and mounted upon a wooden strip, is set up in a horizontal position just below the telescope. The reading telescope with its scale should be set up at a distance of 1 or 2 m. from the galvanometer. The telescope should be focused upon objects at twice that distance. If, then, it be pointed at the mirror from a position such that the image of the scale can be seen, the divisions upon the latter will be nearly in focus.

The construction of such a homemade galvanometer is a matter requiring some little skill and patience, and the mastery of the instrument after it has been constructed is likewise somewhat difficult. To the student of physics, however, there is no instrument more important, and the time required to learn its construction and to gain command of it is well employed. In a laboratory containing a considerable number of students, it is wise to provide several instruments of the type described above. In such a case the services of a carpenter will enable the teacher to secure a valuable equipment at very small expense.

APPENDIX VII

A SIMPLE FORM OF GLASS CELL

To make a glass cell with plane sides suitable for use in the field of the lantern, take two pieces of plate glass about 15 cm. square. Bend a piece of glass tubing into the shape shown in Fig. 408. Over this bent tube, when cool, slip a rubber tube which should fit it rather snugly. The bent tube with its rubber coating may now be laid between the glass plates, and they may be pressed against it so that the rubber will make contact with the glass at every point. A watertight cell is thus produced. The walls of

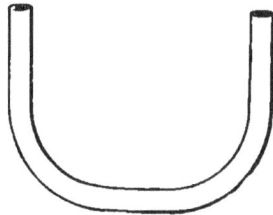

FIG. 408.

the cell may be held in place by means of simple clamps of the form commonly used in such cells, or they may be tied temporarily by means of a strong cord, and the outer space between the tube and the glass plates may be filled in with beeswax or rosin, or other suitable cement. The cord may then be cut away, and the cell will retain its form.

APPENDIX VIII

THE CONSTRUCTION OF AN ELECTROSCOPE

To make a serviceable electroscope, cut out a disk, about 4 cm. in diameter, of sheet metal (copper, brass, or zinc). The metal should be 1 or 2 mm. in thickness. Round off the edges of the disk with a file, and solder to its center a piece of copper wire about 10 cm. long. Over the wire slip a piece of glass tubing 5 cm. long, and seal the glass in place at the middle of the wire by means of melted sulphur, resin, or paraffin. With a pair of pliers bend the free end of the wire into the form shown in Fig. 409, adjusting carefully so that the horizontal portion will be straight and exactly parallel to the plane of the disk.

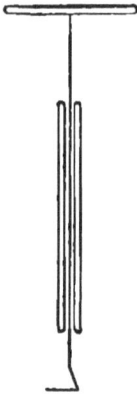

Select a round, short-necked bottle of white glass. It should be 8 to 10 cm. in diameter, and must have a neck wide enough to admit the glass tube freely. Fit the tube to the bottle by means of a cork, a rubber stopper, or a collar made of rubber tubing. Cut from a sheet of gold leaf a strip about 8 cm. long and 0·5 cm. in width. This is readily done with a pair of scissors or a very sharp knife, by allowing the sheet of gold to remain between the adjacent leaves of the "book" and cutting through the leaves and through the layer of gold which they inclose, at a single stroke.

The mounting of the leaves must be performed in a place free from draughts. It is an operation of considerable delicacy, the success of which depends upon the following points: The gold leaf is to be handled only with dry metal tools. To any wet surface the leaf will cling, and cannot be detached without tearing. Dry non-metallic tools become electrified, and attract the foil with a similar result.

By means of a joiner's clamp mount a clean dry steel knitting needle horizontally as in Fig. 410. With one blade of the scissors, or with a knife blade, carefully remove the upper layer of paper from the strip of gold leaf, and the latter from the underlying paper. The tendency of the gold and paper to adhere at the cut edges constitutes the chief difficulty of the entire operation and makes it necessary, sometimes, to cut several strips before one can be detached entire.

FIG. 409.

The strip is to be raised upon the tool and laid over the knitting needle. It is to be adjusted until the ends hang down vertically and to the same length, side by side from the needle, which will then support the strip, folded transversely at its middle. (See Fig. 410.)

The horizontal arm of the wire shown in Fig. 409 is then to be touched along its lower face with shellac or other varnish, after which it is to be brought down care-

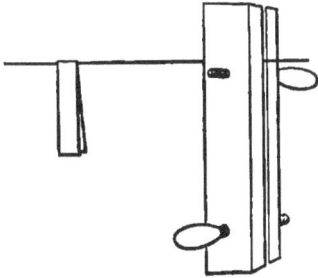

FIG. 410.   FIG. 411.

fully from above until it makes contact with the strip of gold leaf at the fold. The latter will cling to the wire and may then be transferred to the bottle. The electroscope thus constructed will be similar to that depicted in Fig. 176.

A form of electroscope more suitable for use in the field of the lantern is shown in Fig. 411.

The gold leaves hang in the middle of a horizontal cylindrical brass box, 15 cm. in diameter, with glass ends. The beam of light from the lantern passes through the parallel glass plates, which form the ends of this inclosing vessel, without appreciable distortion, and a good image of the gold leaves is projected upon the screen.

APPENDIX IX

THE USE OF THE LANTERN

No instrument is more useful in the physical laboratory than a suitable lantern for projection. It is important that the lantern

should be so constructed as to allow free space between the condensing lens and the objective, so that apparatus of various kinds may be
readily inserted. Many lanterns are intended for use with lantern
slides only. Next of importance in
the selection of the lantern is the
question of the source of light to be
employed. The most powerful light
is that of the electric arc, and this
should be used whenever practicable.
To produce a satisfactory effect, the
carbons of the arc lamp must be so
inclined as to bring the crater of the
positive carbon into view from the center of the condensing lens. (See Fig.
412.)   It is furthermore important
to use carbons of the very best quality.
The incessant flickering and the hissing of most arc lights are due to
the poor quality of the carbon pencils. Arc lamps are to be had
which are constructed especially for use in the lantern, and these may
be employed wherever a *direct-current* incandescent lighting circuit is
at command. Since the potential difference necessary to maintain
the arc is only about 50 volts, it is necessary to place a considerable
resistance in series with the lamp on such a circuit. Resistance boxes
(rheostats) specially designed
for this purpose are easily obtained, or one may be constructed which will answer every
purpose, by taking four full-
sized sheets of tinned iron and
slitting them nearly through
from opposite sides at distances
of about 1 cm. with tinmen's
shears. The method of cutting
is shown in Fig. 413. These may be mounted upon a wooden frame
as shown in Fig. 414, and their adjacent free ends may be attached
together. We thus have a long strip of thin sheet metal capable
of carrying 15 or 20 amperes of current without undue heating. The
apparatus is cheap, serviceable, and easily portable. In laboratories
where electrical work is going on such resistance frames are very use-

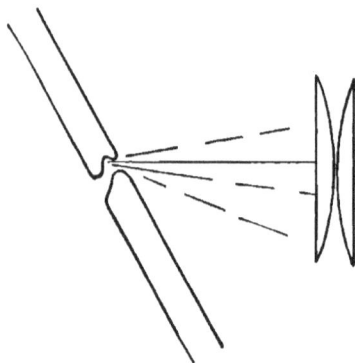

FIG. 412.

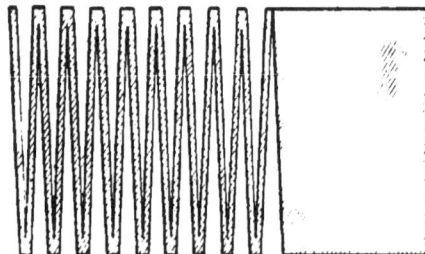

FIG. 413.

ful. One such frame placed in series with the arc lamp is more than sufficient to reduce the current to the suitable size.

Where the arc lamp is not available, the source of light usually employed is the lime light. This is an expensive and inconvenient light; involving as it does the manufacture of oxygen, or the purchase of oxygen and hydrogen in iron cylinders.

It seems likely that the *acetylene flame* will entirely supplant the lime light for use in the lantern. The illumination attained from a suitable acetylene burner is fully equal, both as regards brightness

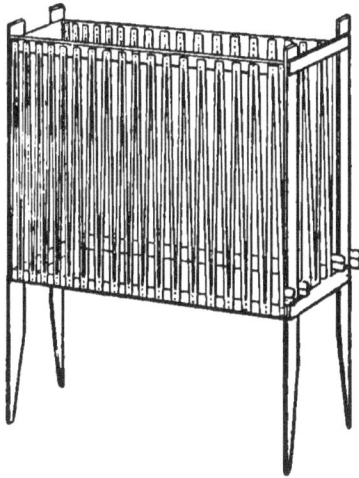

FIG. 414.

and color, to that of the lime light. Generators are obtainable of the manufacturers of lanterns by means of which the acetylene gas can be produced directly from calcium carbide at a sufficient rate to supply the burner. Care should be exercised, as in the case of all illuminating gases, to avoid explosions by mixture with air. Experiment has shown that acetylene is particularly dangerous when stored at considerable pressures. When generated freely, as fast as it is used, there should be no serious danger. Teachers using the lantern will find much that is suggestive and valuable in the following books: Lewis Wright on *Light*; Dolbear, *On the Art of Projecting*; Hopkins, *Experimental Science*, Chapter XXII.

INDEX

Aberration, chromatic, 398;
spherical, 397.
Absolute pitch, 351.
Absorption spectra, 380.
Absorption spectrum of chlorophyl
and potassium permanganate,
388.
Acceleration, 28.
Accommodation, in vision, 413.
Achromatic lenses, 399.
Action of current upon a magnet, 277.
Adhesion, 86.
Air columns, vibration of, 362.
Air thermometer, the, 157.
Ampère's rule, 279.
Analyzer and polarizer, 406.
Angle of contact, 134;
of refraction, 378.
Angles of incidence and reflection, 371.
Aperture of lenses, 395.
Aqueous humor of the eye, 415.
Archimedes, principle of, 116.
Arc light, the, 305.
Atmosphere, nature of, 141.
Attraction, electrostatic, 211.
Attractive and repellent forces be-
tween a magnet and a wire car-
rying current, 325.
Audibility, limits of, 342.
Axle, the wheel and, 70.

Balance, the, 79;
sensitiveness of, 86.
Ballistic curve, 34.
Barometer, construction of, 140.
Battery, plunge, 270.
Beats, graphic representation of, 352;
the method of, 351.

Bells, 345.
Binocular vision, 417.
Boiling point, influence of pressure
upon, 178;
and pressure, relation between, 180.
Bowed strings, motion of, 357.
Boyle's law, 137.
Bridge, Wheatstone's, 297.
Bright-line spectra, 386.
Brightness of image and aperture of
lens, 394.
British units and the metric system, 3.

Calorimeter, the ice-block, 165.
Calorimetry, 161.
Capacity for heat, 161.
Capacity, of water, thermal, 162;
specific inductive, 246.
Cell, bichromic, 269;
simple, 266;
voltaic, forms of, 268.
Cells, closed circuit, properties of, 269.
Centimeter and inch compared, 4;
defined, 2.
Charge, bound and free, 250;
distribution of, 240;
induced, distribution of, 243;
intensity of, 238.
Charges, equal and opposite, 222.
Charging by contact, 217;
by induction, 219.
Charles's law, 157.
Chladni's figures, 343.
Chlorophyl, spectrum of, 388.
Chromatic aberration, 398.
Circular plates, vibration of, 345.
Closed and open pipes, 362.
Closed circuit cells, properties of, 269.

Cohesion, 86.
Coils, induction, 330.
Coincidences, method of, 42.
Color, defined, 417;
　by interference, 409;
　of pigments, 418;
　of soap films, 409;
　purity and saturation of, 418;
　sensations, theory of, 420.
Color blindness, 424.
Concave mirrors, 373.
Condensers, 245;
　capacity of, 249.
Conduction, 194.
Conductivity (thermal) of copper,
　iron, and glass, 194;
　of liquids, 196.
Conductors and non-conductors, 221.
Conjugate foci, 374;
　of a lens, 391.
Conservation of energy, law of, 64.
Contact, angle of, 134;
　charging by, 217.
Continuous spectra, 386.
Contrast, 422.
Convection, 194, 198.
Converging lenses, 391.
Convex mirrors, 373.
Copper, conducting power of, 194;
　electrolysis of, 309.
Copper voltameter, measurement of
　current by, 312.
Current, action of, upon a magnet, 277;
　the electric, 265;
　heating platinum wire by means of,
　302;
　induced by cutting lines of force, 321;
　magnetic effect of, 271;
　measurement of, by voltameter, 312;
　production of, by means of heat, 315;
　transformation of energy by means
　of, 302;
　induced by moving a wire, 321.
Curve, the ballistic, 34;
　of sines, 54.

Dark-line spectra, 386.
Density, defined, 122;
　measurement of, 123;
　of certain substances, 128.

Diamagnetism defined, 289.
Diameters, law of, 132.
Difference of potential of a voltaic
　cell, 265.
Dipping needle, 281. ·
Direction of thermo-electric current in
　antimony and bismuth, 317.
Discharge in vacuo, 258, 260.
Disruptive discharge, 255.
Dispersion defined, 382.
Distance and size, estimation of,
　416.
Distribution of charge, 240.
Diverging lenses, 391.
Double refraction, 404;
　in strained glass, 408.
Dynamos and motors, 326.

Ebullition, 177.
Elasticity, 95;
　limit of, 96;
　of torsion, 105.
Electric current, the, 265.
Electrical machines, 225.
Electricity, defined, 211;
　hypothesis of two fluids, 216;
　quantity of, 237.
Electrification, defined, 213;
　by chemical action, 264.
Electrochemical equivalents (table of),
　314.
Electrolysis, the law of, 307;
　of copper, 309;
　of sodium, 310.
Electromagnets, 289.
Electromotive force and resistance,
　294.
Electrophorus, the, 227.
Electroscope, the, 214.
Electrostatic attraction, 211.
Electrostatic repulsion, 213.
Electrostatic series, 223.
Energy, conservation of, 65;
　heat a form of, 161;
　kinetic, 58;
　of pendulum, 61;
　of the spark, 256;
　potential, 58;
　transformation of, by electric cur-
　rent, 302.

Equilibrium, conditions of, 75;
  stable and unstable, 75.
Equivalents, electrochemical, 314.
Erg, the, 60.
Estimation of tenths, 6.
Evaporation, freezing by, 179;
  influence of, upon temperature, 184.
Expansion of a bar, 148;
  of gases, 151;
  of liquids, 149;
  of liquids (absolute), 150.
Extraordinary ray, the, 405.
Eye, the, 413.
Eyepieces, 393.

Fall of potential, 293.
Falling bodies, laws of, 19.
Faraday's bag, 236;
  ice-pail experiment, 235.
Far and near sightedness, 414.
Field, the magnetic, 271;
  magnetic, of the earth, 280;
  of a coil, influence of iron upon, 287.
Figures, of Lissajous, 354.
Fire syringe, the, 190.
Flame, the sensitive, 334.
Fluids, hypothesis of two, 216.
Focal length of a lens, 391.
Focal length of lens and size of image,
  395.
Foci, conjugate, 374;
  of a lens, 391.
Focus, the principal, 374.
Foot pound, the, 61.
Force, definition of, 10;
  moment of a, defined, 67.
Forces, cohesive, 88;
  composition of, 11;
  graphical representation of, 11;
  in equilibrium, 75;
  molecular, 86;
  parallelogram of, 12;
  polygon of, 13.
Fraunhofer lines, 387.
Friction, 86;
  electrical charging, by, 222;
  sliding, 89;
  starting, 92;
  of a shaft, 94;
  rolling, 94.

Fulcrum, the, 67.
Fundamental tones, 347;
  tone of a pipe, 366.
Fusion, change of volume due to, 174;
  heat of, 169.

Galileo's experiments, 19.
Galvanometer, the, 291.
Galvanometer needles, forms of, 292.
Galvanometers, sensitive, 292.
Gas, definition of, 137;
  measurement of mass of, 137.
Gases, expansion of, 151;
  Pascal's principle for, 139;
  properties of, 137;
  specific heat of, 165.
Gay-Lussac's law, 150.
Geissler tubes, 262.
Glass, conducting power of, 194.
Glow lamps, 303;
  determination of power expended
    upon, 304.
Graduation of a thermometer, 153.
Grain, the, defined, 3.
Gram and grain compared, 4.
Gravitation, 21;
  in combination with other forces, 31;
  measurements of, 55.

Hare's method, 127.
Heat, defined, 178;
  a form of energy, 161;
  and work, relations between, 187;
  capacity for, 161;
  mechanical equivalent of, 187, 192;
  nature and effects of, 148;
  of friction, steam from, 188;
  of fusion, 169;
  of fusion, numerical values of, 170;
  of vaporization, 171;
  production of current by means of,
    315;
  specific, 163;
  specific, of a metal, 164;
  transmission of, 194;
  transmission by radiation, 204;
  units, 163.
Holmgren test for color blindness, 424.
Holtz machines, 233.
Humors of the eye, 415.

Hydraulic press, 110.
Hydrometer, method of, 124;
  Nicholson's, 125.
Hydrometers of constant immersion,
  125.
Hydrostatic pressure, 108.

Ice-block calorimeter, the, 165.
Ice-pail experiment, 235.
Image in plane mirror, 372.
Images, formed by lenses, 391;
  real and virtual, 374.
Incidence, angle of, 371.
Inclined plane, 72;
  method of the, 22;
  relations of screw to, 73.
Index of refraction, 378.
Induced current, 321.
Induced currents by the starting or
  stopping of currents in a neigh-
  boring circuit, 329.
Induction, charging by, 218;
  coils, 330;
  magnetization by, 284.
Inertia, law of, 14;
  moment of, 62.
Influence machines, 229.
Influence of temperature upon resist-
  ance, 300.
Interference colors in nature, 411;
  of light, 408.
Ions, defined, 307.
Iridescence, 411.
Iris, the, 415.
Iron, conducting power of, 194;
  magnetization of, 273.
Isochronism, law of, 41.

Joule's law, 302.

Kilogram, the standard, 3;
  and pound compared, 4.
Kilometer and mile compared, 4.
Kinetic energy, 58;
  measured, 60.
Kirchhoff's law, 207.

Lamp, the glow or incandescent, 303.
Law, Boyle's, 137;
  Charles's, 157.

Law, Gay-Lussac's, 158;
  Joule's, 302;
  Kirchhoff's, 207;
  Mariotte's, 137;
  Ohm's, 294;
  of the conservation of energy, 64;
  of diameters, 132;
  of electrolysis, 307;
  of inertia, 14;
  of isochronism, 41;
  of the lever, 68;
  of the simple pendulum, 38;
  of the simple pendulum, formula
    of, 45.
Laws of falling bodies, 19;
  of friction, 90;
  of motion, 10, 13, 23.
Leakage, measurement of, 145.
Length, measurements of, 1;
  units of, 2.
Lenses, defined, 390;
  converging and diverging, 391.
Leslie's cube, 206.
Lever, the, defined, 66;
  law of the, 68;
  principle of work applied to, 70.
Leyden jar, the, 245.
Light, the arc, 305;
  composition of, 382;
  propagation of, 370;
  reflection of, 371;
  velocity of, 370.
Limits of audibility, 342.
Lines of Fraunhofer, 387.
Liquid, convection currents in, 198.
Liquids, conductivity of, 196;
  density of, 127;
  expansion of, 149;
  properties of, 108;
  properties of surface films of, 130;
  specific heat of, 165.
Lissajous's figures, 354;
  method, 353.
Liter and quart compared, 4.

Machine, definition of a, 65;
  the Holtz, 233;
  the Toepler-Holtz, 230;
  the Wimshurst, 233.
Machines, electrical, 225.

Machines, influence, 229; simple, 65.
Magnet, action of current upon a, 277.
Magnetic effects of the current, 271.
Magnetic field, the, 271; of the earth, 280; of a coil of wire, 272.
Magnetic pole, nature of, 281.
Magnetic saturation, 288.
Magnetization, by induction, 284; by the earth's field, 285; of chromium, 289; of iron by means of a solenoid, 273; of manganese, 289; of nickel, 289; of steel, 275.
Magnifying power, 396.
Manometers, 143.
Manometric flame, the, 368; analysis of sound by, 368.
Mariotte's law, 137.
Mass, measurement of, 1; unit of, 3.
Measurement of resistance by Wheatstone's bridge, 299.
Measurements, physical, 1; of length, 6.
Melting point, influence of pressure on, 176.
Menzel and Raps's experiment, 358.
Metals, resistance of, 296.
Meter, the, 2; and inch compared, 4.
Method of mixtures, 163.
Metric system, 1; units of, 3.
Metronome, the, 5; illustrations of, application of, 5; testing of, 46.
Micrometer, the eyepiece, 401; the stage, 401.
Microscope, the, 400.
Mirrors, convex and concave, 373; plane, 372.
Mixtures, method of, 163.
Modulus, Young's, etc., 100.
Molecular forces, 86.
Motion, characteristics of, 10; energy of, 59; laws of, 10, 13.

Motion, of pendulum, analysis of, 53; simple harmonic, 53; uniform, defined, 10.
Motors and dynamos, 326.
Musical, instrument, parts of a, 365.
tones and noises, 343.

Near and far sightedness, 414.
Needle, dipping, 281.
Neutral point, the, 319.
Newton's laws of motion, 13; rings, 411.
Nicholson hydrometer, the, 125.
Nicol prism, the, 405.
Nodal lines, 344.
Nodes, in a vibrating string, 356.
Noises, defined, 343.

Object lens of telescopes, 394.
Ohm's law, 294.
Open and closed pipes, 362.
Optical study of vibration, 353.
Ordinary ray, the, 405.
Organ pipe, the, 365.
Organ pipes, open and closed, 362.
Oscillation, to find center of, 51.
Overtones, 347; of wind instruments, 367.

Parallelogram of forces, 12.
Pascal's principle applied to gases, 159; vases, 114.
Peltier effect, the, 316.
Pendulum, definition of, terms referring to, 38; analysis of the motion of, 53; energy of, 61; isochronism of, 41; law of, 38; law of equal times, 41; law of relation between force and period, 44; the physical, 53; the reversion, 56; the simple, 37; which beats seconds, length of, 47.
Permeability, bodies of low, 290.
Phenomena accompanying vaporization, 177.
Physical measurements, 161.

Physical pendulum, 51.
Physics, defined, 11;
  branches of, 1.
Pipes, closed and open, 362.
Pitch, defined, 342;
  influence of temperature on, 367;
  of a pipe, 366;
  of a pipe, modified by the contained gas, 367;
  of vibrating strings, 360;
  relative and absolute, 351.
Plane, the inclined, 72.
Plates, vibrating, 341.
Plunge battery, the, 270.
Points, action of, 241.
Polarization, by double refraction, 404;
  by reflection, 403;
  of the voltaic cell, 268.
Polarized light, 403.
Polarizer and analyzer, 406.
Poles, magnetic, 281.
Polygon of forces, the, 13.
Potassium permanganate, spectrum of, 388.
Potential, difference of, 265;
  energy, 58;
  energy of configuration, 60;
  fall of, 293.
Press, the hydraulic, 110.
Pressure, distribution of, 112;
  gauges, 144;
  hydrostatic, 108;
  influence of, upon the electric spark, 257;
  influence of, on the melting point, 176;
  relation between boiling point and, 180;
  transmission of, 109.
Principal focus of a mirror, 374;
  of a lens, 391.
Principle of Archimedes, 116, 122.
Prism, passage of light through a, 379.
Production of current by heat, 315.
Projectiles, defined, 31.
Propagation of light, 370.
Properties of gases, 137;
  of liquids, 108.
Pulley, the, defined, 65;
  considered as a lever, 70.

Pure spectrum, formation of a, 384.
Purity of color, 418.

Quadrant electrometer, the, 252.
Quantity, of electricity, 237.

Radiation, 194, 202;
  and temperature, 210;
  influence of surface upon, 206.
Raps and Menzel's experiment, 360.
Reaction, illustration of, 16.
Real images, 374.
Reflecting telescopes, 377.
Reflection, angle of, 371;
  of sound, 340;
  the law of, 371;
  total, 380.
Refracting telescopes, 393.
Refraction, angle of, 378;
  double, 404;
  index of, 378;
  law of, 377.
Relative pitch, 351.
Repulsion, electrostatic, 213.
Resistance, boxes, 298;
  and electromotive force, 294;
  influence of temperature upon, 300;
  measurement of, 299;
  specific, 296.
Resolution of forces, 11.
Resonance of an air column, 363.
Resonators, 364.
Retina, the, 413.
Reversal of sodium lines, 386.
Rods, torsion of, 106;
  vibrating, 341.
Roentgen rays, 262.
Rosse, the great telescope of, 377.

Saturation, magnetic, 288;
  of color, 418.
Segments, vibrating, 344.
Sensitive flame, the, 334.
Sensitive galvanometers, 292.
Silver voltameter, the, 313.
Simple cell, the, 266.
Sines, curve of, 54.
Size of image and focal length of lens, 395.
Size and distance, estimation of, 416.

www.ingramcontent.com/pod-product-compliance
Lightning Source LLC
Chambersburg PA
CBHW020907210326
41598CB00018B/1799